Signaling and Communication in Plants

Series Editors

František Baluška
Department of Plant Cell Biology, IZMB, University of Bonn, Kirschallee 1, D-53115 Bonn, Germany

Jorge Vivanco
Center for Rhizosphere Biology, Colorado State University, 217 Shepardson Building, Fort Collins, CO 80523-1173, USA

For further volumes:
http://www.springer.com/series/8094

František Baluška • Velemir Ninkovic
Editors

Plant Communication from an Ecological Perspective

Springer

Editors
Dr. František Baluška
Universität Bonn
Inst. Zelluläre und Molekulare
Botanik (IZMB)
Kirschallee 1
53115 Bonn
Germany
baluska@uni-bonn.de

Dr. Velemir Ninkovic
Department of Ecology
Swedish University of Agricultural
Sciences
P.O. Box 7044
SE-750 07 Uppsala
Sweden
Velemir.Ninkovic@ekol.slu.se

ISSN 1867–9048 e-ISSN 1867–9056
ISBN 978-3-642-26366-8 ISBN 978-3-642-12162-3 (eBook)
DOI 10.1007/978-3-642-12162-3
Springer Heidelberg Dordrecht London New York

Cover design: WMXDesign GmbH, Heidelberg, Germany

Printed on acid-free paper

Springer is part of Springer Science+Business Media (www.springer.com)

Preface

Since the concept of allelopathy was introduced almost 100 years ago, research has led to an understanding that plants are involved in complex communicative interactions. They use a battery of different signals that convey plant-relevant information within plant individuals as well as between plants of the same species or different species. The 13 chapters of this volume discuss all these topics from an ecological perspective. Communication between plants allows them to share physiological and ecological information relevant for their survival and fitness. It is obvious that in these very early days of ecological plant communication research we are illuminating only the 'tip of iceberg' of the communicative nature of higher plants. Nevertheless, knowledge on the identity and informative value of volatiles used by plants for communication is increasing with breath-taking speed. Among the most spectacular examples are situations where plant emitters warn neighbours about a danger, increasing their innate immunity, or when herbivore-attacked plants attract the enemies of the herbivores ('cry for help' and 'plant bodyguards' concepts). It is becoming obvious that plants use not only volatile signals but also diverse water soluble molecules, in the case of plant roots, to safeguard their evolutionary success and accomplish self/non-self kin recognition. Importantly, as with all the examples of biocommunication, irrespective of whether signals and signs are transmitted via physical or chemical pathways, plant communication is a rule-governed and sign-mediated process.

The previous volumes focused on signalling molecules and pathways, as well as on communication related to plant sensory biology underlying the emerging concept of plant behaviour. Here, individual chapters deal with diverse aspects of plant communication such as evolution of plant signals and toxins, chemical signals in plant photobiology and 'arms-races' in pathogen defence, allelopathy of exotic plant invasion, volatile chemical interactions between undamaged plants and their effects at higher trophic levels, chemical communication in plant–ant symbioses, as well as effects of global atmospheric changes on plants and their trophic interactions. Finally, two chapters deal with the perspective of exploiting the chemical signals of plant communication for sustainable agriculture, and the technological possibility of monitoring plant volatile signals to obtain information about plant health status in greenhouses.

For many years, plants were placed outside of the communicative and even the sensitive living domain. Immanuel Kant even went so far as to place plants outside the living realm. The vocal-based physical (acoustic) language of humans depends on air vibrations that are decoded in the ears. The volatile-based chemical language of plants is communicated by volatiles decoded via diverse receptors (most of them still unknown). Plants are unique and differ greatly from animals. This makes it very difficult for us, biassed by the human-centric perspective of our world-view, to grasp their whole communicative complexity and to understand the true nature of their communications. The sessile nature of plants and the dual character of plant bodies, with the above-ground autotrophic shoots and the below-ground heterotrophic roots, are further phenomena obscuring the real nature of plant communication. In science, one should try to keep a neutral unbiased position and not exclude any possibility. We can look forward to witnessing the next wave of surprising discoveries.

Bonn, April 2010 František Baluška
Uppsala, April 2010 Velemir Ninkovic

Further Reading

Baluška F, Volkmann D, Mancuso S (2006) Communication in plants: neuronal aspects of plant life. Springer, Berlin

Greenfield MD (2002) Signalers and receivers: mechanisms and evolution of arthropod communication. Oxford University Press, Oxford

Searcy WA, Nowicki S (2005) The evolution of animal communication: reliability and deception in signalling systems. Princeton University Press, Princeton, NJ

Witzany G (2009) Biocommunication and natural genome editing. Springer, Berlin

Contents

Evolutionary Ecology of Plant Signals and Toxins: A Conceptual Framework 1
H. Jochen Schenk and Eric W. Seabloom

The Chemistry of Plant Signalling 21
Michael A. Birkett

Plant Defense Signaling from the Underground Primes Aboveground Defenses to Confer Enhanced Resistance in a Cost-Efficient Manner 43
Marieke Van Hulten, Jurriaan Ton, Corné M.J. Pieterse, and Saskia C.M. Van Wees

Allelopathy and Exotic Plant Invasion 61
Amutha Sampath Kumar and Harsh P. Bais

Volatile Interaction Between Undamaged Plants: A Short Cut to Coexistence 75
Velemir Ninkovic

Volatile Chemical Interaction Between Undamaged Plants: Effects at Higher Trophic Levels 87
Robert Glinwood

Within-Plant Signalling by Volatiles Triggers Systemic Defences 99
Martin Heil

Volatile Interactions Between Undamaged Plants: Effects and Potential for Breeding Resistance to Aphids 113
Inger Åhman and Velemir Ninkovic

Communication in Ant–Plant Symbioses 127
Rumsaïs Blatrix and Veronika Mayer

Photosensory Cues in Plant–Plant Interactions: Regulation and Functional Significance of Shade Avoidance Responses 159
Diederik H. Keuskamp and Ronald Pierik

Global Atmospheric Change and Trophic Interactions: Are There Any General Responses? 179
Geraldine D. Ryan, Susanne Rasmussen, and Jonathan A. Newman

Exploiting Plant Signals in Sustainable Agriculture 215
Toby J.A. Bruce

Plant Volatiles: Useful Signals to Monitor Crop Health Status in Greenhouses 229
R.M.C. Jansen, J. Wildt, J.W. Hofstee, H.J. Bouwmeester, and E.J. van Henten

Index 249

Evolutionary Ecology of Plant Signals and Toxins: A Conceptual Framework

H. Jochen Schenk and Eric W. Seabloom

Abstract Plants are capable of acquiring information from other plants, but are they able to send signals and communicate with them? Evolutionary biologists define biological communication as information transmission that is fashioned or maintained by natural selection and signals as traits whose value to the signaler is that they convey information to receivers. Plants, then, can be said to communicate if the signaling plant derives a fitness benefit from conveying information to other plants. Examples for interplant communication that fit these definitions potentially include territorial root communications, self/non-self recognition between roots and associated with self-incompatibility, volatile signals that induce defenses against herbivores, signals from ovules to mother plants, signals associated with root graft formation, and male to female signals during pollen competition. Natural selection would favor signals that are costly to the signaler and therefore are likely to convey reliable information because they cannot be easily faked. Toxins in low concentrations may commonly act as signals between plants rather than as inhibitory allelochemicals. This explains why toxic concentrations of plant allelochemicals are rarely found in natural coevolved systems.

1 Introduction

Do plants communicate with other plants? To many readers, this would appear to be a redundant question in a volume devoted to plant communication from an ecological perspective. However, anyone even vaguely familiar with the voluminous

H.J. Schenk (✉)
Department of Biological Science, California State University Fullerton, Fullerton, CA 6850, USA
e-mail: jschenk@fullerton.edu

E.W. Seabloom
Department of Zoology, Oregon State University, Corvallis, OR, USA
e-mail: seabloom@science.oregonstate.edu

F. Baluška and V. Ninkovic (eds.), *Plant Communication from an Ecological Perspective*, Signaling and Communication in Plants,
DOI 10.1007/978-3-642-12162-3_1, © Springer-Verlag Berlin Heidelberg 2010

literature on human communication (e.g., Littlejohn and Foss 2008; Watzlawick et al. 1967) and animal communication (e.g., Dawkins and Krebs 1978; Maynard Smith and Harper 2003; Otte 1974; Searcy and Nowicki 2005; Zahavi and Zahavi 1997) will know that this is by no means an easy question to answer, the answer depending very much on one's definition of biological communication. The purpose of this chapter is to review the pertinent biological communications literature and apply it to communication between plants.

2 What Is Communication?

The Merriam–Webster dictionary defines communication as *a process by which information is exchanged between individuals through a common system of symbols, signs, or behavior*. For biological communication, this definition would have to be expanded to include information exchanges between any kind of signaler and receiver, e.g., within organisms, among organs, or cells. (To avoid confusion, we will use the terms signaler and receiver throughout this chapter instead of the synonymous terms emitter, agent, actor, source, or sender on one side and target, reactor, and recipient on the other.) Following common usage in biological sciences, it is also useful to replace the terms *symbols, signs, or behavior* with *signal*, which Webster's defines as *a detectable physical quantity or impulse by which messages or information can be transmitted*. This gives us the following general definition: *Communication is a process by which information is exchanged between a signaler and a receiver through a common system of signals*. Definitions similar to this one have been widely used in studies of human communications (Watzlawick et al. 1967).

2.1 What Is Information?

If communication is information exchange, what exactly is information? That turns out to be a surprisingly difficult question to answer, and interested readers are referred to the voluminous literature on information theory starting with Shannon (1948) and Wiener (1948). The most helpful and most memorable definition was offered by Gregory Bateson (2000, p. 381): Information is *any difference which makes a difference in some later event*. Information comes in the form of answers to binary questions such as self or non-self, male or female. Continuous information can be expressed as a series of binary choices. This means that the amount of information can be measured in bits (Bradbury and Vehrencamp 1998).

2.2 What Is Biological Communication?

The definition of communication as information exchange, however, is not the one used by most evolutionary biologists, for whom it is important to adopt a pragmatic view that distinguishes between evolved functions and incidental effects. Pragmatic definitions of the terms *signal* and *communication* in evolutionary biology, then, should be restricted to behavioral, physiological, or morphological information transmission that is fashioned or maintained by natural selection (Dicke and Sabelis 1988; Maynard Smith and Harper 2003; Otte 1974). Otte (1974) suggested using the term *cue* for information exchanges that have not been under selection to inform, and this usage, which is widely accepted in animal communication studies (Bradbury and Vehrencamp 1998; Maynard Smith and Harper 2003), will be adopted in this chapter. For example, a plant detecting the presence of a neighbor through alterations in the red/far-red light ratio is considered to have received a cue rather than a signal.

2.3 What Is a Signal?

Unfortunately, there are several colloquial meanings of the word signal, referring either to the physical quantity or impulse being transmitted – such as quanta of light or molecules –, the signal's meaning, or to the emitting source, such as a pattern of skin pigments, a vibrating vocal cord, or the biochemical pathway that creates a signaling molecule. Most animal communication researchers adopt the third usage and define signals as genetically encoded traits that are under natural selection (Maynard Smith and Harper 2003; Otte 1974; Wilson 1975; Zahavi and Zahavi 1997). There is almost universal agreement that a biological signal benefits the sender (i.e., increases its fitness) by altering the likelihood that the receiver will respond in a certain way (Bradbury and Vehrencamp 1998; Dawkins and Krebs 1978). Some have argued that the receiver has to benefit from the information for "true communication" to occur (Dusenbery 1992; Marler 1977). This, however, would exclude deceptive signaling from biological communication, and few biologists appear to have adopted this very restrictive definition (Bradbury and Vehrencamp 1998). A receiver has to benefit, on the average, from responding to a certain type of signal in a certain way. For example, a male insect benefits from responding to a pheromonal signal that is most likely to originate from a female, even though it may be tricked occasionally into responding this way by an orchid that mimics the signal. For the remainder of this chapter we will adopt the pragmatic definition of biological signals from Zahavi and Zahavi (1997): "Signals are traits whose value to the signaler is that they convey information to those who receive them," which is a more generalized version of an earlier definition by Otte (1974). As customary in evolutionary biology, terms such as "value" and "benefit" are understood to mean adaptive value or benefit, on average, a positive effect on fitness.

Pragmatic definitions of biological communication focus on the evolutionary aspects of communication and sidestep other aspects of communication, such as the nature of information transmission (syntactics) and of meaning (semantics) (Watzlawick et al. 1967). The downside of the evolutionary approach, of course, is that we do not actually know whether most traits are under natural selection (Gould and Lewontin 1979). In many cases, it will be relatively easy to determine whether signaler and/or receiver on average benefit from a signal transmission, but in many other cases this will be less apparent. Consider, for example, altruistic signal exchange, that appears to benefit only the receiver or even a group of receivers. Even in more clear-cut cases of signal exchange between two individuals, the advantage to sender or receiver must often be assumed to exist rather than empirically demonstrated (Slater 1983). Defining a process by its supposed function can invite adaptationist reasoning when natural selection has not in fact been involved in the shaping of the process (Gould and Lewontin 1979). It is important to keep in mind that whenever we speak of biological communication we are in fact formulating a hypothesis about the adapativeness of a process (see chapter "Volatile Interaction between Undamaged Plants: A Short Cut to Coexistence").

Plants have innumerable ways of gathering information from cues received from their environment, including from other plants, but, as the discussion above has made clear, information gathering from incidental cues, while immensely important to organisms, is not considered biological signaling or communication (Bradbury and Vehrencamp 1998; Maynard Smith and Harper 2003; Otte 1974) and therefore is not further discussed here.

2.4 What Is Allelopathy?

For most of its history, botany has had its own concepts and terms that often were quite different from those used in other areas of biology. A good example is the term allelopathy, coined by Hans Molisch (1937) to refer to "the influence of one plant on another," i.e., *all kinds* of stimulatory and inhibitory interactions between plants. Allelopathy today is normally used in a much more restricted meaning to denote chemical inhibition - an understanding that may have originated from translating the two Greek words that make up the term, *allēlōn* as "one another other" and *pathē* as "suffering." In fact, *pathē* also has a more general meaning, "subject to, experience," and this is obviously what Molisch (1937) had in mind, because his research in allelopathy largely concerned the volatile plant hormone ethylene, not a toxin at concentrations normally found in plants. Rice (1984) and Einhellig (1995) used the term allelopathy in a slightly narrower meaning to include only *chemical* interactions: communication, as well as inhibitory and stimulatory (e.g., nutritional) ones. Because these are very different kinds of interactions, none of which are unique to plants, there is really no reason, other than deference to history, to retain this broad concept of allelopathy. Previously, accepting the current usage of terms, we adopted the view that chemical communication should be distinguished from allelopathy, which would be defined as chemical interactions that involve toxic allelochemicals (Schenk et al. 1999). However, as we will see, it is often extremely difficult to determine whether chemicals act as toxins, signals, or both. Moreover, words are powerful in directing thoughts, and retaining the word allelopathy for plants brings with it the powerful suggestion that chemical interactions in plants are somehow fundamentally different from those in bacteria, fungi, protists, or animals. For these reasons, the term allelopathy has outlived its usefulness and, in the interest of integrating general concepts across all of biology, should be retired (Schenk 2006).

2.5 *What Is the Difference Between a Toxin and a Signal?*

The distinction between transmission of energy and transmission of information is vital in studies of organismal interactions (Dawkins and Krebs 1978; Dicke and Sabelis 1988; Dusenbery 1992; Wiley 1994). Expenditure of energy on aggressive interactions is costly and surprisingly rare in animals which have evolved communication mechanisms. Probabilistic information about the fighting ability of an adversary allows many organisms to exchange signals and avoid the costly fight. Calls and songs of birds, insects, frogs, and toads all around us attest to the evolutionary power of signaling over aggression (Krebs and Davies 1997; Wilson 1975). Obviously aggressive behavior can contain important information, but conceptually the information contained in a blow to the head of a rival male bighorn sheep is quite distinct from the damage or backward movement caused by the blow. The distinction between energy and information exchange was reflected in Wiley's (1994) definition of biological communication (which did not include criteria of adaptation): "A signal is any pattern of energy or matter produced by one individual (the signaler) and altering some property of another (the receiver) *without providing the power to produce the entire response* (p. 162, author's italics)."

An important difference between signaling and energy exchange is that the receiver has full physiological control over its response; it can respond or ignore the signal depending on the circumstances or the nature of the signal (Dusenbery 1992). In the case of energy transmission (including toxins), the energy source has the physiological control over the response and the receiver does not have the option of ignoring the transmission. Obviously, the ability to potentially ignore a signal will usually be an advantage for a receiver. For example, it was found that male mice of low body weight tend to avoid territories scent-marked by another male, while heavy mice with higher competitive ability are more likely to ignore such signals (Gosling et al. 1996). Larger frogs and toads are more likely to ignore high-frequency calls from smaller competitors than low-frequency calls from larger ones (Arak 1983; Wagner 1989). Similarly, in plants, the ability to ignore root signals from a competitor (Schenk 2006; Schenk et al. 1999) may be an advantage for a strong competitor, while an inefficient competitor, such as a seedling, may benefit from avoiding soil volume occupied by other roots. Signalers can also benefit from the receivers' ability to ignore their signals, as indiscriminate responses from all potential receivers are unlikely to benefit a signaler. In contrast, a powerful toxin could potentially harm a large variety of other organisms, including some that could be beneficial to the emitter of the toxin.

It clearly is important for organisms to be able to ignore a signal, but this ability brings up an interesting conundrum for researchers. It is universally agreed upon that for pragmatic reasons biological communication can only be said to have occurred when a response of the receiver is observed (Searcy and Nowicki 2005). Yet, in the case of a potential receiver that does not respond to a signal it is often impossible to know if the signal was received. In the case of acoustic communication, as in the frog and toad studies mentioned above (Arak 1983; Wagner 1989), it may be safe to assume that receivers heard a call, but in the case of chemical communication the distinction between not perceiving or ignoring a signal will be almost impossible to make. This creates a special problem for plant researchers, who typically face signals that are difficult to observe.

2.6 Differences Between Plant and Animal Communication

So far, much of our discussion has been about animal communication. The idea that plants may possibly communicate was controversial until quite recently. Reports in the early 1980s of pheromonal signal exchange among trees (Baldwin and Schultz 1983; Rhoades 1985) were much debated, heavily criticized on methodological and analytical grounds, and ridiculed as "talking trees" (Fowler and Lawton 1985). Silvertown and Gordon (1989) stated that visual and olfactory signals transmitted from plants are exclusively directed at animals. Since then, a wealth of information on signal exchange and chemical interactions among plants, and among plants and other organisms, including microbes, fungi, and animals has accumulated, forcing a re-evaluation of the nature of plant interactions (Baluška 2009; Baluška and Mancuso 2009b, this volume). Already it seems hard to believe that plants used to be singled out as the only group of organisms not thought to be able to exchange chemical signals – an ability easily acceded to bacteria, fungi, protists, and animals. The book by Zahavi and Zahavi (1997) on biological signaling, for example, did not include a single reference to plants, even though one of the authors was a plant physiologist. Plant communications research clearly has come a long way since then. However, the question remains: are there important or even fundamental differences between communication in plants and in other groups of organisms?

The main trait that sets plants apart from other organisms is the rigid cellulose cell wall that restricts their movement to relatively slow rates. The modular nature of plants is not unique to them, but it certainly sets them apart from unitary animals. Does either of these traits affect the abilities of plants to communicate? The modular nature of all plants and the clonal nature of about 40% of all plants (Tiffney and Niklas 1985) certainly has interesting implications for the evolution of plant signals through individual selection (more on that below). Rigid cell walls generally do not allow plants to send and perceive signals that require rapid movement of organs or cells. However, plants clearly emit and perceive visual cues, better called radiational cues, as plants do not have eyes, and nobody seriously disputes the ability of plants to produce and perceive chemical cues. Plants also create and respond to electrical fields (Baluška and Mancuso 2009a; Davies 2004; Fromm and Lautner 2007; Lund 1947), and are able to perceive tactile information (Chehab et al. 2009). As far as we know, plants do not appear to have evolved the ability to produce or perceive sound, but this statement has to be qualified by noting that outside pseudoscientific, unreplicated experiments (Retallack 1973), reactions of plants to sound do not appear to have been studied, and that plants are known to produce sounds in the acoustic and ultrasonic range as byproducts of physiological processes (Ritman and Milburn 1988; Zweifel and Zeugin 2008). Thus, the main difference between plant and animal communication is that plants lack complex sensory organs and signals that require rapid movement. Most communication between plants is likely to be chemical or possibly electrochemical – unfortunately the most difficult types of communication to observe.

Thus, other than in animals, where many signals such as calls or visual displays are easily observed, the study of plant signals typically requires specialized equipment and complex analytical procedures. Frequently, the existence of signals is only inferred from observations of a plant's response to a neighbor, and the actual signal may never be identified (e.g., Mahall and Callaway 1991, 1996). This of course makes it impossible to determine whether a signal was received when no response is observed. Thus plant communication is much more difficult to study than animal communication, and this likely has been the reason for the long-held, tacit assumption that plants do not communicate.

3 How Can Communication Between Plants Evolve?

Research on plant communication is still in its infancy compared to animal communication, and an evolutionary biology of plant signals is still lacking. The key evolutionary question that must be asked about any hypothesized communication between organisms is: Who benefits from the interaction? Individual selection is the major driving force of evolution, so a signal exchange that does not benefit the signaler would seem to be impossible to evolve (Dawkins and Krebs 1978). However, individuality in plants is a much less clear concept than it is in unitary animals. All plant ancestors were clonal, all plants are modular, and about 40% of all plants today are still clonal (Tiffney and Niklas 1985). Adding to that the observation that many plant species have poor long-distance dispersal abilities, one has to conclude that a sizable proportion of plants, perhaps even the majority, will have some long-term neighbors, which are either genetically identical or closely related. This would suggest that evolutionary pathways of traits involved in plant interactions may differ substantially from those in unitary animals, and that evolution of cooperative signaling that benefits a conspecific neighbor may not be unusual in plants. Moreover, plants tend to live in extraordinary stable groups of neighbors, which create conditions that allow for group selected traits to evolve under certain circumstances (Dudley and File 2007; Goodnight 1985; Tuomi and Vuorisalo 1989; Wilson and Sober 1994; Wilson 1987).

3.1 Evolution of Signaling Through Individual Selection

Signal reliability has been the major focus of biological signaling theory for the last three decades (Searcy and Nowicki 2005), but with the exception of deceptive plant signaling to pollinators, the topic has not received much attention by researchers who study signaling between plants. Yet the subject is of vital importance, because signals that provide false information about the signaler are not evolutionarily stable unless the deception only occurs in a small proportion of instances (Searcy and Nowicki 2005). Thus, receivers will respond only in a fashion that, on the average, benefits the signaler if the signal has a high probability of being reliable (Zahavi and Zahavi 1997). After much initial debate and controversy, the theory that signals have to be costly to the signaler (Zahavi 1975, 1977; Zahavi and Zahavi 1997) has been largely supported by the evidence from a multitude of studies, both modeling and experimental (Bradbury and Vehrencamp 1998; Grafen 1990; Johnstone 1997; Searcy and Nowicki 2005). Signal costs may include direct and indirect costs, such as the metabolic energy to produce a toxin and the costs for the biochemical machinery to prevent autotoxicity, as well as ultimately the fitness costs for producing the signal (Searcy and Nowicki 2005). Costly signals are unlikely to be faked and therefore will tend to be reliable (Zahavi and Zahavi 1997). Some researchers continue to maintain that there is a separate category of signals that are inherently reliable and come at no cost to the signaler (Maynard Smith and Harper 2003). An example would be claw marks made by an animal in the bark of a tree that indicate the true height of the animal. However, in practice it turns out that there are hardly any kinds of signals that are truly impossible to fake – imaging an animal jumping up the tree to make the claw marks – (Searcy and Nowicki 2005), which suggests that the handicap principle (Zahavi and Zahavi 1997) of high signal cost is essentially the only way through which signaling can evolve by individual selection. Unless it is in their own benefit to respond, receivers would not continue to respond to a signal that comes with little cost to the signaler and therefore is easily faked.

It is surprising to note that to date only a single paper on the subject of signaling between plants (Zhang and Jiang 2000) – a modeling study of sibling rivalry among ovules – appears to have invoked the handicap principle. The idea of signaling cost still appears to be foreign to the debate about plant communication. This puts the field at a huge disadvantage, because signaling systems continue to be proposed without reference to whether or not they benefit the signaler and convey reliable information to a receiver and thus could possibly evolve. An example will help to make the point (see Box 1): roots of the desert shrub *Ambrosia dumosa* have been found to cease growth after contact with other roots belonging to conspecifics of the same population (Mahall and Callaway 1991, 1992, 1996). This has been attributed to signals received from the neighbor's roots. It seems intuitively clear in this example that the hypothesized signaler would benefit from the self-curtailing behavior of a potential competitor, but why would the receiver respond in this fashion? A modeling study (see Box 1) of root competition for water between plants with *Ambrosia*-type behavior suggests that plants could benefit from sensing the presence of competing roots and reallocating root growth to parts of the soil that are not occupied by competing roots. However, this would only be true for root competition for relatively immobile resources. Allowing higher rates of soil water conductivity eliminated the advantage of root territoriality (Box 1, Fig. 1f). Moreover, the advantage of intraspecific root territoriality also disappears in the presence of a nonterritorial competitor (Box 1, Fig. 1e), such as desert annuals that normally compete with *Ambrosia dumosa* shrubs (Holzapfel and Mahall 1999). And here lies the problem: root signals that are produced by a signaler regardless of whether soil resources are depleted or available do not provide reliable information to receiver roots and therefore would appear to be unlikely to evolve. The alternative, evolution of such signals by kin or group selection is discussed below.

To take this example further, *Ambrosia dumosa* roots have also been found to cease growth when approaching roots of the much larger desert shrub *Larrea tridentata* (Mahall and Callaway 1991, 1992), with which *A. dumosa* is co-dominant over huge areas of the Sonoran and Mojave Deserts of North America. In this case, kin or group selection cannot be invoked to explain the existence of a signaling system, which suggests that *Ambrosia* roots either respond to a costly and reliable root signal from *Larrea* or that *Larrea* roots exude an unidentified toxin that cannot be ignored (Schenk et al. 1999). *Larrea* roots also cease growth when approaching other *Larrea* roots (Mahall and Callaway 1991, 1992). The modeling study presented in Box 1 found that the self-curtailing root behavior of an *Ambrosia*-like plant in competition with a *Larrea*-like plant could also benefit the "*Ambrosia*" if soil resources were immobile (Box 1, Fig. 1e) and if therefore the presence of the competitor's root reliably indicated local resource depletion. However, in nature, *Larrea* roots are just as unlikely as *Ambrosia* roots to deplete local soil resources continuously to such an extent that the mere presence of a *Larrea* root would reliably indicate resource depletion (Box 1). Interestingly, in the modeling study, *Larrea*-type plants only benefited from self-curtailing root behavior of competitors when these competitors also behaved like *Larrea* roots (Box 1, Fig. 1e). These examples show that benefits and costs for signalers and receivers of root signals are not easily determined, thereby leaving it open to question how they could evolve.

The alternative idea that *Larrea* produces root toxins in sufficient quantities to poison the roots of a coevolved competitor seems exceedingly unlikely. In fact, there are rather few documented cases of toxic root exudates that are exuded in such large quantities that they can affect competing roots before being absorbed by soil particles or broken down by oxidation or by microbes (Cheng 1995; Newman 1978). Yet toxic root exudates undoubtedly exist (Inderjit and Weston 2003).

So why would plants produce root toxins that cannot poison the roots of their neighbors? An answer to this puzzling question is provided by Zahavi's handicap principle (Zahavi and Zahavi 1997): a toxin is a powerful and reliable signal because it comes at a substantial cost to the signaler for production and autotoxicity prevention. If only the most active fine roots produced it then the toxin would be a reliable signal to roots of coevolved competitors of the presence of an active root that belongs to a competitor strong enough to produce such a costly signal. Thus in

Box 1 A Cellular Automaton Model of Root Territoriality

This model (Fig. 1) was developed to explore the potential benefits to plants of root signaling systems associated with root territoriality (Schenk et al. 1999). The spatially-explicit root model is run within a 100 by 100 cell two-dimensional grid, in which each cell represents 1 cm^3 of soil. Simulations are run for 50 time steps of 4 days each (200 days total). All carbon costs are converted to a common currency of water units (375 mg H_2O/mg C) for production (60 mg H_2O cm^{-1}) and maintenance of roots (0.75 mg H_2O cm^{-1} day^{-1}), for associated shoots (2.5 shoot/root ratio), and production of root signals (1.75 mg H_2O cm^{-1} day^{-1}). Each cell in the grid is initialized with 150 mg of H_2O, with no replenishment, as might occur in a desert following a saturating rain. Initially ten plants are placed randomly in the grid, each starting with enough resources to produce four initial root nodes. During each time step, the following actions are applied in random order to each plant in the grid:

1. Pay maintenance costs in water for the total roots system
2. Extract up to 15 mg of water units per day from each cell of soil contacted by the roots
3. Produce a new root growing in a random direction starting at a node, the location where growth stops at the end of the previous time step

Roots may grow into any unoccupied cell of the nine grid cells adjacent to a node, and each new root can grow up to 1 cm per day. Root growth continues in a straight line within a time step until the plant is out of resources, the root encounters a root that it cannot cross, as determined by its territorial behavior (see below), or the root is 4 cm long. Following root growth, all water in the system diffuses to neighboring cells based on an exponential probability density function. The model outputs total root length and water uptake of each plant at each step in the simulation, produces maps of roots and water content of each cell in the grid.

Root behavior is determined by two variables that determine whether a root can cross another root of the same species or of another species. No roots are allowed to cross their own roots. We set combinations of these two variables to establish three species with different territorial behaviors: "Non-territorial" (no inter-or intraspecific root territories), *Ambrosia*-type (intraspecific root territories only), and *Larrea*-type (intra- and interspecific root territories).

In our simulations, we ran a full factorial combination of all six unique pairs of the three species (including monocultures) at each of two water conductivities (ϕ = 1 cm and ϕ = 80 cm) for a total of 12 unique treatments. All treatments were replicated ten times for a total of 120 simulations. Note that in the high conductivity treatment, water is redistributed nearly evenly across the entire grid, as the mean diffusion distance (80 cm) is nearly the maximum grid dimension (100 cm).

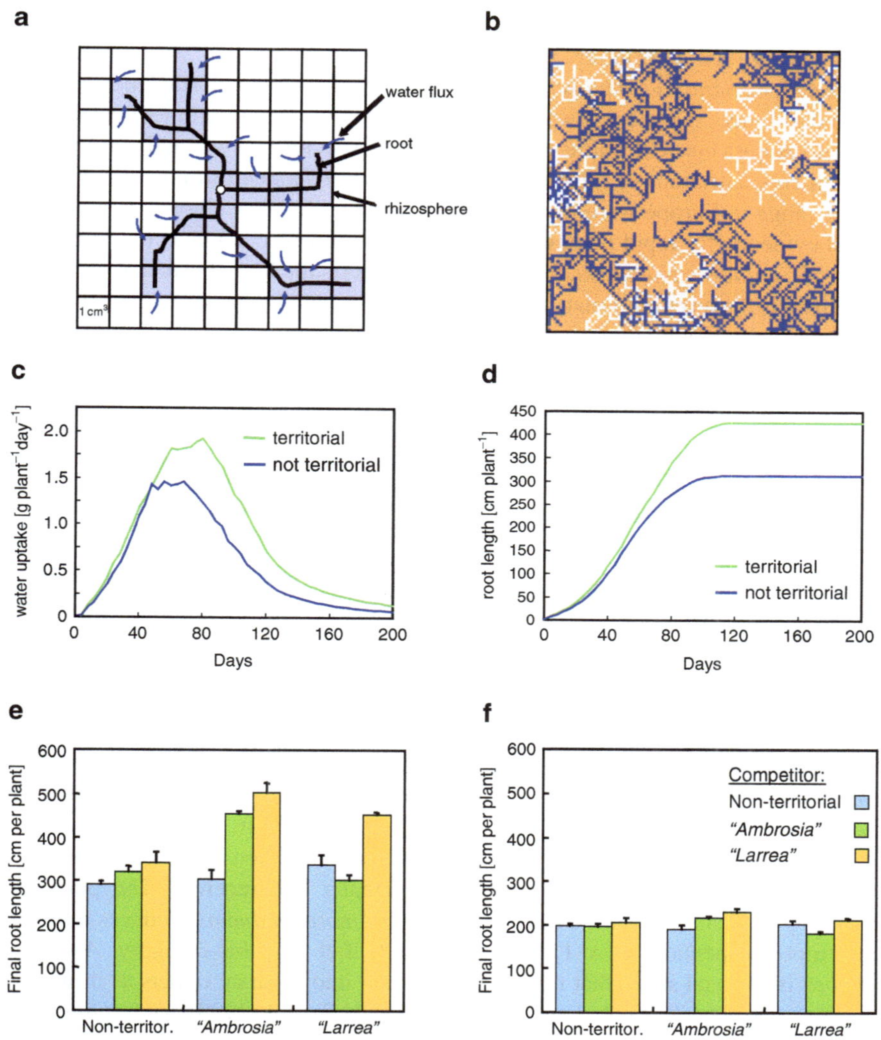

Fig. 1 (**a**) Basic structure of the cellular automaton model. (**b**) A root map at the end of a 200 day simulation. (**c**) Mean water uptake per plant for non-territorial and territorial plants. (**d**) Total root length per plant for non-territorial and territorial plants. Because of the structure of the model, cumulative water uptake is closely correlated with cumulative root length. (**e**) Final root lengths per plant at the end of ten 200 day simulations at low soil water conductivity. (**f**) Final root lengths per plant at the end of ten 200 day simulations at high soil water conductivity

coevolved systems one would not expect to find production of root toxins at levels high enough to actually poison a neighbor's roots. However, toxin-producing plant species outside their native range can encounter new neighbors that do not recognize the signal. In that case, natural selection would either favor elimination of the signal or an increase in its production to a level where it actually does poison neighbors' roots. The latter case is exactly what was found with the spotted knapweed, *Centaurea maculosa*, which is invasive in western North America. In its nonnative range, the species was found to produce the phytotoxin (±)-catechin in the field at high concentrations that inhibit native species' growth and germination, but soil concentrations of the phytotoxin in *Centaurea maculosa* populations in its native range in Europe were much lower (Bais et al. 2003). Callaway et al. (2005) found evidence for rapid natural selection for tolerance of (±)-catechin in competitors of *Centaurea maculosa*, which further supports the hypothesis that poisoning neighbors is not an evolutionary stable strategy. The handicap principle, on the other hand, can explain why toxic substances, including reactive oxygen species (del Río and Puppo 2009) and nitric oxide (Tuteja and Sopory 2008), are common signaling molecules both within and between plants.

3.2 Evolution of Signaling Through Kin or Group Selection

Evolution of signaling between a signaler and a receiver can be explained without recourse to the handicap principle, if the interests of both participants overlap and both benefit from the information exchange. Unfortunately, the history of biological communications research is rife with examples of studies where common interests have been assumed rather than tested (Dawkins and Krebs 1978). Because plants appear to lack social behavior, cases of common interests between individual plants are likely to be restricted to interactions between genetically identical or related plants and potentially to close mutualistic associations between plant species. Because many plants are clonal and/or lack long-distance dispersal mechanisms, they are likely to interact with genetically related neighbors, and this would create conditions in which "true communication" can evolve that benefits both signaler and receiver. The purported signaling mechanism by which the desert shrub *Ambrosia dumosa* reduces intraspecific root competition (Box 1) would appear to fall into this category. *Ambrosia dumosa* is a clonal shrub that normally fragments into separate ramets as it matures (Espino and Schenk 2009; Jones and Lord 1982; Schenk 1999), and competition among these ramets would create costs with no benefits to the genetic individual. Interestingly, *Ambrosia dumosa* ramets segregate their root systems only when they are disconnected from each other and they also segregate root systems from those of other ramets from the same population (Mahall and Callaway 1996). This suggests that root communication that leads to root segregation in this species may have evolved by a combination of individual, kin, and group selection, which may not be uncommon in plants (Goodnight 1985; Tuomi and Vuorisalo 1989).

Volatile "alarm calls" between conspecific plants in response to herbivore attack may offer other examples for kin- or group-selected signaling systems (Baldwin and Schultz 1983; Dolch and Tscharntke 2000; Farmer and Ryan 1990), but in clonal plants these could also evolve by individual selection (Karban et al. 2006; Shiojiri and Karban 2006, 2008). The common interest between communication partners in this case could be the use of induced chemical defenses to deter herbivores from a whole plant neighborhood and thereby reduce the risk of further attack for all plants in that neighborhood. An alternative explanation for "alarm calls" is that they evolve through individual selection and are directed at predators (Zahavi and Zahavi 1997), informing them of defense induction or that they are directed at a predator's predator (Kessler and Baldwin 2001).

The animal communications literature holds many examples for communication between related organisms, some of which may also occur in plants. For example, begging for food from a parent is a common behavior in birds and many other animals with parental care. The plant equivalent for this type of sibling rivalry is signaling associated with competition between ovules for resources from the maternal plant. Interestingly, research in plants has focused mostly on the maternal regulation of ovule abortion in plants (Bañuelos and Obeso 2003; Ganeshaiah and Shaanker 1988; Korbecka et al. 2002; Shaanker et al. 1996), but the animal literature suggests that offspring may be more likely to affect the outcome of sibling rivalry than the mother (Mock and Parker 1998; Searcy and Nowicki 2005). Conflicts between selfish interests of ovules and interests of the mother plant were addressed in a modeling study by Zhang and Jiang (2000) that explicitly included the costs of signals produced by ovules.

Although there are many examples for positive interactions between plants (Callaway 2007), there is little evidence for mutualistic associations between plant species that are so close that signaling may be involved in forming the association. Graft formation between root systems (Graham and Bormann 1966) may fall into this category, as graft formation involves signaling between the graft partners (Pina and Errea 2005; Yeoman 1984). However, the costs and benefits of natural root grafts are poorly understood, and it remains to be seen whether they can be truly mutualistic (Loehle and Jones 1990).

3.3 Evolution of Signaling Through Sexual Selection

Sexual signaling in plants has been thought to be directed exclusively at animal pollinators (Silvertown and Gordon 1989), but a wealth of recent information on pollen competition and pollen-pistil interactions (Aizen and Harder 2007; Cruzan 1993; Erbar 2003; Herrero and Hormaza 1996; Lankinen et al. 2009; Nakamura and Wheeler 1992; Ruane 2009; Snow and Spira 1991) forces a re-evaluation of this view. Sexual selection associated with mate choice involves an abundance and variety of conspicuous signaling systems in animals (Wilson 1975), and there is no *a priori* reason to think that processes that are such powerful selective forces in animals would not be equally powerful in plants. Sexual signaling between males and females involves diverging interests between signaler and receiver, including high fitness benefits to females if they can detect high-quality males and high fitness benefits to low-quality males if they can deceive females into mating with them

(Searcy and Nowicki 2005). Male gametophytes in plants would appear to lack the resources for a plant equivalent to the male peacock's tail. Instead, male competition (Ruane 2009; Snow and Spira 1991) and female choice (Cruzan 1993; Herrero and Hormaza 1996) take place hidden from sight at the stigmatic surface or in the pollen-tube transmitting tissue (Erbar 2003). Signaling between males and females associated with sexual selection in plants has been discussed in great detail by Skogsmyr and Lankinen (2002), and readers are referred to that review.

4 A Conceptual Framework for the Evolutionary Ecology of Plant Signals

In plant literature, the term signaling has mostly been used for plant-internal signals (Baluška and Mancuso 2009b) or for interactions between plants and their environment (Baluška 2009). Consistency in terminology with other scientific literature in biology would exclude from signaling any information gathering from the abiotic or biotic environment that does not benefit a signaler. While acknowledging the separate traditions, we argue that there is much to be gained from adopting consistent terms and concepts across all of biology. Plant biology can benefit from the accumulated knowledge of many decades of research on communications in other organisms by looking for similarities and differences between communication in plants and communication in animals, bacteria, protists, and fungi. Certain categories of interactions among individuals – including territorial defense, mate choice, parent-offspring, and kin interactions – have produced a wealth of signaling systems in other organisms and are likely to have produced signaling in plants as well. Evolution of biological signals is likely to differ greatly between systems where the interests of signalers and receivers overlap, diverge, or oppose (Searcy and Nowicki 2005). Table 1 presents a conceptual framework of plant signals grouped into these three categories and further divided into specific types of interactions.

5 Conclusions

The history of animal communications research provides some useful lessons to researchers engaged in the emerging field of plant communications research. For some of the last three decades, progress in the understanding of animal communications had been hampered by conflicting uses of concepts and terms and by fundamental disagreements about the processes that underlie the evolution of animal signals. Conflicts and disagreements are important parts of the scientific process, but it is even more important for that process to learn both from past mistakes and advances in understanding. There is now an emerging consensus that

Table 1 Different types of biological communication that have been observed to occur or could potentially occur between plants, grouped by the relationship between the interests of signaler and receiver. Interest here refers to potential fitness benefits resulting from the signal exchange. References cited are only meant to cite examples and more citations may be found in the text

Relationship between signaler and receiver	Roles of signaler and receiver	Examples in plants
Interests oppose	Competitors	Territorial root communications (Schenk 2006; Schenk et al. 1999)
	Host and parasite	Signals from potential hosts that warn off parasites?
Interests overlap	Male and female gametes of the same plant	Self/non-self recognition during self-incompatibility (Haring et al. 1990; Rea and Nasrallah 2008)
	Ramets	Self/non-self recognition in roots (Falik et al. 2003; Holzapfel and Alpert 2003); "Alarm calls": volatile signals that induce defenses against herbivores (Karban et al. 2006)
	Kin	"Begging calls": Sibling rivalry between ovules (Bañuelos and Obeso 2003; Ganeshaiah and Shaanker 1988); "Alarm calls": volatile signals that induce defenses against herbivores (Farmer and Ryan 1990)
	Mutualists	Root graft formation? (Loehle and Jones 1990)
Interests diverge	Male and female gametes of different plants	Pollen competition (Ruane 2009; Snow and Spira 1991); "Female choice" of pollen (Cruzan 1993; Herrero and Hormaza 1996)
"Signaler" has no interest in signal exchange	Various	This is not biological communication and signaling, but information gathering from cues. Examples too numerous to list.

signaling costs are vital for the evolution of many, if not most, biological signaling systems. Except for communications between genetically related individuals, only costly and therefore reliable signals are likely to evolve by individual selection, and this is likely to be true also for plants.

The handicap principle that led to the understanding of the importance of signaling costs may also throw new light on the role of phytotoxins in plant interactions. The ecological roles of allelochemical toxins have been puzzling to plant ecologists for a long time, because such toxins rarely occur in concentrations large enough to actually poison a competitor. Reinterpreting toxins as costly, and therefore reliable, signals provides a new explanation for a long-standing mystery in plant ecology.

Finally, we argue that the term and concept of allelopathy are much less useful than the more consistent and integrative term and concept of plant communications. Communication and chemical inhibition are very different concepts, but molecules may commonly serve both as toxins and as signals; therefore, these two concepts cannot be relegated to separate fields of inquiry and instead should all be part of plant interactions research. Moreover, communication and inhibition are universal processes across all of biology, and maintaining separate terminologies for different biological disciplines would only serve to obscure the commonalities. Adopting some concepts and terms from animal research will allow plant behavioral researchers to build on knowledge and understanding gained from the longer and more productive history of animal behavioral ecology and perhaps to avoid some of its pitfalls and mistakes.

References

Aizen MA, Harder LD (2007) Expanding the limits of the pollen-limitation concept: effects of pollen quantity and quality. Ecology 88:271–281
Arak A (1983) Sexual selection by male-male competition in natterjack toad choruses. Nature 306:261–262
Bais HP, Vepachedu R, Gilroy S, Callaway RM, Vivanco JM (2003) Allelopathy and exotic plant invasion: from molecules and genes to species interactions. Science 301:1377–1380
Baldwin IT, Schultz JC (1983) Rapid changes in tree leaf chemistry induced by damage: evidence for communication between plants. Science 221:277–279
Baluška F (ed) (2009) Plant-environment interactions: from sensory plant biology to active plant behavior. Springer, Berlin
Baluška F, Mancuso S (2009a) Plant neurobiology: from stimulus perception to adaptive behavior of plants, via integrated chemical and electrical signaling. Plant Signal Behav 4:475–476
Baluška F, Mancuso S (eds) (2009b) Signaling in plants. Springer, Berlin
Bañuelos MJ, Obeso JR (2003) Maternal provisioning, sibling rivalry and seed mass variability in the dioecious shrub *Rhamnus alpinus*. Evol Ecol 17:19–31
Bateson G (2000) Steps to an ecology of mind. University of Chicago Press, Chicago
Bradbury JW, Vehrencamp SL (1998) Principles of animal communication. Sinauer Associates, Sunderland, MA
Callaway RM (2007) Positive interactions and interdependence in plant communities. Springer, Berlin
Callaway RM, Ridenour WM, Laboski T, Weir T, Vivanco JM (2005) Natural selection for resistance to the allelopathic effects of invasive plants. J Ecol 93:576–583
Chehab EW, Eich E, Braam J (2009) Thigmomorphogenesis: a complex plant response to mechano-stimulation. J Exp Bot 60:43–56
Cheng HH (1995) Characterization of the mechanisms of allelopathy: modeling and experimental approaches. In: Inderjit, Dakshini KMM, Einhellig FA (eds) Allelopathy: organisms, processes, and applications, vol 582. American Chemical Society, Washington, DC, pp 132–141
Cruzan MB (1993) Analysis of pollen-style interactions in *Petunia hybrida*; the determination of variance in male reproductive success. Sex Plant Reprod 6:275–281
Davies E (2004) New functions for electrical signals in plants. New Phytol 161:607–610
Dawkins R, Krebs JR (1978) Animals signals: information or manipulation? In: Krebs JR, Davies NB (eds) Behavioural ecology – an evolutionary approach. Blackwell Scientific Publications, Oxford, pp 282–309
del Río LA, Puppo A (2009) Reactive oxygen species in plant signaling. Springer, Berlin
Dicke M, Sabelis MW (1988) Infochemical terminology: based on cost-benefit analysis rather than origin of compounds? Funct Ecol 2:131–139
Dolch R, Tscharntke T (2000) Defoliation of alders (*Alnus glutinosa*) affects herbivory by leaf beetles on undamaged neighbours. Oecologia 125:504–511
Dudley SA, File AL (2007) Kin recognition in an annual plant. Biol Lett 3:435–438
Dusenbery DB (1992) Sensory ecology. W.H. Freeman and Company, New York
Einhellig FA (1995) Allelopathy: current status and future goals. In: Inderjit, Dakshini KMM, Einhellig FA (eds) Allelopathy: organisms, processes, and applications, vol 582. American Chemical Society, Washington, DC, pp 1–24

Erbar C (2003) Pollen tube transmitting tissue: place of competition of male gametophytes. Int J Plant Sci 164:S265–S277
Espino S, Schenk HJ (2009) Hydraulically integrated or modular? Comparing whole-plant-level hydraulic systems between two desert shrub species with different growth forms. New Phytol 183:142–152
Falik O, Reides P, Gersani M, Novoplansky A (2003) Self/non-self discrimination in roots. J Ecol 91:525–531
Farmer EE, Ryan CA (1990) Interplant communication: airborne methyl jasmonate induces synthesis of proteinase inhibitors in plant leaves. Proc Natl Acad Sci USA 87:7713–7716
Fowler SV, Lawton JH (1985) Rapidly induced defenses and talking trees: the devil's advocate position. Am Nat 126:181–195
Fromm J, Lautner S (2007) Electrical signals and their physiological significance in plants. Plant Cell Environ 30:249–257
Ganeshaiah KN, Shaanker RU (1988) Seed abortion in wind-dispersed pods of *Dalbergia sissoo*: maternal regulation or sibling rivalry. Oecologia 77:135–139
Goodnight CJ (1985) The influence of environmental variation on group and individual selection in a cress. Evolution 39:545–558
Gosling LM, Atkinson NW, Collins SA, Roberts RJ, Walters RL (1996) Avoidance of scent-marked areas depends on the intruder's body size. Behaviour 133:491–502
Gould SJ, Lewontin RC (1979) The spandrels of San Marco and the Panglossian paradigm: a critique of the adaptationist programme. Proc R Soc Lond B 205:581–598
Grafen A (1990) Biological signals as handicaps. J Theor Biol 144:517–546
Graham BF Jr, Bormann FH (1966) Natural root grafts. Bot Rev 32:255–292
Haring V, Gray JE, McClure BA, Anderson MA, Clarke AE (1990) Self-incompatibility: a self-recognition system in plants. Science 250:937–941
Herrero M, Hormaza J (1996) Pistil strategies controlling pollen tube growth. Sex Plant Reprod 9:343–347
Holzapfel C, Alpert P (2003) Root cooperation in a clonal plant: connected strawberries segregate roots. Oecologia 134:72–77
Holzapfel C, Mahall BE (1999) Bidirectional facilitation and interference between shrubs and annuals in the Mojave Desert. Ecology 80:1747–1761
Inderjit, Weston LA (2003) Root exudates: an overview. In: de Kroon H, Visser EJW (eds) Root ecology, vol 168. Springer-Verlag, Berlin, pp 235–255
Johnstone RA (1997) The evolution of animal signals. In: Krebs JR, Davies NB (eds) Behavioural ecology. Blackwell Publishing, Oxford, pp 155–178
Jones CS, Lord EM (1982) The development of split axes in *Ambrosia dumosa* (Gray) Payne (Asteraceae). Bot Gaz 143:446–453
Karban R, Shiojiri K, Huntzinger M, McCall AC (2006) Damage-induced resistance in sagebrush: volatiles are key to intra- and interplant communication. Ecology 87:922–930
Kessler A, Baldwin IT (2001) Defensive function of herbivore-induced plant volatile emissions in nature. Science 291:2141–2144
Korbecka G, Klinkhamer PGL, Vrieling K (2002) Selective embryo abortion hypothesis revisited - A molecular approach. Plant Biol 4:298–310
Krebs JR, Davies NB (eds) (1997) Behavioural ecology: an evolutionary approach, 4th edn. Blackwell Science, London
Lankinen A, Maad J, Armbruster WS (2009) Pollen-tube growth rates in *Collinsia heterophylla* (Plantaginaceae): one-donor crosses reveal heritability but no effect on sporophytic-offspring fitness. Ann Bot 103:941–950
Littlejohn SW, Foss KA (2008) Theories of human communication. Thomson, Belmont, CA
Loehle C, Jones RH (1990) Adaptive significance of root grafting in trees. Funct Ecol 4:268–271
Lund EJ (1947) Bioelectric fields and growth. The University of Texas Press, Austin, TX
Mahall BE, Callaway RM (1991) Root communication among desert shrubs. Proc Natl Acad Sci USA 88:874–876

Mahall BE, Callaway RM (1992) Root communication mechanisms and intracommunity distributions of two Mojave Desert shrubs. Ecology 73:2145–2151
Mahall BE, Callaway RM (1996) Effects of regional origin and genotype on intraspecific root communication in the desert shrub *Ambrosia dumosa* (Asteraceae). Am J Bot 83:93–98
Marler P (1977) The evolution of communication. In: Sebeok TB (ed) How animals communicate. Indiana University Press, Bloomington, IN, pp 45–70
Maynard Smith J, Harper D (2003) Animal signals. Oxford University Press, Oxford
Mock DW, Parker GA (1998) Siblicide, family conflict and the evolutionary limits of selfishness. Anim Behav 56:1–10
Molisch H (1937) Der Einfluss einer Pflanze auf die andere - Allelopathie. Gustav Fischer, Jena, Germany
Nakamura RR, Wheeler NC (1992) Pollen competition and paternal success in Douglas-Fir. Evolution 46:846–851
Newman EI (1978) Allelopathy: adaptation or accident? In: Harborne JB (ed) Biochemical aspects of plant and animal coevolution. Academic Press, London, pp 327–342
Otte D (1974) Effects and functions in the evolution of signaling systems. Ann Rev Ecol Syst 4:385–417
Pina A, Errea P (2005) A review of new advances in mechanism of graft compatibility-incompatibility. Sci Hortic 106:1–11
Rea AC, Nasrallah JB (2008) Self-incompatibility systems: barriers to self-fertilization in flowering plants. Int J Dev Biol 52:627–636
Retallack DL (1973) The sound of music and plants. DeVorss, Santa Monica, CA
Rhoades DF (1985) Pheromonal communication between plants. In: Cooper-Driver GA, Swain T, Conn EE (eds) Chemically mediated interactions between plants and other organisms, vol 19. Plenum Press, New York, pp 195–218
Rice EL (1984) Allelopathy, 2nd edn. Academic Press, Orlando
Ritman KT, Milburn JA (1988) Acoustic emissions from plants: ultrasonic and audible compared. J Exp Bot 39:1237–1248
Ruane LG (2009) Post-pollination processes and non-random mating among compatible mates. Evol Ecol Res 11:1031–1051
Schenk HJ (1999) Clonal splitting in desert shrubs. Plant Ecol 141:41–52
Schenk HJ (2006) Root competition: beyond resource depletion. J Ecol 94:725–739
Schenk HJ, Callaway RM, Mahall BE (1999) Spatial root segregation: Are plants territorial? Adv Ecol Res 28:145–180
Searcy WA, Nowicki S (2005) The evolutionof animal communication: reliability and deception in signaling systems. Princeton University Press, Princeton, NJ
Shaanker RU, Ravishankar KV, Hegde SG, Ganeshaiah KN (1996) Does endosperm reduce intra-fruit competition among developing seeds? Plant Syst Evol 201:263–270
Shannon CE (1948) A mathematical theory of communication. The Bell Syst Tech J 27:379–423
Shiojiri K, Karban R (2006) Plant age, communication, and resistance to herbivores: young sagebrush plants are better emitters and receivers. Oecologia 149:214–220
Shiojiri K, Karban R (2008) Vascular systemic induced resistance for *Artemisia cana* and volatile communication for *Artemisia douglasiana*. Am Midl Nat 159:468–477
Silvertown J, Gordon DM (1989) A framework for plant behavior. Annu Rev Ecol Syst 20:349–366
Skogsmyr I, Lankinen A (2002) Sexual selection: an evolutionary force in plants. Biol Rev 77:537–562
Slater PJB (1983) The study of communication. In: Halliday TR, Slater PJB (eds) communication, vol 2. Blackwell Scientific Publications, Oxford, pp 9–42
Snow AA, Spira TP (1991) Pollen vigour and the potential for sexual selection in plants. Nature 352:796–797

Tiffney BH, Niklas KJ (1985) Clonal growth in land plants: a paleobotanical perspective. In: Jackson JBC, Buss LW, Cook RE (eds) Population biology and evolution of clonal organisms. Yale University Press, New Haven, pp 35–66
Tuomi J, Vuorisalo T (1989) Hierarchical selection in modular organisms. Trends Ecol Evol 4:209–213
Tuteja N, Sopory SK (2008) Chemical signaling under abiotic stress environment in plants. Plant Signal Behav 3:525–536
Wagner WE (1989) Fighting, assessment, and frequency alteration in Blanchard's cricket frog. Behav Ecol Sociobiol 25:429–436
Watzlawick P, Beavin JH, Jackson DD (1967) Pragmatics of human communication: a study of interactional patterns, pathologies, and paradoxes. W.W.Norton, New York
Wiener N (1948) Cybernetics or control and communication in the animal and the machine. Wiley, New York
Wiley RH (1994) Errors, exaggeration, and deception in animal communication. In: Real L (ed) Behavioral mechanisms in evolutionary ecology. University of Chicago Press, Chicago, pp 157–189
Wilson EO (1975) Sociobiology: the new synthesis. The Belknap Press of Harvard University Press, Cambridge, MA
Wilson JB (1987) Group selection in plant populations. Theor Appl Genet 74:493–502
Wilson DS, Sober E (1994) Group selection: the theory replaces the bogey man. Behav Brain Sci 17:639–654
Yeoman MM (1984) Cellular recognition systems in grafting. In: Linskens HF, Heslop-Harrison J (eds) Cellular interactions, vol 17. Springer-Verlag, Berlin, pp 453–472
Zahavi A (1975) Mate selection - a selection for a handicap. J Theor Biol 53:205–214
Zahavi A (1977) The cost of honesty (further remarks on the handicap principle). J Theor Biol 67:603–605
Zahavi A, Zahavi A (1997) The handicap principle: a missing piece of Darwin's puzzle. Oxford University Press, Oxford
Zhang DY, Jiang XH (2000) Costly solicitation, timing of offspring conflict, and resource allocation in plants. Ann Bot 86:123–131
Zweifel R, Zeugin F (2008) Ultrasonic acoustic emissions in drought-stressed trees - more than signals from cavitation? New Phytol 179:1070–1079

The Chemistry of Plant Signalling

Michael A. Birkett

Abstract This chapter highlights the contribution that chemical sciences, i.e. analytical and synthetic organic chemistry, has made to the understanding of plant–insect interactions from an ecological perspective. This includes a general overview of the approaches and techniques used in the isolation of natural products that play a role in mediating such interactions and recent examples of the important role that chemical techniques have played. It covers plant-derived signals that are both constitutively produced and those induced in response to defence signalling stimuli, including insect attack. It also includes insect-derived elicitors of plant defence. Finally, future prospects of the role of chemical sciences in plant–insect interaction studies are discussed.

1 Introduction

The study of plant–insect interactions comprises a vast range of disciplines, including behavioural and chemical ecology, organic chemistry, neurophysiology, biochemistry, molecular biology and field behaviour. Whilst any or all of these could rightly claim to be of the utmost importance from a scientific perspective, it is the chemical sciences, i.e. analytical and synthetic organic chemistry, which hold most weight from a fundamental and applied perspective, by providing new tools for studying plant responses at the chemical level in genomically sequenced plants and for the manipulation of organisms that have a negative impact on the performance of arable crops and other ecosystems. Nevertheless, chemists working in the field of plant–insect interactions are fully aware that their role must fit in seamlessly with those around them to enhance the prospects of elucidating new pathways or

M.A. Birkett
Biological Chemistry Department, Rothamsted Research, Harpenden, Herts AL5 2JQ, UK
e-mail: mike.birkett@bbsrc.ac.uk

F. Baluška and V. Ninkovic (eds.), *Plant Communication from an Ecological Perspective*, Signaling and Communication in Plants,
DOI 10.1007/978-3-642-12162-3_2, © Springer-Verlag Berlin Heidelberg 2010

products and exploit new developments from areas of chemical sciences to this scientific area.

This chapter is not designed to be a comprehensive review of every published paper where the chemistry of plant–insect interactions is mentioned. It is intended to provide the reader with examples of how chemical techniques that have been, or are currently being, applied to plant–insect interaction studies. There is also a focus on recent examples of how the areas of analytical and synthetic organic chemistry have played a crucial role in elucidating plant–insect interactions. Finally, the chapter describes how the chemical sciences will play a role in future plant–insect interaction studies, with an emphasis on new and emerging chemical techniques.

2 Approaches to the Isolation and Identification of Plant and Insect-Derived Signals

2.1 *Collection of Biological Samples for Analysis*

The collection of biological material from plants, and the approaches to be used in that process, depends on the chemical nature of the component or components to be studied. Typically, plant–insect interactions are mediated by small lipophilic molecules (SLMs) that are either emitted as volatile organic compounds (VOCs), present within plant tissue, or deposited on the plant surface. Interactions at a distance are mediated by olfactory perception of VOCs, and this phenomenon has been exploited through the development of electrophysiological recordings from insect antennae (Pickett et al. 2009) for the identification of host attractants (kairomones). At close distances, or once contact has been made, interactions are influenced by the detection of toxic/antifeedant plant compounds. There are many examples in the literature referring to the identification of toxic and antifeedant plant natural products, but these are beyond the scope of this chapter and the reader can refer to alternative literature (e.g. Gordon-Weeks and Pickett 2009). Instead, this chapter is generally restricted to the SLMs that are generated upon insect herbivory. In many cases, but not all, such molecules are generated via oxidative metabolism of the polyunsaturated fatty acid (PUFA) and isoprenoid pathways.

2.1.1 Dynamic Headspace Collection

The chemical composition and intensity of plant VOCs carry much information on plant status (D'Alessandro and Turlings 2006), and indeed on the identity of the insect involved when attacked (e.g. Du et al. 1998; Dicke 1999). A vast amount of knowledge has been generated on the range of VOCs emitted by plants, with over 1,000 VOCs having been identified at present. These belong to several different class of compound (isoprenoids, fatty acid derived, amino-acid derived, aromatic compounds, and compounds arising through oxidative stress). The most common technique for VOC collection used is that of dynamic headspace collection, otherwise known as air entrainment. This technique provides the ability to capture VOCs from plants enclosed in purified air chambers, using porous polymeric materials such as Porapak Q, Super Q, TENAX TA and activated charcoal (see D'Alessandro and Turlings 2006 and references therein; for a specific example, see Agelopoulos et al. 1999). VOCs are then desorped either thermally or by elution using a high purity solvent. Thermal desorption, when performed in the inlet of a GC injector port, provides the advantage of whole sample analysis, thus increasing the prospect of compound detection through enhanced sensitivity. However, such samples can only be considered as "one–offs", whereas liquid desorption provides the capability of using the same sample to link biological and chemical studies – a crucial step in defining the role of a natural product as a semiochemical. This was exemplified in the discovery of *cis*-jasmone as an insect semiochemical and as a plant activator, where coupled GC-electrophysiology (GC-EAG) was used to identify this compound within a blend of blackcurrant, *Ribes nigrum*, VOCs (Birkett et al. 2000), and in the identification of the VOC blend emitted by faba beans, *Vicia faba,* used by black bean aphids, *Aphis fabae*, in host location (Webster et al. 2008). A note of caution to the reader is that samples collected using air entrainment are effectively "snapshots" of the VOCs being emitted, i.e. they are "averaged blends". Therefore, although information on VOC production can be generated, information on temporal dynamics of VOC emission is incomplete. Furthermore, in many cases, VOCs are collected separately from a behavioural assay, and so it is difficult to directly link behavioural analysis and VOC production. Turlings et al (2004) developed a six-arm olfactometer which allows simultaneous behavioural testing and collection of plant VOCs.

2.1.2 Solid Phase Microextraction (SPME)

Solid phase micro extraction (SPME) includes the use of a small fibre coated with an adsorbent material, typically polydimethylsiloxane (PDMS). This technique has been used in studying plant–insect interactions, but suffers from the fact that samples are again "lost" through thermal desorption, and cannot be linked to biological studies. Furthermore, SPME appears to suffer from being selective in its ability to trap a range of VOCs (Agelopoulos and Pickett 1998). This appears to sideline its use in studies where VOC blends are known to play a crucial role in the plant–insect interactions. SPME has been used for studying belowground interactions, e.g. measurements of uptake of allelochemicals (Loi et al. 2008). Here, SPME was used to measure uptake of exogenously applied 1,8-cineole by tomato plants, by insertion of a SPME fibre into the stem of test plants at a height of 6 cm above the soil, with the fibre being preconditioned in phosphate-buffered saline (PBS) for 30 min. After 1 h, the fibre was removed and then subjected to GC-MS analysis. The authors claim that this technique provides a means of tracking compounds within target plants. SPME was also used in the first identification of an insect-induced belowground plant signal, (*E*)-caryophyllene (Rasmann et al. 2005). Here, roots damaged by *Diabrotrica virgifera* were frozen in liquid nitrogen, ground to a powder, and VOCs collected by SPME. The VOCs were analysed by thermal desorption directly into a GC-MS instrument. Farag et al. (2006) used SPME in conjunction with GC-MS to profile rhizobacterial volatiles that induce systemic resistance and growth in *Arabidopsis thaliana* (Ryu et al. 2003, 2004; Ping and Boland 2004). Here, bacteria grown on medium were sealed in glass vials, kept at 50°C, and were sampled for 30 min prior to thermal desorption and GC-MS analysis.

2.1.3 Vacuum Distillation

Vacuum distillation involves the distillation and trapping of volatile-laden air or plant/insect extracts in liquid nitrogen-cooled traps under high vacuum (Griffiths and Pickett 1980). Although this technique has been used in the identification of insect pheromones (Al Abassi et al. 1998; Griffiths and Pickett 1980), it has not been used extensively in plant–insect interaction studies.

2.1.4 Liquid–Liquid Extraction

Biological samples can be prepared for analysis by partitioning between aqueous and organic phases which are less polar and immiscible. There are no fixed rules for choosing the solvent system to be used in the partitioning process, but guidelines have been published elsewhere (e.g. Millar and Haynes 1998). This technique forms the basis of extracting compounds from plant tissue, whether they are volatile or involatile. For plant–insect interaction studies, it has been used for extraction and analysis of oxylipins in plants (e.g. Schulze et al. 2006). Recently, the technique has also been used to study other plant signalling mechanisms, e.g. to evaluate the impact of the naturally-occurring plant activator, *cis*-jasmone, on the secondary metabolism of *Triticum aestivum* (Moraes et al. 2008). Here, the use of liquid phase extraction allowed measurement of levels of benzoxazinoids and phenolic acids, which are known to have allelopathic effects on competitive weeds, pests and diseases. Significantly higher levels of DIMBOA and phenolic acids were found in aerial and root parts of *cis*-jasmone treated plants. These results showed for the first time that *cis*-jasmone induces production of secondary metabolites capable of direct control over pests, diseases and weeds.

2.1.5 Solid Phase Extraction (SPE)

Solid-phase extraction (SPE) is a convenient process that involves the concentration of analytes from dilute samples, and is particularly useful when targeting a specific group of natural products. Phases for "normal" SPE include silica, alumina and florisil, which are used to retain unwanted polar compounds, whereas phases for reverse SPE (e.g. C_{18}) are used to trap wanted lipophilic compounds. SPE has also been used as part of a strategy for purifying oxylipins from plant tissue. Here, aminopropyl cartridges were used to remove interference analytes that interfere with the derivatisation process (Schulze et al. 2006). This purification procedure was used in studies that showed conversion of the oxylipin 12-oxophytodienoic acid (OPDA) to the isomeric *iso*-OPDA (Dabrowska and Boland 2007).

2.1.6 Stir Bar Sorptive Extraction (SBSE)

Stir bar sorptive extraction (SBSE) was developed as a rapid technique for extracting organic chemicals from very dilute aqueous media (Soini 2005). A wide range of volatile and semi-volatile substances (from aqueous and gaseous media) can be retained on a PDMS-coated magnetic bar (TwisterTM). SBSE-based extractions have been described in a number of applications. For plant studies, stir bars coated with PDMS were used as probes to assess the production of sorgoleone in the rhizosphere of sorghum-sudangrass plants (Weidenhamer 2005). Compounds were eluted from stir bars by solvent desorption using acetonitrile, followed by HPLC. SBSE has also been used to study the release of defence VOCs by cabbage plants upon herbivory by caterpillars, and the attraction of *Cotesia* spp. parasitoids. SBSE was used to collect solvent extracts of damaged plants that could be used for both bioassays and chemical analysis (Scascighini et al. 2005).

2.1.7 Other PDMS Materials

Other types of PDMS materials have been used to study plant signalling processes. PDMS-coated optical fibres and PDMS tubing have been used in addition to PDMS-coated stir bars to study the dynamics of allelochemical production in the rhizosphere, specifically the production of sorgoleone over time (Weidenhamer 2007).

2.1.8 Vapour Phase Extraction (VPE)

Vapour phase extraction (VPE) was first reported as a new method for the easy, sensitive and reproducible quantification of both jasmonic and salicylic acid in plant defence responses (Engelberth et al. 2003). The method is based on a one-step extraction, phase partitioning, methylation with HCl/methanol, and collection of methylated, and thus, volatilised compounds on Super Q filters, thereby omitting further purification steps. Eluted samples are analysed and quantified by GC-MS using chemical ionisation (GC-CI-MS). Using authentic samples of jasmonic and salicylic acid, recovery rates were estimated between 90–100% and 70–90% respectively. The limits of detection were about 500 femtograms (fg) by using GC-MS in SIM mode. This technique is described as being highly efficient, allowing for reliable quantification of small levels of compounds from small amounts of plant material (5–400 mg). The technique was slightly modified, using trimethylsilyldiazomethane instead of HCl/methanol, and applied to the simultaneous analysis of phytohormones, phytotoxins and VOCs in *A. thaliana* following *Pseudomonas syringae* infection, *Zea mays* herbivory by *Helicoverpa zea*, *Nicotiana tabacum* after mechanical damage and *Lycopersicon esculentum* during drought stress in plants (Schmelz et al. 2003). The numerous complex changes led the authors to propose that this technique can facilitate simple quantification of plant signalling cross talk that occurs at the level of synthesis and accumulation. The same authors extend the use of VPE to include unsaturated fatty acids and OPDA (Schmelz et al. 2004), and phytohormone mapping of insect–herbivore produced elicitors (Schmelz et al. 2009). The technique has also been used to evaluate the impact of *cis*-jasmone on the secondary metabolism of wheat, in conjunction with liquid–liquid extraction as described above (Moraes et al. 2008). Here, the use of VPE allowed measurement of levels of benzoxazinoids and phenolic acids, with levels of HBOA in aerial parts and roots being higher in *cis*-jasmone treated plants.

2.1.9 In-Situ Derivatisation

Comprehensive details of derivatising agents are published elsewhere (Millar and Haynes 1998), but examples relating to plant–insect interactions are mentioned briefly here. The collection of samples for analysis can be enhanced through the use of derivatising agents designed for specific functional groups. Examples include the use of 2,4-dinitrophenylhydrazine (DNPH)-coated filters to facilitate collection of short-chain unstable aldehydes ((*Z*)-3-hexenal, (*E*)-2-hexenal) (D'Alessandro and Turlings 2006), and the use of pentafluorobenzyl hydroxylamine (PFBHA) to facilitate collection of labile oxylipin compounds produced in plant tissues (Schulze et al. 2006). Fatty acid analysis is facilitated by the formation of fatty acid methyl esters (FAMes) via the use of reagents such as HCl/methanol or diazomethane (e.g. Engelberth et al. 2003; Schulze et al. 2006), whereas lipid analysis can be achieved through transesterification using sodium methoxide. Involatile secondary metabolites such as glycosides can be permethylated using sodium hydride and methyl iodide to aid analysis by mass spectrometry, whereas benzoxazinoids and phenolic acids and higher oxidised oxylipins can be converted to trimethylsilyl ethers using reagents such as MSTFA, and thus become suitable for GC-MS analysis (Moraes et al. 2008; Schulze et al. 2006).

3 Recent Advances

The aim of this section is to explain briefly how mass spectrometry (MS), nuclear magnetic resonance (NMR) spectroscopy and synthetic organic chemistry can be applied to plant–insect interactions studies. It is not the intention of this section to provide details of how each of these approaches operates, and the reader is advised to consult alternative literature which will explain these techniques in a clear and precise manner. Nevertheless, the importance of each of these aspects cannot be overestimated. Examples of how these approaches have been used to facilitate important recent breakthroughs in plant–insect interaction studies are provided.

As stated above, in many cases, plant–insect interactions are characterised by increased oxidative metabolism which generates small molecular weight lipophilic compounds, for example, oxidation products from the unsaturated fatty acid and isoprenoid pathways. These compounds are often produced in vanishingly small amounts (sub-nanogram) in complex mixtures, and therefore require the use of highly sensitive analytical equipment. Thus, mass spectrometers (magnetic sector, ion trap, quadruple, time-of-flight) are the natural and logical choice for identifications. They are able to generate stable and reproducible physical data at the nanogram level, and can also be hyphenated to chromatography systems, i.e. GC-MS and HPLC-MS. Despite the challenge of working at low levels of material, identifications of plant and insect-derived semiochemicals are facilitated by the use of biological detectors which are able to operate at levels of material much lower than those used by analytical systems, e.g. coupled GC-electrophysiology (GC-EAG), which exploits the olfactory sensilla located on insect antennae (Pickett et al. 2009). However, the pace of development of modern mass spectrometers, where instruments are increasingly sensitive and accurate, are able to detect broad spectra of molecules with diverse chemical and physical properties, and are generally easier to operate and handle, is now such that identifications should, in theory become easier, assuming that the underlying ecological aspects are fully understood, and consequent semiochemical collection and detection is straightforward. Such instruments are now being employed heavily in modern metabolomic and metabolite profiling strategies.

Mass spectrometry is the predominant technique for structure elucidation, due to the higher degree of sensitivity that such instruments possess. However, developments in NMR instrumentation in recent years are now enabling its application in similar studies, with increased ability to generate NMR data on small amounts of material in a short space of time. GC-MS is the approach used almost universally to study VOC-mediated plant–insect interactions, but GC-Fourier Transformed Infrared Spectroscopy (GC-FTIR) has also been used (see later). Involatile plant compounds, and more recently, insect-derived elicitors, can be characterised using soft ionisation, i.e. electrospray or atmospheric pressure chemical ionisation MS coupled to HPLC, e.g. benzoxazinoids and flavonoids (Bonnington et al. 2003; Cuyckens and Claeys 2004; March et al. 2006). Both GC and HPLC have the potential to be coupled to NMR, but no such examples have been presented in the literature at this point.

Synthetic organic chemistry is one of the key tenets of natural products chemistry, and has been used to confirm the structure of plant natural products in a vast number of studies conducted since the last century. In the context of plant signalling, chemical synthesis continues to play a vital role through the provision of authentic samples for structure confirmation, and intermediate and large-scale synthesis of materials for field testing. Synthesis has also been applied to newly-identified elicitors of plant defence that originate from insects.

3.1 Plant Derived Chemical Signals

3.1.1 *cis*-Jasmone

The isolation and identification of *cis*-jasmone as a plant-derived insect semiochemical and activator of plant defence is a classic example of where mass spectrometry has a key role in the identification of plant-derived signals. In many cases, such molecules are produced and emitted in vanishingly small amounts in highly complex blend, and present a real challenge to the chemist. However, in Birkett et al. (2000), the use of ultra-sensitive magnetic sector instrumentation, closely allied to GC-EAG using recordings from the antennae of aphids enabled the identification of the minor component *cis*-jasmone with high EAG activity. In this case, as with all tentative identifications made by MS, the identification was confirmed by GC peak enhancement using an authentic sample obtained from a commercial supplier (Pickett 1990). Since the seminal publication, further chemical studies have started to provide an understanding of the mechanisms by which *cis*-jasmone activates indirect and direct plant defence. Induction of defence VOC production has been demonstrated for *A. thaliana* (Bruce et al. 2008), soybean, *Glycine max* (Moraes et al. 2009) and for cotton, *Gossypium hirsutum* (Birkett, unpublished data). In each of these cases, production of the plant stress semiochemical (*E,E*)-4,8,12-trimethyltrideca-1,3,7,11-tetraene (TMTT) has been reported. Activation of direct defence pathways has also been demonstrated for wheat, *T. aestivum*, with *cis*-jasmone treatment leading to enhanced levels of benzoxazinoids and phenolic acids (Moraes et al. 2008). The levels of these compounds were investigated using a combination of liquid–liquid extraction and VPE. The latter has also been applied to study defence induction in faba beans, *Vicia faba*, following *cis*-jasmone treatment, with the data suggesting enhanced levels of defence compounds (Moraes and Birkett, unpublished data).

3.1.2 Oxylipins and Phytohormones

Plant compounds derived from the family of unsaturated C_{18} fatty acids play an important role in plant–insect interactions. Jasmonic acid (JA) and other members of the jasmonate family, along with its early precursor OPDA, and other fatty acid-derived compounds, all appear to play a role in plant defence. Following an oxidative burst associated with plant stress response, fatty acid hydroperoxides are generated, which are then further processed into oxylipins. Many of these compounds, however, are unstable, as a consequence of the presence of unsaturated ketones and aldehyde moieties. Therefore, an accurate assessment of their production upon herbivory is difficult to generate. Several different methods for derivatising oxylipins for their extraction from plant tissue have been devised, mostly based on methyl ester formation for GC analysis (see for example, Mueller et al. 2006). However, in most cases, the extraction process is selective and fails to prevent unstable oxylipin degradation, which in the case of compounds containing an α,β-unsaturated ketone or aldehyde moiety, arises through conjugation with active agents such as glutathione during the extraction process. A number of studies have attempted to overcome these problems. VPE was developed for the simultaneous monitoring of phytohormones JA, SA, abscisic acid, VOCs and low oxidised oxylipins, i.e. OPDA, from plant tissue (Schmelz et al. 2003, 2004, 2009; Engelberth et al. 2003), but this approach is viewed in some quarters as not appropriate for analysis of higher oxidised oxylipins. For these compounds, Schulze et al (2006) developed a new approach for in situ trapping and extraction based on the immediate conversion of oxo-derived compounds into stable O-2,3,4,5,6-pentafluorobenzyloximes (PFB oximes), thereby preventing any oxylipin degradation and isomerisation. Detection and identification of the derivatised compounds was best achieved using negative ion GC-CI-MS due to the characteristic MS fragments of the different oxylipins. Thus, accurate profiles of fatty acid and oxylipin levels could be generated and investigated following insect herbivory. This approach for oxylipin analysis was used to demonstrate that OPDA undergoes rapid isomerisation to *iso*-OPDA following exposure to insect gut enzymes (Schulze et al. 2007), and in studies which showed that *iso*-OPDA is a natural precursor for *cis*-jasmone (Dabrowska and Boland 2007).

Although the pathway of jasmonic acid biosynthesis was established in the 1980s, studies on oxylipin pathways have been hindered in certain areas due to synthesis of small amounts of material and at high cost. Nevertheless, synthesis for the provision of commercially unavailable compounds has been reported. Tetrahydrodicraneone B (*iso*-OPDA) was synthesised by Lauchli and Boland (2003). 11-Oxoundec-9-enoic acid was synthesised as described in Schulze et al (2006). 13-Oxotrideca-9,11-dienoic acid was obtained as described by Adolph et al. (2003). 13-HOTE and 13-KOTE were obtained from 13-hydroperoxyoctadeca-9,12,15-trienoic acid by Koch et al (2002). A mixture of 9-hydroxy-10-oxo-stearic acid and 10-hydroxy-9-oxostearic acid can be obtained by oxidising *threo*-9, 10-dihydroxystearic acid with Bobbits reagent (Schulze et al. 2006). *cis*-OPDA, 13-hydroxy-12-oxooctadeca-9,15-dienoic acid and 9-hydroxy-12-oxooctadeca-10,15-dienoic acid have been synthesised using a modified Zimmermann–Feng approach (Schulze et al. 2007). A mixture of *cis*-and *trans*-OPDA isomers can be obtained by treating *cis*-OPDA with DBN (1:1 molar ratio of OPDA and DBN for 2 h at room temperature (Schulze et al. 2007). Recently, the production of optically pure enantiomers of octadecanoids in high amounts in a cost- and time-efficient manner has been described, with the key step being the expression and purification of allene oxide synthase (AOS) and allene oxide cyclase (AOC) enzymes, and their coupling to solid matrices (Zerbe et al. 2007).

Recently a new class of unique oxylipins has been reported from *A. thaliana*. These compounds, termed Arabidopsides, are monogalactosyl diacyl glycerides containing OPDA and/or dinor-OPDA. Arabidopsides A, B, C, D, E and F have all been isolated from the aerial parts of *A. thaliana* and characterised (Hisamatsu et al. 2003, 2005; Andersson et al. 2006; Nakajyo et al. 2006). Later studies have shown that induction of defence in *A. thaliana* leads to the production of Arabidopside E

(Andersson et al. 2006; Kourtchenko et al. 2007). Although the authors conclude that these compounds are specifically generated in *A. thaliana* following pathogen infection, it is possible that they may play a role in plant–insect interactions.

3.2 Insect-Derived Chemical Signals

3.2.1 Bruchins

Bruchins are long-chain α,ω-diols, esterified at one or both oxygens with 3-hydroxypropionic acid. They were identified from both cowpea weevils, *Callosobruchus maculates*, and pea weevils, *Bruchus pisorum*, with the authors reporting these compounds to be the first natural products to induce neoplasm formation applied to intact plants (Doss et al. 2000). Extraction and isolation of bruchins was accomplished through bioassay-guided normal and reverse-phase low pressure liquid chromatography. Final separation from inactive fatty acids was achieved through reaction with 2-bromoacetophenone. Alternatively, HPLC was also used instead of low pressure liquid chromatography. Prior to analysis, further microchemistry was applied, with compounds being hydrolysed and converted to trimethylsilylethers using BSTFA, and subjected to ozonolysis. Synthesis of bruchins was accomplished by standard routes involving acetylene alkylations and semihydrogenations and/or Wittig condensations. The (3-hydroxypropyl) esters were initially prepared by oxidative desilylation of 3-(phenyldimethylsilyl) propionates as described for bruchin A ((*Z*)-9-docosene-1,22-diol, 1-(3-hydroxypropanoate) ester) (Oliver et al. 2000). Initially, 9-decyn-1-ol was deprotonated with butyllithium in THF and alkylated with 12-bromododecanol THP ether. The product was hydrogenated using Lindlar catalyst and the olefinic alcohol esterified with the acid chloride obtained by treating 3-(phenyldimethylsilyl) propanoic acid with oxalyl chloride. Removal of the THP group and treatment of the resulting monoester with fluoroboric acid etherate in dichloromethane, followed by flash chromatography yielded the mono 3-(fluorodimethylsilyl)propanoate. Stirring in methanol–THF containing sodium bicarbonate, potassium fluoride and hydrogen peroxide, followed by flash chromatography, yielded the desired Bruchin A.

3.2.2 Volicitin and Related Compounds

The oral secretion of beet armyworm caterpillars (BAW), *Spodoptera exigua*, when applied to damaged tissues of maize, induces the production of VOCs that attract the natural enemies of the caterpillars. Alborn et al. (1997, 2000) and Turlings et al. (2000) reported the identification of the key elicitor present in BAW oral secretions as *N*-[17-hydroxylinolenoyl]-L glutamine (volicitin). Analysis of the oral secretion showed that it also contained *N*-[17-hydroxyolinoleoyl]-L-glutamine, free

17-hydroxylinolenic and 17-hydroxylinoleic acids, the glutamine conjugates of linolenic and linoleic acid as well as free linolenic and linoleic acid. Isolation of the active components included initial centrifugation, filtration of the supernatant and precipitation of proteinaceous material by treatment with citric acid, followed by SPE and further fractionation using reverse-phase HPLC. At each stage, extracts and fractions were tested for biological activity by addition to *Z. mays* plants in water and monitoring VOC production and parasitoid wind tunnel bioassays. Final purification was achieved using further SPE. Characterisation of volicitin was achieved through fast atom bombardment mass spectroscopy (FABMS) and FABMSMS, giving information on the molecular weight, and revealing the possible presence of a glutamine unit. Acid methanolysis followed by GC-CI-MS confirmed the presence of glutamine. GC-EI-MS suggested a straight-chain unsaturated hydrocarbon, consistent with a methyl ester of an 18-carbon hydroxy acid.

Microhydrogenation of the methyl ester over PdO/H_2, followed by GC-MS indicated that more than 1 double bond was present in the side chain. GC-FTIR confirmed the presence of a hydroxyl group, indicated non-conjugation in the unsaturated side chain, and no presence of *trans* double bonds. The methyl ester of the hydroxy C_{18} acids was subjected to further microdegradative analysis to determine the positions of the double bonds and the hydroxyl group. Partial reduction resulted in both cases in a mixture of monoand diunsaturated products as established by GC-MS analysis. The mixtures were then ozonised, with GC-CI-MS analysis showing the presence of three diagnostic GC peaks, which was similar to that for methyl linolenate. EI mass spectra of a pyrrolidide derivative of the reduced products confirmed the C-17 location of the hydroxyl group. Alborn et al. (2000) synthesised racemic 17-hydroxylinolenic acid starting from the ethoxyethyl ester of 3,6-heptadiyn-1-ol, followed by coupling with the *p*-nitrobenzyl ester of L-glutamine using a method developed for peptide synthesis. Since the initial identification of volicitin, synthesis has enabled the elucidation of the absolute stereochemistry of volicitin (Sawada et al. 2006; Pohnert et al. 1999b).

Following the initial reports of volicitin as an insect-derived elicitor from *S. exigua* (Alborn et al. 1997), further fatty acid – amino acid conjugates were identified from the oral secretions of other freshly harvested Lepidopteran species by Pohnert et al. (1999a) using an APCI LC-MS method to analyse oral secretions. The compounds present in regurgitates were identified as a structurally diverse group of conjugates of glutamine and glutamic acid linked *via* an amide bond to saturated and unsaturated C_{14}, C_{16} and C_{18} fatty acids, with proportions being species specific. Dihydroxy and epoxy fatty acid – glutamine conjugates were later isolated from the regurgitant of *S. exigua* and *S. frugiperda*, using LC-MS, in conjunction with methanolysis and derivatisation with MSTFA to determine the positions of the hydroxy groups by GC–MS (Spiteller and Boland 2003). The synthesis of volicitin and analogues has since been published in a number of studies (see for example Hansen and Stenstrom 2000; Itoh et al. 2002; Wei et al. 2003; Krishnamachari et al. 2007), which highlights its suitability as a natural product target for synthesis chemists.

3.2.3 Caeliferins

Caeliferins were isolated from the regurgitant of the grasshopper, *Schistocerca americana* (Alborn et al. 2007). These novel natural products comprise saturated and unsaturated sulphated α-hydroxy fatty acids in which the ω carbon is functionalised with either a sulphated hydroxyl group or a carboxyl conjugated with a glycine unit via an amide bond. The predominant compound possessed a 16-carbon chain and appeared to have most biological activity. Isolation and identification were achieved in an analogous manner to that of volicitin, but with negative and positive ion electrospray LC/MS and LC-MSMS used as the soft ionisation technique rather than FABMS to provide initial information of the molecular mass. GC-CI-MS analyses of methanolysed material suggested methyl esters, and GC-EI-MS indicated two alcohols and a carboxylic acid methyl ester. On the basis of MS data and NMR analyses of intact molecules, compounds were shown to be 2, 16-dihydroxy C_{16} fatty acids with the addition of two unknown 80-amu groups. The only difference between the two compounds was the presence of a double bond, explaining its weak UV absorption. To test the one double-bond hypothesis, the methyl ester was subjected to hydrogenation, which, as expected, gave a GC/MS peak identical to that of the other methyl ester. The presence of two alcohols was confirmed by acetylation that resulted in the expected increase in molecular weight. Finally, the presence of only one carboxylic acid was confirmed by ethanolysis and GC-CI-MS that, for both compounds, gave an ethyl ester with $M + 1$ ions 14 amu higher than for the corresponding methyl esters. GC-CI-MS analyses of ozonised acetylated methyl ester confirmed the double bond was located between carbon 6 and 7 in the 16-carbon chain. GC-FTIR confirmed two non-identical alcohols and the presence of a *trans* double bond. NMR analysis of the original intact material did not indicate the presence of any other organic structure than a di-O(H) substituted C_{16} fatty acid. The consecutive loss of 80 amu in LC/MS analyses was explained by the loss of sulphate esters, which also explained the high water solubility of the natural products. Thus, two compounds were identified as 2, 16-disulfooxy-(*E*)-6-hexadecenoic acid and 2, 16-disulfooxyhexadecanoic acid, which were named caeliferin A16:1and caeliferin A16:0, respectively. Both proposed (racemic forms of) dihydroxy acids were synthesised and transformed to disulfate esters.

4 Structure–Activity Relationships

A classical approach used by chemists working on biologically active natural products is to generate analogous structures and investigate their biological activity relative to the parent compound. This approach has been used traditionally by the pharmaceutical and agrochemical industry in the development of new active agents with activity and stability greater than the lead compound, e.g. the development of the synthetic pyrethroids (Elliott et al. 1973). Such a strategy is difficult to envisage in plant–insect ecological interactions, as the systems have evolved to respond specifically to the actual signal, i.e. the effective "lead" compound. However, the production and/or testing of analogues of plant signals, activators and insect-derived elicitors can potentially be of great value in investigating structural requirements for their biological activity.

Jasmonic acid, and 66 structurally related compounds, were tested by Miersch et al. (1999) to find the structural requirements which induce the expression of jasmonate-responsive genes in *H. vulgare*. Studies showed that an intact cyclopentanone ring, as well as a pentenyl side chain exhibiting only minor alterations, were necessary for this activity. The (−)-enantiomeric and the (+)-7-iso-enantiomeric structure increased activity of jasmonyl compounds. Earlier studies by the same group (Kramell et al. 1997) showed that in response to jasmonic acid, methyl jasmonate and its amino acid conjugate, (−)-enantiomers were more active, and that conjugates were more active if they carry an L-amino acid moiety. Mithofer et al. (2005) reported evidence showing that common and different biological responses are mediated by OPDA and / or jasmonic acid, suggesting the existence of at least two separate structure–activity groups. Based on the structure of a bacterial phytotoxin, coronatine, with similar biological activities as jasmonates, Lauchli et al. (2002) designed indanoyl isoleucine conjugates as functional synthetic mimics of octadecanoid-derived signals. Zhang et al. (1997) explored the structural requirements for jasmonic acid activity in *N. sylvestris*. The authors examined jasmonates, mimics and a biosynthetic precursor for nicotine-inducing activity (NIA). They examined the importance of the keto group on the five-membered ring and the double bond in the *n*-pentenyl side chain by comparing the activity of methyl jasmonate with that of closely related structures. From this work, they suggested that the keto functional group on the five-membered ring and the double bond in the *n*-pentenyl side chain are crucial components of jasmonic acid for activity.

Compounds containing α,β-unsaturated carbonyl groups are reactive electrophile species and have been implicated as potent regulators of gene expression in plants (Vollenweider et al. 2000; Weber et al. 2004), with some compounds being powerful cytotoxins that accumulate at the site of lesion. Almeras et al. (2003) quantified the level of defence gene expression in *A. thaliana* with a variety of lipid derivatives. Small α, β-unsaturated compounds (i.e. acrolein, methyl vinyl ketone) were shown to be potent stimulators of gene expression and far more potent than larger alkenyl homologues such as (*E*)-2-hexenal, e.g. treatment with reactive electrophile species massively increased the levels of unesterified jasmonate production/perception.

Heil et al. (2008) searched for a structural motif that characterises VOCs which elicit defensive responses in neighbouring plants. *P. lunatus* plants were exposed to (*Z*)-3-hexenyl acetate, which is known to be released from damaged plants and known to induce extra floral nectar (EFN) secretion, and to several structurally related compounds, including (*E*)-3-hexenyl acetate, (*E*)-2-hexenyl acetate, 5-hexenyl acetate, (*Z*)-3-hexenyl isovalerate and (Z)-3-hexenyl butyrate. All compounds elicited significant increases in EFN secretion, demonstrating that neither the (*Z*)-configuration nor the position of the double bond nor the size of the acid moiety are critical for the EFN-inducing effect. The authors postulated that physicochemical processes, including interactions with odorant binding proteins and resulting in changes in transmembrane potentials, can underlie VOC-mediated signalling processes.

Synthetic analogues of bruchins were prepared by Oliver et al. (2002) and examined to evaluate structural requirements for inducing the unusual neoplastic response that is induced in plants. The authors concluded that chain length (optimum length C_{22}–C_{24}) is important, whereas unsaturation within the chain is relatively unimportant. Difunctionality is required for maximum activity, but the free diols themselves are inactive. The most critical aspect is the ester portion(s) of the molecules, with 3-hydroxypropanoate esters being far more active than any analogues examined.

5 Biosynthetic Studies

Isotopically labelled VOCs, oxylipins, phytohormones and their putative biosynthetic precursors have been used in a number of studies relating to plant–insect interactions, with mass spectrometry being the route for detection and characterisation of labelled, intact or metabolised products following incubation with plant material and subsequent extraction. Labelled materials are also used to validate and quantify the recovery of plant metabolites using the extraction techniques such as those described above, e.g. VPE (Schmelz et al. 2003) and the PFBHA approach described by Schulze et al. (2006). In the latter, synthesis of deuterium-labelled *iso*-OPDA and incubation with various plant species was critical to its discovery as a biosynthetic precursor for *cis*-jasmone (Lauchli and Boland 2003; Dabrowska and Boland 2007).

Similarly, elegant studies on the biosynthesis of the acyclic homoterpenes (*E*)-4,8-dimethylnona-1,3,7-triene (DMNT) and TMTT have shown that these plant defence VOCs are biosynthesised from nerolidol and geranyllinalool respectively by oxidative fragmentation resulting in loss of 3-buten-2-one. Synthesis of ^{2}H-nerolidol and ^{2}H-geranylinalool, incubation with various plant species and GC-MS analysis of the VOCs showed the presence of isotopically labelled DMNT and TMTT (Boland and Gabler 1989). Further studies on this pathway using labelled geranylacetone showed that this can also be a precursor for DMNT (Gabler et al. 1991; Boland et al. 1998). By using chiral labelled intermediates, the stereochemistry of the intermediates of TMTT biosynthesis was revealed (Donath and Boland 1994), and by using labelled enantiomers of nerolidol, the stereochemistry of the starting material for DMNT biosynthesis was revealed (Donath and Boland 1995). Together, these studies represent a classic example of elucidating a biosynthetic pathway using chemical approaches.

Isotopically labelled carbon dioxide (CO_2) has also been used to investigate biosynthetic pathways linked to plant–insect interactions. Pare et al. (1998) fed *Z. mays* seedlings labelled with labelled CO_2 to beet armyworms, and obtained chemical evidence that the caterpillars acquire linolenic acid from the plants prior to its hydroxylation and conjugation with glutamine. The authors surmised that this modification provided a distinct chemical cue that allows the plant to distinguish between herbivore damage and other stresses.

6 Future Prospects

Due to the evolving nature of chemical sciences technology, there is an ever increasing prospect of understanding the chemistry underlying plant–insect interactions. As described above, there is a diverse range of approaches available for the isolation and analysis of plant and insect chemistry (SPE, SPME, VPE, SBSE etc.), which is aimed at reducing sample size and sampling time. However, there is a drawback to all of these approaches, namely that they collect samples at specific time points. In the real world, plant–insect interactions are dynamic processes; there is a continuum in place, which current sampling techniques cannot deal with. Thus, analytical-based approaches which allow "real-time" sampling are required. In recent years, there have been rapid improvements in mass spectrometry instrumentation, which includes the development of in situ analysis of trace level components present in air or water, and, of particular relevance to chemical marker identification, the development of miniaturised lab scale instruments for rapid, portable use and which may or may not involve pre-concentration of samples. A method which has been developed for rapid real-time analysis is proton-transfer-reaction mass spectrometry (PTR-MS) (Rieder et al. 2001). This technique has already been applied to plant–insect interactions, e.g. for the analysis of root-secreted VOCs by *A. thaliana* (Steeghs et al. 2004) and the detection of methanol emitted by *Nicotiana* plants when attacked by *Manduca sexta* (von Dahl et al. 2006). A drawback of this technique however is that empirical identification of VOCs is not possible, as it cannot distinguish between different compounds with the same molecular mass. A similar method which could be exploited involves membrane introduction (inlet) mass spectrometry (MIMS), which includes a miniature ion trap mass spectrometer coupled with PDMS membranes used in SPME (Riter et al. 2003). New ionisation methods are also being developed, which have the potential to be applied to portable systems. These include desorption electrospray ionisation (DESI) (Takats et al. 2004), which has already been applied to the analysis of plant alkaloids (Talaty et al. 2005). Alongside those systems, current space research programmes are also providing new leads in mass spectrometer miniaturisation. These include the Ptolemy GC-MS instrument developed for the European Space Agency's Rosetta mission, which includes a miniature GC coupled to an ion trap mass spectrometer (Todd et al. 2007).

A number of physical sensors (biosensors) can potentially be deployed in plant–insect interaction studies. Where there are examples of biosensors being developed for detecting plant VOCs, these rely on previous knowledge of the appropriate markers. These include a portable handheld minaturised GC system (zNose) (Kunert et al. 2002), which has been used in real-time analysis of aphid alarm pheromone production (Schwartzberg et al. 2008). However, these systems – in particular the simpler artificial nose systems – can suffer from not being able to discriminate very low levels of key markers in the presence of large amounts of irrelevant but related compounds. Thus, new, highly tuned physical sensors appear to be highly suited, in particular those miniature systems that utilise mass spectrometry (MS). Of these, the most effective will be those that incorporate gas chromatography (GC) and mass spectrometry (MS), e.g. the Ptolemy instrument mentioned above. The physical devices developed will need to be capable of rapid-throughput screening of airborne and aqueous-derived volatile samples, possibly using multi-location devices.

From the applied perspective, there is potential to exploit the chemistry of plant defence in novel plant protection strategies. Here, chemicals can be selected as markers for the presence of insects at an early stage. However, for the development of rapid detection systems, there are several challenges that need to be addressed. These relate to (1) the identification of chemical markers that are specific to particular insects and can be clearly resolved from the "normal" i.e. healthy plant situation and (2) the development of advanced physical sensors which can detect the identified markers. For (1), techniques for rapid collection of airborne and aqueous-derived volatiles from plant are already well established. However, the complex nature and high variability of plant chemical profiles, implies that insect marker detection through direct comparison of profiles from "healthy" and "infested" individuals is a challenge that needs to be overcome. For (2), any potential detector system that will be employed will have to be able to 'see' the chemical markers, once they have been identified, effectively through extraneous material not related directly to the specific disease to be detected. A number of physical devices have been devised in an attempt to satisfy this need. Biosensors for detection of pathogen infection and insect infestations via detection of induced plant volatiles have been reported (e.g. Schultz et al. 1996). These include a system for detecting fungal-infected plants (Schultz et al. 1999) and beetle-damaged plants (Schultz et al. 2000).

For a greater understanding of plant–insect interactions, the chemical sciences community has to link in much more with expertise in other scientific disciplines, including plant biology, biochemistry and molecular biology, insect behaviour, neurophysiology, morphology and molecular biology. A specific opportunity includes exploiting the growing field of chemical genetics, whereby molecular probes can be designed and exploited to study gene expression in plant systems, particularly for genomically-sequenced plants that can be used in plant–insect-interactions studies. There is also an opportunity to study the olfactory mechanisms underlying the interaction between insects and their host plants (Pickett et al. 2009), specifically the design of molecular probes to study the "chemical space" surrounding antennal perception of plant VOCs. Finally, an understanding of the mechanisms underlying plant "perception" of plant activators (including phytopheromones) and insect-derived elicitors is an area which will also be explored in more depth, again using molecular probes (e.g. Schuler et al. 1999). Alongside these opportunities will be continuing work to identify elicitors of plant defence from other insect classes.

In summary, the chemical sciences have a central role in the understanding of plant–insect interactions, by providing approaches for the collection, analysis and synthesis of plant natural products, and appear to continue to do so for the foreseeable future.

References

Adolph S, Poulet SA, Pohnert G (2003) Synthesis and biological activity of α, β, γ, δ unsaturated aldehydes from diatoms. Tetrahedron 59:3003–3008

Agelopoulos N, Pickett JA (1998) Headspace analysis in chemical ecology: effects of different sampling methods on ratios of volatile compounds present in headspace samples. J Chem Ecol 24:1161–1172

Agelopoulos N, Hooper AM, Maniar SM, Pickett JA, Wadhams LJ (1999) A novel approach for isolation of volatile chemicals released by individual leaves of a plant in situ. J Chem Ecol 25:1411–1425

Al Abassi S, Birkett MA, Pettersson J, Pickett JA, Woodcock CM (1998) Ladybird beetle odour identified and found to be responsible for attraction between adults. Cell Mol Life Sci 54:876–879

Alborn HT, Turlings TCJ, Jones TH, Stenhagen G, Loughrin JH, Tumlinson JH (1997) An elicitor of plant volatiles from beet armyworm oral secretion. Science 276:945–949

Alborn HT, Jones TH, Stenhagen GS, Tumlinson JH (2000) Identification and synthesis of volicitin and related components from beet armyworm oral secretions. J Chem Ecol 26:203–220

Alborn HT, Hansen TV, Jones TH, Bennett DC, Tumlinson JH, Schmelz EA, Teal PEA (2007) Disulfooxy fatty acids from the American bird grasshopper *Schistocerca americana*, elicitors of plant volatiles. Proc Natl Acad Sci USA 104:12976–12981

Almeras E, Stolz S, Vollenweider S, Reymond P, Mene-Saffrane L, Farmer EE (2003) Reactive electrophile species activate defense gene expression in Arabidopsis. Plant J 34:205–216

Andersson MX, Hamberg M, Kourtchenko O, Brunnstrom A, McPhail KL, Gerwick WH, Gobel C, Feissner I, Ellerstrom M (2006) Oxylipin profiling of the hypertensive response in Arabidopsis thaliana. J Biol Chem 281:31528–31537

Birkett MA, Campbell CAM, Chamberlain K, Guerrieri E, Hick AJ, Martin JL, Matthes M, Napier J, Pettersson J, Pickett JA, Poppy GM, Pow EM, Pye BJ, Smart LE, Wadhams GE, Wadhams LJ, Woodcock CM (2000) New roles for *cis*-jasmone as an insect semiochemical and in plant defence against insects. Proc Natl Acad Sci USA 97:9329–9334

Boland W, Gabler A (1989) Biosynthesis of homoterpenes in higher plants. Helv Chim Acta 72:247–253

Boland W, Gabler A, Gilbert M, Feng Z (1998) Biosynthesis of C11 and C16 homoterpenes in higher plants: stereochemistry of the C–C-bond cleavage reaction. Tetrahedron 54: 14725–14736

Bonnington LS, Barcelo D, Knepper TP (2003) Utilisation of electrospray time of flight mass spectrometry for solving complex fragmentation patterns: application to benzoxazinone derivatives. J Mass Spectrom 38:1054–1066

Bruce TJ, Matthes M, Chamberlain K, Woodcock CM, Mohib A, Webster B, Smart LE, Birkett MA, Pickett JA, Napier JA (2008) Multitrophic interactions involving *Arabidopsis thaliana* investigated by means of *cis*-jasmone defence activation. Proc Natl Acad Sci USA 105:4553–4558

Cuyckens F, Claeys M (2004) Mass spectrometry in the structural analysis of flavonoids. J Mass Spectrom 39:1–15

D'Alessandro M, Turlings TCJ (2006) Advances and challenges in the identification of volatiles that mediate interactions among plants and arthropods. Analyst 131:24–32

Dabrowska P, Boland W (2007) iso-OPDA: an early precursor of cis-jasmone in plants? Chembiochem 8:2281–2285

Dicke M (1999) Specificity of herbivore-induced plant defences. In: Chadwick DJ, Goode JA (eds) Insect–plant interactions in induced plant defence. Wiley, New York, pp 43–59

Donath J, Boland W (1994) Biosynthesis of acyclic homoterpenes in higher plants parallels steroid hormone metabolism. J Plant Physiol 143:473–478

Donath J, Boland W (1995) Biosynthesis of acyclic homoterpenes: enzyme selectivity and absolute configuration of the nerolidol precursor. Phytochemistry 39:785–790

Doss RP, Oliver JE, Proebsting WM, Potter SW, Kuy S, Clement SL, Williamson RT, Carney JR, DeVilbiss ED (2000) Bruchins: insect-derived plant regulators that stimulate neoplasm formation. Proc Natl Acad Sci USA 97:6218–6223

Du YJ, Poppy GM, Powell W, Pickett JA, Wadhams LJ, Woodcock CM (1998) Identification of semiochemicals released during aphid feeding that attract parasitoid *Aphidius ervi*. J Chem Ecol 24:1355–1368

Elliott M, Farnham AW, Janes NF, Needham PJ, Pulman DA (1973) Potent pyrethroid insecticides from modified cyclopropane acids. Nature 244:456–457

Engelberth J, Schmelz EA, Alborn HT, Cardoza Y, Huang J, Tumlinson JH (2003) Simultaneous quantification of jasmonic acid and salicylic acid in plants by vapour-phase extraction and gas chromatography-chemical ionization-mass spectrometry. Anal Biochem 312:242–250

Farag MA, Ryu CM, Sumner LW, Pare PW (2006) GC-MS SPME profiling of rhizobacterial volatiles reveals prospective inducers of growth promotion and induced systemic resistance in plants. Phytochemistry 67:2262–2268

Gabler A, Boland W, Preiss U, Simon H (1991) Stereochemical studies on homoterpene biosynthesis in higher plants – mechanistic, phylogenetic and ecological aspects. Helv Chim Acta 74:1773–1789

Gordon-Weeks R, Pickett JA (2009) Role of natural products in nature: plant–insect interactions. In: Osborn AE, Lanzotti V (eds) Plant-derived natural products. Springer, Berlin, pp 321–347

Griffiths DC, Pickett JA (1980) Composition of aphid alarm pheromones. J Chem Ecol 6:349–360

Hansen TV, Stenstrom Y (2000) A facile formal synthesis of volicitin. Synth Commun 30:2549–2557

Heil M, Lion U, Boland W (2008) Defense-inducing volatiles: in search of the active motif. J Chem Ecol 34:601–604

Hisamatsu Y, Goto N, Hasewaga K, Shigemori H (2003) Arabidopsides A and B, two new oxylipins from *Arabidopsis thaliana*. Tetrahedron Lett 44:5553–5556

Hisamatsu Y, Goto N, Sekiguchi M, Hasewaga K, Shigemori H (2005) Oxylipins Arabidopsides C and D from *Arabidopsis thaliana*. J Nat Prod 68:600–603

Itoh S, Kuwahara S, Hasegawa M, Kodama O (2002) Synthesis of the (17*R*)- and (17*S*)-isomers of volicitin, an elicitor of plant volatiles contained in the oral secretion of the beet armyworm. Biosci Biotechnol Biochem 66:1591–1596

Koch T, Hoskovec M, Boland W (2002) Efficient syntheses of (10E, 12Z, 15Z)-9-oxo- and (9Z, 11E, 15E)-13-oxo-octadecatrienoic acids; two stress metabolites of wounded plants. Tetrahedron 58:3271–3274

Kourtchenko O, Andersson MX, Hamberg M, Brunnstrom A, Gobel C, McPhail KL, Gerwick WH, Fuessner I, Ellerstrom M (2007) Oxo-phytodienoic acid-containing galactolipids in Arabidopsis: jasmonate signalling dependence. Plant Physiol 145:1658–1669

Kramell R, Miersch O, Hause B, Ortel B, Parthier B, Wasternack C (1997) Amino acid conjugates of jasmonic acid induce jasmonate-responsive gene expression in barley (*Hordeum vulgare*) leaves. FEBS Lett 414:197–202

Krishnamachari VK, Xie X, Zhu S, Wei HX, Pare PW (2007) Efficient synthesis of the insect elicitor volicitin and biologically active analogues. Nat Prod Commun 2:1019–1023

Kunert M, Biedermann A, Koch T, Boland W (2002) Ultrafast sampling and analysis of plant volatiles by a hand-held miniaturised GC with pre-concentration unit: kinetic and quantitative aspects of plant volatile production. J Sep Sci 25:677–684

Lauchli R, Boland W (2003) Efficient synthesis of [$^2H_2$]-tetrahydrodicranenone and a 3-oxa-analogue resistant against β-oxidation. Tetrahedron 59:149–153

Lauchli R, Schuler G, Boland W (2002) Selective induction of secondary metabolism in *Phaseolus lunatus* by 6-substituted indanoyl isoleucine conjugates. Phytochemistry 61:807–817

Loi RX, Solar MC, Weidenhamer JD (2008) Solid-phase microextraction method for in vivo measurement of allelochemical uptake. J Chem Ecol 34:70–75

March RE, Lewars EG, Stadley CJ, Miao XS, Zhao X, Metcalfe CD (2006) A comparison of flavonoid glycosides by electrospray tandem mass spectrometry. Int J Mass Spectrom 248:61–85

Miersch O, Kramell R, Parthier B, Wasternack C (1999) Structure–activity relations of substituted, deleted or stereospecifically altered jasmonic acid in gene expression of barley leaves. Phytochemistry 50:353–361

Millar JG, Haynes KF (1998) Methods in chemical ecology volume 1: chemical methods. Kluwer, Massachusetts

Mithofer A, Maitrejean M, Boland W (2005) Structural and biological diversity of cyclic octadecanoids, jasmonates and mimetics. J Plant Growth Regul 23:170–178

Moraes MCB, Birkett MA, Gordon-Weeks R, Smart LE, Martin JL, Pye BJ, Bromilow R, Pickett JA (2008) cis-Jasmone induces accumulation of defence compounds in wheat, *Triticum aestivum*. Phytochemistry 69:9–17

Moraes MCB , Laumann RA, Pareja M, Sereno FTPS, Michereff MFF, Birkett MA, Pickett JA, Borges M (2009) Attraction of the stink bug egg parasitoid *Telenomus podisi* to defence signals from soybean activated by treatment with cis-jasmone. Entom Exp et Appl 131:178–188

Mueller MJ, Mene-Saffrane L, Grun C, Karg K, Farmer EE (2006) Oxylipin analysis methods. Plant J 45:472–489

Nakajyo H, Hisamatsu Y, Sekiguchi M, Goto H, Hasewaga K, Shigemori H (2006) Arabidopside F, a new oxylipin from *Arabidopsis thaliana*. Heterocycles 69:295–301

Oliver JE, Doss RP, Williamson RT, Carney JR, DeVillbriss ED (2000) Bruchins – mitogenic 3-(hydroxypropanoyl) esters of long chain diols from weevils of the Bruchidae. Tetrahedron 56:7633–7641

Oliver JE, Doss RP, Marquez B, DeVillbris ED (2002) Bruchins, plant mitogens from weevils: structural requirements for activity. J Chem Ecol 28:2503–2513

Pare PW, Alborn HT, Tumlinson JH (1998) Concerted biosynthesis of an insect elicitor of plant volatiles. Proc Natl Acad Sci USA 95:13971–13975

Pickett JA (1990) Gas chromatography-mass spectrometry in insect pheromone identification: three extreme case histories. In: McCaffrey ID, Wilson AR (eds) Chromatography and isolation of insect hormones and pheromones. Plenum, New York, pp 281–288

Pickett JA, Birkett MA, Woodcock CM, Zhou JJ (2009) Scents and sex. The Biochemist, 1–6

Ping L, Boland W (2004) Signals from the underground: bacterial volatiles promote growth in Arabidopsis. Trends Plant Sci 9:263–266

Pohnert G, Jung V, Haukioja E, Lempa K, Boland W (1999a) New fatty acid amides from regurgitant of Lepidopteran (Noctuidae, Geometridae) caterpillers. Tetrahedron 55: 11275–11280

Pohnert G, Koch T, Boland W (1999b) Synthesis of volicitin: a novel three-component\wittig approach to chiral 17-hydroxylinolenic acid. Chem Commun 12:1087–1088

Rasmann S, Kollner TG, Degenhardt J, Hiltpold I, Toepfer S, Kuhlmann U, Gershenzon J, Turlings TCJ (2005) Recruitment of entomopathogenic nematodes by insect-damaged maize roots. Nature 434:732–737

Rieder J, Lirk P, Ebenbichler C, Gruber G, Prazeller P, Lindinger W, Amman A (2001) Analysis of volatile organic compounds: possible applications in metabolic disorders and cancer screening. Wien Klin Wochenschr 13:181–185
Riter LS, Meurer EC, Cotte-Rodriguez I, Eberlin MN, Cooks RG (2003) Solid phase micro-extraction in a miniature ion trap mass spectrometer. Analyst 128:1119–1122
Ryu CM, Farag MA, Hu CH, Reddy MS, Wei HX, Pare PW, Kloepper JW (2003) Bacterial volatiles promote growth in Arabidopsis. Proc Natl Acad Sci USA 100:4927–4932
Ryu CM, Farag MA, Hu CH, Reddy MS, Kloepper JW, Pare PW (2004) Bacterial volatiles induce systemic resistance in Arabidopsis. Plant Physiol 134:1017–1026
Sawada Y, Yoshinaga N, Fujisaki K, Nishida R, Kuwahara Y, Mori N (2006) Absolute configuration of volicitin from the regurgitant of Lepidopteran caterpillers and biological activity of volicitin-related compounds. Biosci Biotech Biochem 70:2185–2190
Scascighini N, Mattiacci L, D'Alessandro M, Hern A, Rott A, Rott AS, Dorn S (2005) New insights in analysing parasitoid attracting synomones: early volatile emission and use of stir bar sorptive extraction. Chemoecology 15:97–104
Schmelz EA, Engelberth J, Alborn HT, O'Donnell P, Sammons M, Toshima H, Tumlinson JH (2003) Simultaneous analysis of phytohormones, phytotoxins and volatile organic compounds. Proc Natl Acad Sci USA 100:10552–10557
Schmelz EA, Engelberth J, Tumlinson JH, Block A, Alborn HT (2004) The use of vapour phase extraction in metabolic profiling of phytohormones and other metabolites. Plant J 39:790–808
Schmelz EA, Engelberth J, Alborn HT, Tumlinson JH, Teal PEA (2009) Phytohormone-based activity mapping of insect herbivore-produced elicitors. Proc Natl Acad Sci USA 106: 653–657
Schuler G, Wasternack C, Boland W (1999) Synthesis of 6-azido-1-oxo-indan-4-oyl isoleucine; a photoaffinity approach to plant signalling. Tetrahedron 55:3897–3904
Schultz S, Weissbecker B, Hummel HE (1996) Biosensor for volatiles released by damaged plants. Biosens Bioelectron 11:427–433
Schultz S, Weissbecker B, Koch UT, Hummel HE (1999) Detection of volatiles released by diseased potato tubers using a biosensor on the basis of intact insect antennae. Biosens Bioelectron 14:221–228
Schultz S, Schoning MJ, Schroth P, Malkoc U, Weissbecker B, Kordos P, Luth H, Hummel HE (2000) An insect-based BioFET as a bioelectronic nose. Sensors Actuators B 65:291–295
Schulze B, Lauchli R, Mekem Sonwa M, Schmidt A, Boland W (2006) Profiling of structurally labile oxylipins in plants by in situ derivatisation with pentafluorobenzyl hydroxylamine. Anal Biochem 348:269–283
Schulze B, Dabrowska P, Boland W (2007) Rapid enzymatic isomerisation of 12-oxophytodienoic acid in the gut of Lepidopteran larvae. Chembiochem 8:208–216
Schwartzberg EG, Kunert G, Stephan C, David A, Rose USR, Gershenzon J, Boland W, Weisser W (2008) Real-time analysis of alarm pheromone emission by the pea aphid (*Acyrthosiphon pisum*) under predation. J Chem Ecol 34:76–81
Soini HA, Bruce KE, Wiesler D, David F, Sandra P, Novotny M (2005) Stir bar sorptive extraction: a new quantitative and comprehensive sampling technique for determination of chemical signal profiles biological media. J Chem Ecol 31:377–392
Spiteller D, Boland W (2003) *N*-(15, 16-Dihydroxylinoleoyl)-glutamine and *N*-(15, 16-epoxylinoleoyl)-glutamine isolated from oral secretions of lepidopteran larvae. Tetrahedron 59:135–139
Steeghs M, Pal Bais H, de Ouw J, Goldan P, Kuster W, Northway M, Fall R, Vivanco JM (2004) Proton-transfer-reaction mass spectrometry as a new tool for real time analysis of root-secreted volatile organic compounds in Arabidopsis. Plant Physiol 135:47–58
Takats Z, Wisman JM, Gologan B, Cooks RG (2004) Mass spectrometry sampling under ambient conditions with desorption electrospray ionization. Science 306:471–473
Talaty N, Takats Z, Cooks RG (2005) Rapid in situ detection of alkaloids in plant tissue under ambient conditions using desorption electrospray ionization. Analyst 130:1624–1633

Todd JF, Barber SJ, Wright IP, Morgan GH, Morse AD, Sheridan S, Leese MR, Maynard J, Evans ST, Pillinger CT, Drummond DL, Heys SC, Huq SE, Kent BJ, Sawyer EC, Whalley MS, Waltham NR (2007) Ion trap mass spectrometry on a comet nucleus: the Ptolemy instrument and the Rosetta space mission. J Mass Spectrom 42:1–10

Turlings TCJ, Alborn HT, Loughrin JH, Tumlinson JH (2000) Volicitin, an elicitor of maize volatiles in oral secretion of *Spodoptera exigua*: isolation and bioactivity. J Chem Ecol 26:189–202

Turlings TCJ, Davison AC, Tamo C (2004) A six-arm olfactometer permitting simultaneous observation of insect attraction and odour trapping. Physiol Entomol 29:45–55

Vollenweider S, Weber H, Stolz S, Chetelat A, Farmer EE (2000) Fatty acid ketodienes and fatty acid ketotrienes: Michael addition acceptors that accumulate in wounded and diseased Arabidopsis leaves. Plant J 24:467–476

Von Dahl CC, Havecker M, Schlog R, Baldwin IT (2006) Caterpiller-elicited methanol emission: a new signal in plant–herbivore interactions? Plant J 46:948–960

Weber H, Chetelat A, Reymond P, Farmer EE (2004) Selective and powerful stress gene expression in Arabidopsis in response to malondialdehyde. Plant J 37:877–888

Webster B, Bruce T, Dufour S, Birkemeyer C, Birkett M, Hardie J, Pickett JA (2008) Identification of volatile compounds used in host location by the black-bean aphid, Aphis fabae. J Chem Ecol 34:1153–1161

Wei HX, Truitt CL, Pare PW (2003) Synthesis of hydroxyl-substituted unsaturated fatty acids and the amino-acid insect derivative volicitin. Tetrahedron Lett 44:831–833

Weidenhamer J (2005) Biomimetic measurement of allelochemical dynamics in the rhizosphere. J Chem Ecol 31:221–236

Weidenhamer JD (2007) New approaches to analyse allelochemicals in the soil. Allelopath J 19:135–142

Zerbe P, Weiler EW, Schaller F (2007) Preparative enzymatic solid phase synthesis of *cis*(+)-12-oxo-phytodienoic acid – physical interaction of AOS and AOC is not necessary. Phytochemistry 68:229–236

Zhang ZP, Krumm T, Baldwin IT (1997) Structural requirements of jasmonates and mimics for nicotine induction in *Nicotiana sylvestris*. J Chem Ecol 23:2777–2789

Plant Defense Signaling from the Underground Primes Aboveground Defenses to Confer Enhanced Resistance in a Cost-Efficient Manner

Marieke Van Hulten, Jurriaan Ton, Corné M.J. Pieterse, and Saskia C.M. Van Wees

Abstract Plants can be induced to develop below and aboveground enhanced resistance to pathogens and herbivorous insects by root-colonizing beneficial micro-organisms. The resistance induced is broad-spectrum and can be long lasting. The enhanced resistance is based at least partially on priming of defense responses, leading to a more rapid or more intense mobilization of defense responses upon encounter with harmful organisms. Several molecular players in local and systemic tissues of plants treated with resistance-inducing microbes have been identified and are reviewed in this chapter. We also discuss the ecological consequences of expression of induced resistance through a primed defense response.

1 Introduction

Below the soil surface, interactions between plants and microbes take place. Plant roots are quickly colonized by members of the indigenous microflora. Colonization by pathogens could have deleterious effects on the plant, but interactions between plants and microbes can also be advantageous for both the plant and the microbe. A well-known example of symbiosis between plants and soil-borne micro-organisms is that between plants and arbuscular mycorrhizal fungi, where the fungus aids the plant in the uptake of water and mineral nutrients such as phosphate by enhancing its absorbance surface through the fungal mycelium, while the plant provides carbohydrates to the fungus (Harrison 2005). Another classical example of symbiosis is the interaction between legume plants and *Rhizobium* spp. bacteria, in which the bacteria induce the formation of root nodules where they fix atmospheric nitrogen to convert it into organic nitrogenous compounds that become available

M. Van Hulten, J. Ton, C.M.J. Pieterse, and S.C.M. Van Wees (✉)
Faculty of Science, Plant–Microbe Interactions, Department of Biology, Utrecht University, Utrecht 3508 TB, The Netherlands
e-mail: s.vanwees@uu.nl

F. Baluška and V. Ninkovic (eds.), *Plant Communication from an Ecological Perspective*, Signaling and Communication in Plants,
DOI 10.1007/978-3-642-12162-3_3,

for the plant, while the bacteria obtain various organic photosynthetic compounds from the plant (Spaink 2000). Plant growth-promoting rhizobacteria (PGPR) form yet another class of beneficial soil-borne micro-organisms. PGPR such as *Pseudomonas* spp. and *Baccilus* spp. colonize the rhizosphere, where they live off nutrients exuded by plant roots. They can improve plant growth either directly by augmenting photosynthesis (Zhang et al. 2008) or indirectly by suppressing plant diseases (Van Loon et al. 1998; Bloemberg and Lugtenberg 2001). Disease suppression can be established through direct effects on soil-borne pathogens, such as competition for nutrients or secretion of toxic compounds (Van Loon et al. 1998; Weller et al. 2002; De Bruijn et al. 2007). However, root colonization by PGPR was also documented to suppress diseases caused by foliar pathogens in aboveground tissue (Kloepper et al. 2004; Van Loon and Bakker 2006; Van Wees et al. 2008). This plant-mediated effect of PGPR on pathogens is dependent on activation of the host's immune response and is commonly referred to as induced systemic resistance (ISR; Van Loon 2000). In addition to PGPR, soil-borne beneficial fungi can also trigger an ISR response in plants – for instance, plant growth-promoting fungi (PGPF) that include members of *Trichoderma* spp. (De Meyer et al. 1998; Harman et al. 2004; Shoresh et al. 2005; Vinale et al. 2008; Segarra et al. 2009) and *Piriformospora* spp. (Waller et al. 2005). Moreover, mycorrhizal association has also been reported to protect systemic plant tissues (Pozo and Azcón-Aguilar 2007).

Here, the molecular mechanisms of ISR triggered by PGPR are reviewed in light of the ecological perspective of the costs and benefits that are associated with plant defense.

2 Perception of PGPR by the Plant

Typically, ISR is effective against a broad range of taxonomically different pathogens (Van Loon et al. 1998; Van Wees et al. 2008) and also against herbivorous insects (Zehnder et al. 2001; Van Oosten et al. 2008). ISR induction is dependent on the combination of the plant and the beneficial micro-organism. PGPR strains that induce ISR in one species may not do so in another species and vice versa, suggesting host specificity in PGPR detection. For example, *Pseudomonas putida* WCS358 induces ISR in Arabidopsis (*Arabidopsis thaliana*), but not in its closely related crop relative radish (Van Peer et al. 1991; Van Peer and Schippers 1992; Leeman et al. 1995; Van Wees et al. 1997). Conversely, *Pseudomonas fluorescens* WCS374 is capable of inducing ISR in radish but not in Arabidopsis (Leeman et al. 1995; Van Wees et al. 1997).

The establishment of a symbiotic interaction requires a complex dialog between the plant and the micro-organism. The plant can detect microbe-associated molecular patterns (MAMPs) of beneficial micro-organisms, such as flagellin and lipopolysaccharides (LPS), which is in analogy to the detection of pathogen-associated molecular patterns (PAMPs) of pathogenic microbes (Nürnberger et al. 2004). In support of this, purified flagellin and LPS of beneficial rhizobacteria are reported to elicit ISR. That bacterial mutants lacking one of these determinants are still capable of protecting plants suggests that multiple MAMPs are involved in the induction of ISR (Bakker et al. 2007). However, while PAMP detection triggers a primary defense response in plants, called PAMP-triggered immunity (PTI), which keeps nonadapted pathogens at bay (Jones and Dangl 2006; Schwessinger and Zipfel 2008), perception of beneficial microbes does not trigger such a substantial defense response (Verhagen et al. 2004; Van Wees et al. 2008), or to a much lesser extent (Liu et al. 2007), and the benefactor remains accommodated by the plant.

3 ISR Signal Transduction

ISR induced by several beneficial *Pseudomonas* strains was shown to function independent of the plant defense hormone salicylic acid (SA; Pieterse et al. 1996). This is in contrast to another well-studied form of systemically induced resistance, namely systemic acquired resistance (SAR), which is triggered upon infection by pathogens (Durrant and Dong 2004). However, the use of signaling mutants indicated that components of the signaling pathways controlled by the hormones jasmonic acid (JA) and ethylene (ET) response are required for ISR (Pieterse et al. 1998; Van Wees et al. 2008). This indicates that distinct signaling cascades underlie each form of systemically induced resistance. Concordantly, SAR is predominantly effective against biotrophic pathogens that are resisted through SA-dependent defenses, while ISR is most efficient against necrotrophic pathogens and insects (Fig. 1), which are susceptible to JA-dependent defenses. For instance, SAR was shown to protect *Arabidopsis* plants against turnip crinckle virus in *Arabidopsis*, while ISR did not (Ton et al. 2002). Conversely, ISR induced by *P. fluorescens* WCS417 was shown to be effective in *Arabidopsis* against the necrotrophic pathogens *Alternaria brassicicola* (Ton et al. 2002), *Botrytis cinerea* (Van der Ent et al. 2008), and *Plectosphaerella cucumerina* (Segarra et al. 2009), while pathogen-induced SAR was not. ISR was reported to be effective against *Spodoptora exigua*, a generalist herbivore on *Arabidopsis*; SAR could also induce resistance against this insect (Van Oosten et al. 2008). Both ISR and SAR led to protection against the biotrophic oomycete *Hyaloperonospora arabidopsidis* and the (hemi-)biotroph *Pseudomonas syringae* (Ton et al. 2002). However, this resistance was shown to be accomplished through activation of distinct signaling pathways in SAR and ISR (Cao et al. 1994; Lawton et al. 1995; Pieterse et al. 1998; Van Wees et al. 1999, 2000; Van der Ent et al. 2009b). In agreement with this, SAR and ISR have been shown to have an additive effect on the level of induced resistance against *P. syringae* (Van Wees et al. 2000). For multiple plant–benefical microbe interactions, the involvement of JA and/or ET signaling components has been reported, indicating that ISR signaling is the common route to induce systemic resistance (Van Wees et al. 2008). However, several examples of PGPR and PGPF that trigger SA-dependent SAR signaling leading to enhanced systemic resistance have been documented as well (Van Loon and Bakker 2005; Van Wees et al. 2008).

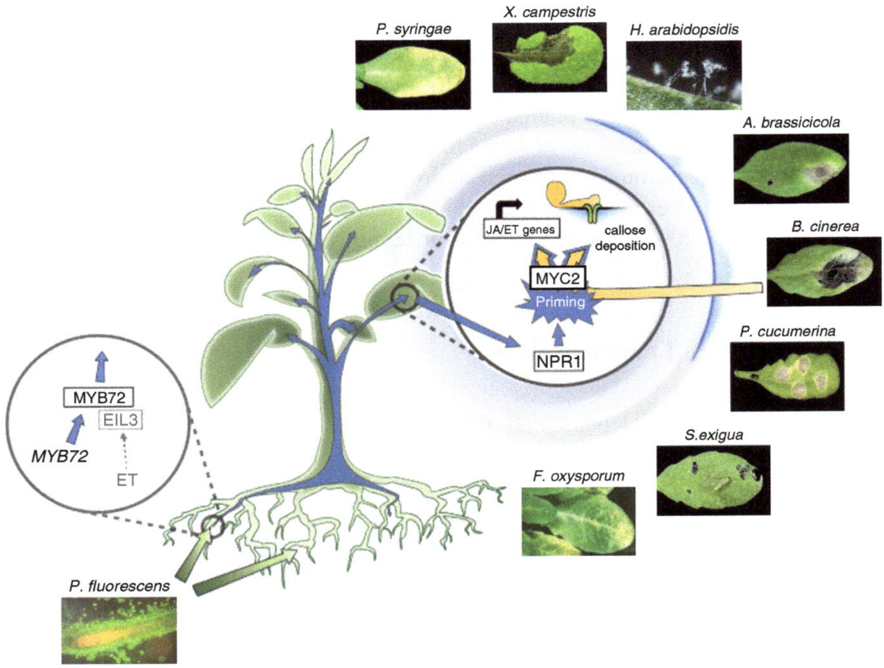

Fig. 1 Spectrum of effectiveness of *Pseudomonas fluorescens* WCS417-mediated ISR in Arabidopsis. ISR induced by the beneficial rhizobacterium *P. fluorescens* WCS417 is effective against the hemibiotrophic bacterial pathogens *Pseudomonas syringae* (Pieterse et al. 1996) and *Xanthomonas campestris* (Ton et al. 2002), the biotrophic oomycete *Hyaloperonospora arabidopsidis* (Ton et al. 2002), the necrotrophic fungal pathogens *Alternaria brassicicola* (Ton et al. 2002), *Botrytis cinerea* (Van der Ent et al. 2008) and *Plectosphaerella cucumerina* (Segarra et al. 2009), the fungal root pathogen *Fusarium oxysporum* (Van Wees et al. 1997), and the insect herbivore *Spodoptora exigua* (Van Oosten et al. 2008). The TF MYB72 is involved in ISR locally in the roots (formation or translocation of the ISR signal; Van der Ent et al. 2008) and the TF MYC2 is required for the priming response in systemic ISR-expressing tissue (Pozo et al. 2008). Adapted from Van Wees et al. (2008)

Microarray analysis of Arabidopsis root tissue identified a differential expression of 94 genes during rhizosphere colonization by *P. fluorescens* WCS417 (Verhagen et al. 2004). One of these genes was the R2R3-MYB-like transcription factor (TF) gene *MYB72* (Verhagen et al. 2004). Interestingly, *myb72* knockout mutants were incapable of mounting ISR, indicating that local *MYB72* induction is required for ISR (Van der Ent et al. 2008; Fig. 1). Interestingly, in *Arabidopsis*, MYB72 is also essential for the induction of ISR by the beneficial fungus *Trichoderma asperellum* T34 (Segarra et al. 2009), suggesting a regulating role for MYB72 in the induction of ISR by taxonomically different organisms. MYB72 may be an early point of convergence in ISR signaling elicited by different MAMPs (Segarra et al. 2009). *MYB72* overexpressors do not show enhanced levels of disease resistance (Van der Ent et al. 2008), indicating that although *MYB72* induction is required, it is not sufficient for ISR induction, suggesting the involvement of an additional signal. It was demonstrated that MYB72 binds to the ETHYLENE INSENSITIVE3-LIKE (EIL3) TF *in vitro*, linking MYB72 function to the ET response pathway. In this respect it is noteworthy that the Arabidopsis *ethylene insensitive root 1* (*eir1*) mutant, which is insensitive to ET in the roots but not in the shoot (Roman et al. 1995), is incapable of mounting an ISR response after root colonization by WCS417r, while leaf infiltration with WCS417r still triggered ISR in this mutant (Knoester et al. 1999). These results indicate that ISR requires an intact ET responsiveness at the site of elicitation (Knoester et al. 1999).

While most SA signaling mutants of *Arabidopsis* are still able to express beneficial microbe-induced ISR, the SA-nonresponsive mutant *npr1 (nonexpressor of PR genes 1*; Cao et al. 1994; Delaney et al. 1995; Shah et al. 1997) of Arabidopsis is disturbed in WCS417-ISR (Pieterse et al. 1998). SA triggers the reduction of inactive NPR1 oligomers into active monomers, which subsequently translocate to the nucleus (Mou et al. 2003). In the nucleus NPR1 can interact with different TFs to regulate the expression of downstream genes, like *PR-1* (Fan and Dong 2002; Wang et al. 2006). NPR1 functions in the ISR signal transduction pathway likely downstream of the JA- and ET-dependency and does not activate *PR* gene expression (Pieterse et al. 1998). Evidence is accumulating that the role of NPR1 in ISR is connected to a cytosolic function of NPR1 (Stein et al. 2008), which is in line with a role of NPR1 in the cytosol in cross-talk between SA and JA signaling (Spoel et al. 2003; Leon-Reyes et al. 2009). Several examples of NPR1-dependency of ISR triggered by PGPR and PGPF in different plant species have been documented (Van der Ent et al. 2009a).

4 Priming for Enhanced Defense

Large-scale transcriptome analysis of ISR-expressing leaves in plants of which the roots were treated with WCS417 or other beneficial microbes revealed that there is no or only a weak direct induction of gene expression in systemic tissue (Verhagen et al. 2004; Liu et al. 2007; Van Wees et al. 2008). However, subsequent infection with a pathogen led to an augmented expression of a large number of genes in ISR expressing plants compared to control plants (Van Wees et al. 1999; Verhagen et al. 2004; Ahn et al. 2007; Cartieaux et al. 2008). In analogy to a similar phenomenon in animals, the enhanced defensive capacity without direct induction of defense responses in the absence of pathogens is called priming (Conrath et al. 2002, 2006). The set of genes that showed WCS417-primed induction after *P. syringae* infection of Arabidopsis was particularly enriched in JA/ET-regulated genes (Verhagen et al. 2004) that are responsive to JA/ET-inducing pathogens and insects, like *P. syringae*, *A. brassicicola*, *Pieris rapae*, and *Frankliniella occidentalis* (De Vos et al. 2005; Van der Ent et al. 2009a). Indeed, WCS417-ISR against the bacterial pathogen *P. syringae* and the insect *Spodoptera exigua* is associated with primed expression of the JA/ET-dependent genes *VSP2* and

PDF1.2, respectively (Van Wees et al. 1999; Hase et al. 2003; Pozo et al. 2008; Van Oosten et al. 2008). The specific priming of JA/ET-dependent defense responses during WCS417-mediated ISR fits with the dependency of ISR on JA/ET-dependent signaling pathways (Pieterse et al. 1998). Recently, Pozo et al. (2008) found an overrepresentation of MYC2 TF binding sites in the promoters of priming-responsive genes in ISR-expressing plants. MYC2 is a well-known player in JA-regulated signaling (Lorenzo et al. 2004). Mutant *jin1* which is impaired in MYC2 was incapable of mounting WCS417-ISR against *P. syringae* and *H. arabidopsidis*, indicating a central role for the MYC2 TF in WCS417-ISR (Pozo et al. 2008; Fig. 1). Also ISR induced by the beneficial fungus *Piriformospora indica* against *Golovinomyces orontii* in *Arabidopsis* was demonstrated to depend on MYC2 (Stein et al. 2008), supporting a role for MYC2 as an important regulator of priming during ISR induced by different microbes.

Besides priming of certain JA/ET-dependent responses, WCS417 also primes the plant to reinforce the cell wall at the site of pathogen attack. An enhanced deposition of callose-rich papillae is observed upon infection by the oomycete *H. arabidopsidis* in WCS417-pretreated plants (Van der Ent et al. 2008). In addition to forming a physical barrier for pathogen penetration, callose depositions are also considered to be a matrix for the accumulation of defense compounds such as H_2O_2, phenolics, and various proteins and glycoproteins with hydrolytic and antifungal properties (Zeyen et al. 2002). Priming for enhanced deposition of callose-containing papillae during WCS417-ISR in Arabidopsis is dependent on the ISR regulators MYB72, NPR1, and MYC2 (Pozo et al. 2008; Van der Ent et al. 2008, 2009b). Moreover, this ISR response is dependent on the phosphoinositide (PtdIns)- and abscisic acid (ABA)-dependent signaling components IBS2 and IBS3 (Van der Ent et al. 2009b), which were previously identified to be required for primed callose deposition induced by the priming agent β-aminobutyric acid (BABA; Zimmerli et al. 2001; Van der Ent et al. 2009b).

Priming of defense responses is also characteristic of other induced resistance phenomena. For instance, mycorrhizal fungi can prime plants for enhanced JA-regulated defense activation (Pozo and Azcón-Aguilar 2007). Insect herbivory can prime plants for a faster and stronger defense response to subsequent stresses in both systemic tissue (De Vos et al. 2006) and in neighboring plants by means of the production of volatile organic compounds (Engelberth et al. 2004; Ton et al. 2007). Although pathogen-induced SAR is accompanied by a large-scale transcriptional reprogramming of the cell (Maleck et al. 2000) and direct accumulation of PR proteins (Van Loon 1997), priming of certain defense responses also occurs during SAR (Cameron et al. 1999; Van Wees et al. 1999). While some PR proteins are thought to contribute to resistance because several of them possess antimicrobial activity (Van Loon and Van Strien 1999; Van Loon et al. 2006), direct activation of these *PR* genes alone seems not sufficient to explain the broad range of protection (Van Loon 1997). The fact that priming can be induced in various ways and has been observed in different plant species ranging from monocots to dicots, conferring protection against a wide variety of pathogens, insects, and abiotic stresses (Conrath et al. 2006) suggests that priming appears to be a common feature of the plant's immune system.

5 Mechanisms of Priming of Defense Responses

Although plants do not possess an adaptive immune system, priming may be considered as a form of immunological memory in plants. The molecular mechanisms underlying priming are the subject of research by us and other scientists. Since defense genes are not activated directly in primed plants, it was hypothesized that the primed state is based on accumulation or posttranslational modification of signal molecules that remain inactive until a subsequent stress stimulus is perceived (Conrath et al. 2006). Due to the enhanced level of signaling components, subsequent pathogen attack would lead to an increased activation of the appropriate defense pathway and thus to a potentiated activation of defense-related genes. TFs are plausible candidates to contribute to the onset of priming. Using reverse transcription quantitative polymerase chain reaction (RTq-PCR) technology, Van der Ent et al. (2009b) recently analyzed the expression of all putative *Arabidopsis* TF genes during WCS417-ISR. The expression of 121 different types of TF genes was induced or repressed, including induction of the *MYC2* gene, which had previously been identified as a player in priming during ISR (Pozo et al. 2008). Interestingly, the AP2/ERF (APETALA2/ETHYLENE-RESPONSIVE FACTORS) TF family was notably overrepresented among the upregulated TFs in ISR-expressing plants. MYC2 and several AP2/ERF TFs are implicated in the regulation of JA/ET-dependent defenses (Lorenzo and Solano 2005; Pré et al. 2008).

Also priming induced by the nonprotein amino acid BABA is accompanied by direct upregulation of TF genes (Van der Ent et al. 2009b). However, while several TFs that had previously been implied in regulation of JA/ET-dependent signal transduction pathways were upregulated during ISR, pretreatment with BABA induced the expression of 22 out of the 72 known WRKY genes in the *Arabidopsis* genome (Van der Ent et al. 2009b). WRKY TF genes have previously been implicated in the regulation of several SA-dependent defense-related genes (Dong et al. 2003). The BABA-responsiveness of a subset of the SA-regulated WRKYs is in agreement with BABA-induced priming of SA-dependent defenses (Zimmerli et al. 2000). WCS417- and BABA-induced priming is associated with induced expression of divergent sets of TF genes, which are in accordance with the defense responses that are primed by these inducers. However, the exact role of the priming-related TFs in the regulation of priming remains to be elucidated. Transcriptome analyses of pathogen-induced SAR expressing tissue demonstrated that TF genes are induced (Maleck et al. 2000; Wang et al. 2006), but it is unknown whether this is related to priming or to direct induction of defense responses.

There is no significant activation of defense-related genes during priming induced by treatment with WCS417 or BABA, which suggests that the accumulating TFs remain inactive until the perception of a subsequent stress signal. One way to activate TFs posttranslationally is via phosphorylation. For instance, phosphorylation of a bZIP TF is crucial for abscicic acid (ABA)-induced transcriptional activity (Kagaya et al. 2002). Interestingly, mitogen activated protein kinases (MPKs), which can phosphorylate proteins, were also recently shown to be primed by low concentrations of the SA analog benzothiadiazole (BTH; Beckers et al. 2009). Inactive MPK3 and MPK6 proteins accumulated in response to BTH and only subsequent exposure to pathogens led to activation of these primed MPKs (Beckers et al. 2009). Epigenetic regulation forms another possible mechanism for the priming phenomenon (Bruce et al. 2007). An altered methylation status or modification of nucleosomal histones could ensure a more accessible chromatin structure for activation of TF genes or defense-related genes, which could facilitate a quicker or more potent transcriptional response to subsequent pathogen attack. The SA-dependent SAR response is documented to be associated with epigenetic regulation as well (Mosher et al. 2006). Recently, Jung et al. (2009) discovered that the metabolite azeleic acid seems an important signal molecule in the establishment of pathogen-induced SAR. Azeleic acid is translocated in the vascular sap from local pathogen-infected tissue to systemic tissue. Application of azeleic acid confers enhanced disease resistance in Arabidopsis which is associated with priming for enhanced accumulation of SA upon pathogen challenge (Jung et al. 2009).

6 Costs of Induced Defenses

While some defense compounds are constitutively present such as toxic compounds that form a pre-existing chemical barrier against pathogens (Osbourn 1996), others are expressed only upon attack by a pathogen or herbivore. The inducible responses can be subjected to priming. Two prominent hypotheses have been proposed to explain the spatial and temporal variation in plant defense. These are the optimal defense theory (ODT), which predicts that plant parts with high fitness value will be highly defended, and the growth-differentiation balance hypothesis (GDBH), which assumes that a balance must be maintained between resources used for growth and defense (Barto and Cipollini 2005). It is assumed that inducible defenses are too costly to be expressed constitutively under enemy free conditions. In agreement with this, the constitutive SA-dependent defense expressing *Arabidopsis* mutant *cpr1* (*constitutive expressor of PR genes 1*) is severely compromised in growth and seed production compared to wild-type plants (Bowling et al. 1994; Heidel et al. 2004; Van Hulten et al. 2006). This indicates a severe fitness penalty for the constitutive expression of SA-inducible defenses, which may explain why SA-dependent defenses are not expressed constitutively (Bowling et al. 1994; Heidel et al. 2004). In agreement with this, Heidel et al. (2004) observed that *cpr1* also displays a decreased fitness under field conditions, in spite of its enhanced resistance. Interestingly, in the same field experiment, SA-insensitive *npr1* mutants that are defected in plant defense exhibited a decreased fitness as well, suggesting that there is a delicate balance between the costs and the benefits of inducible plant defense. Mutants that constitutively express JA- and ET-dependent defenses, such as *cev1* (*constitutive expression of VSP1*), also exhibit undersized measures (Ellis and Turner 2001). Moreover, several studies have demonstrated a fitness reduction upon direct induction of defenses by exogenous application of SA or (Me)JA

(Baldwin 1998; Agrawal et al. 1999; Heil et al. 2000; Van Dam and Baldwin 2001; Cipollini 2002; Heidel et al. 2004).

Fitness costs can arise from various processes (Heil 2002; Heil and Baldwin 2002; Walters and Heil 2007). Allocation costs occur when limited resources are allocated to resistance traits and not to growth and reproduction. In agreement with this, several studies have reported that photosynthesis is repressed during pathogen infection, presumably to free resources needed for defense (Berger et al. 2007). One can predict that plants experience more costs of defense-related traits under low-nutrient conditions. Conversely, resistance levels may be impaired due to limiting resources. Both predictions have been empirically confirmed (Cipollini 2002; Dietrich et al. 2004, 2005).

Ecological costs occur when defense expression affects other organisms besides the challenging pathogen. Activation of certain defenses may for instance have a negative effect on interactions with plant-beneficial organisms, such as mycorrhizal fungi (Glandorf et al. 1997). Moreover, resistance against one pathogen may result in enhanced susceptibility to another pathogen or insect. There is ample evidence of cross communication between the SA and JA/ET defense pathways, which can act both synergistically or antagonistically (Reymond and Farmer 1998; Rojo et al. 2003; Bostock 2005; Beckers and Spoel 2006; Koornneef and Pieterse 2008; Pieterse et al. 2009). For instance, Spoel et al. (2007) recently showed that SA-mediated defenses triggered in *Arabidopsis* upon infection with *P. syringae* rendered the infected tissue more susceptible to *A. brassicicola*. Pathway crosstalk is thought to be a mechanism for fine-tuning defense responses by prioritizing which defensive strategy to employ to cope with the different organisms that (simultaneously) interact with the plant. Koornneef et al. (2008) demonstrated that timing of elicitation of SA and JA signaling pathways is crucial for determining which defense pathway to prioritize, suggesting that there is a window of opportunity during which JA- and SA-regulated defense responses can have cross-effects on organisms other than the attacker. Therefore, laboratory studies that concentrate on single plant-attacker combinations may not take all ecological costs into account. The challenge lies in unraveling the costs of defense mechanisms in a multitrophic environment, such as appears in nature.

7 Fitness Benefits of Priming Under Disease Pressure

While the inducibility of defenses may save resources under enemy-free conditions compared to constitutive activation of defenses, it also causes a time slot between attack and the expression of inducible defenses in which the plant is vulnerable to the attacker. Priming may be a mechanism to ease the trade-off dilemma between costly defense activation and effective protection against harmful organisms (Goellner and Conrath 2008), since primed plants do not activate defenses directly upon induction treatment but activate their defenses faster and stronger when subsequently attacked by pathogens or insects (Conrath et al. 2002, 2006).

Most studies on the costs and benefits of plant defense have concentrated on direct activation of defenses, rather than on priming. Moreover, putative benefits of defenses under disease pressure were often not taken into account (Walters and Boyle 2005). We recently demonstrated in Arabidopsis that priming induced by BABA has clear benefits under conditions of disease pressure (Van Hulten et al. 2006). In the absence of pathogens, priming had no or only marginal effect on the relative growth rate (RGR) and seed production of the plant, whereas there were large effects when defenses were directly activated. In the presence of pathogens, a clear fitness advantage was observed for primed plants over non-primed plants and also over plants in which defenses were activated already before pathogen challenge (Van Hulten et al. 2006). These results clearly indicate that under conditions of disease pressure, the benefits of priming outweigh the costs. In agreement with our study, Walters et al. (2009) recently demonstrated in barley (*Hordeum vulgare*) that priming induced by saccharin, a metabolite of the synthetic SAR-inducer probenazole, had no significant effect on plant growth rate and grain yield in the absence of pathogen infection. However, priming significantly increased plant fitness under high inoculum pressure by the hemibiotrophic fungus *Rhynchosporium secalis*. To take unforeseen ecological costs into account, putative fitness costs of saccharin treatment under natural conditions were also evaluated in a field study. Saccharin treatment increased grain yield slightly but not significantly compared to untreated controls, indicating that priming for enhanced defense responses induced by saccharin did not incur allocation costs in barley under field conditions (Walters et al. 2009).

The studies of Van Hulten et al. (2006) and Walters et al. (2009) are the only ones to date that describe the fitness costs and fitness benefits associated with priming of defense. In these studies priming had been induced either chemically or genetically by the *edr1* mutation, which affects a MPKKK (MPK kinase kinase) that is a negative regulator of SA-inducible defense responses in Arabidopsis (Frye and Innes 1998; Frye et al. 2001). No study has yet been designed to elucidate fitness effects of ecologically more relevant induction of priming, like that induced by beneficial micro-organisms. However, Raupach and Kloepper (1998) reported that in two field trials, conducted in separate years, seed treatment of cucumber with PGPR *Baccilus* spp. increased plant growth and reduced disease severity against different pathogens. In another field study, *Bacillus* pretreatment protected cucumber plants from cucumber beetles and the beetle-transmittable bacterial wilt disease (Zehnder et al. 2001). This was accompanied by significant yield increases. Also field-grown tomato plants were protected by *Baccilus* spp. against the cucumber mosaic virus and tomato mottle virus, which was associated with increased plant yield compared to untreated plants (Zehnder et al. 2001). An increase in yield was not observed during a similar field trial conducted in the consecutive cropping season, but viral titers were also not affected by the bacteria in this second year (Zehnder et al. 2001). These results demonstrate fitness benefits for plants that interact with PGPR. These interactions do not seem to incur costs on the plant. Although the priming phenomenon was not the emphasis of these studies, priming is often found to be the underlying mechanism of systemic plant protection by PGPR (Van Wees et al. 2008; Van der Ent et al. 2009a).

8 To What Extent are Plants in the Field Already Primed?

Priming has predominantly been studied under tightly controlled growth conditions with optimum temperature, light intensity, humidity, and nutrient availability for plant growth. In natural environments the above mentioned parameters could be less favorable for the plant. Furthermore, in nature plants interact with many organisms, such as pathogens, herbivores, other plants, and beneficial microorganisms (Pieterse and Dicke 2007). All these parameters may exert an effect on the plant and consequently on the priming response. Conversely, under natural conditions the priming response may have different effects on plant fitness or the plant's ability to respond to stresses than under controlled conditions, because of the different environment of the plant. This way, effects of priming may become apparent that would not be detected under controlled greenhouse conditions. Studies on the fitness consequences of plants that exhibit a primed defense state in the field will be instrumental to better understand the ecological impact of priming of defense responses. In earlier studies priming-inducing agents have shown to protect plants in the field (Beckers and Conrath 2007; Goellner and Conrath 2008). However, in these studies, the primed state was not verified, which is crucial to impute any protective effect of the agents to a primed state because most chemical priming agents can also activate defenses directly when applied at higher doses (Kohler et al. 2002; Van Hulten et al. 2006). Therefore, from these field studies, it cannot be concluded that the induced protection is caused by priming because contribution of direct activation of defenses can not be excluded. Also in field studies with PGPR and PGPF that can induce resistance through priming of defenses, additional protective mechanisms can not be excluded because many of these beneficials are capable of exerting direct effects on pathogenic organisms through competition for nutrients or production of antibiotics.

Since priming can be induced by a plethora of organisms, whether beneficial or harmful, it is possible that plants in the field are already primed to some extent through their continual interaction with the biotic (and abiotic) environment. Walters (2009) recently suggested that in several published field studies, defenses had already been induced in plants to some extent prior to induction treatment. For example, treatment of field-grown wheat with BTH, a functional analog of SA, did not induce SAR-related genes, which may be due to the already high expression levels of these genes before treatment (Pasquer et al. 2005). However, Herman et al. (2007) reported that tomato plants in the field responded to treatment with ASM, a functional analog of SA, with significant induction of defense-related gene expression, despite their enhanced basal level of expression in the field compared to in the greenhouse. Moreover, a second treatment with ASM resulted in a significantly stronger expression response relative to non-pretreated plants. This latter finding implies that although plants may be already primed, this does not compromise their ability to express even higher levels of induced resistance upon subsequent induction.

Priming of defense responses is an inducible phenomenon, indicating that by default the plant is in a noninduced or nonprimed physiological state. It would be interesting to investigate whether plants that grow in their natural habitat, in which they coevolved with indigenous micro-organisms and are well adapted to the local environment, have naturally acquired the primed state. For this purpose, proper controls need to be included (e.g., mutant plants that are affected in priming only) and markers strictly correlated with priming should be assessed. These tools are not known at the moment.

9 Outlook

In recent years, knowledge on resistance induced by beneficial root-colonizing microbes has greatly expanded. There is ever-growing information available on microbes with resistance-inducing activity, plant species that are perceptive to resistance-inducing microbes, and pathogens/insects to which the induced resistance is effective. The picture is emerging that the plant defense signaling pathway that is triggered by ISR-inducing beneficials depends on responsiveness to the plant hormones JA and ET. ISR is not accompanied by massive changes in gene expression. Instead, ISR is established through priming the plant for enhanced defense responses upon encountering a pathogen or insect. Priming seems to be a cost-efficient defense mechanism. Hence, beneficial microbes-induced priming may be an ecologically relevant feature of plants.

Priming by benefical microbes may be a valuable tool for sustainable crop protection. For effective use of priming agents in agriculture, it is critical to investigate whether and to what extent the specific crops in the field may be already primed by their interacting environment and whether this could be further enhanced by application of priming agents. However, the fact that JA-dependent ISR and SA-dependent SAR can have additive effects (Van Wees et al. 2000) implies that different defense pathways can be primed simultaneously, leaving room for enhancement of naturally primed defenses. PGPR can present an attractive alternative to chemical pesticides for protection against pathogens and insects. A major advantage of PGPR is that once systemic resistance is induced, the natural defense mechanisms of the plant are operative for prolonged periods (Van Loon et al. 1998). However, complete disease control is rarely provided by resistance-inducing agents. Research aimed at determining the factors that influence the success rate of the PGPR is necessary for broad implementation of biocontrol agents that consistently provide acceptable levels of disease control in crop protection programs. Besides inducing of resistance PGPR can also directly control soil-borne pathogens, through competition of nutrients or production of antibiotics. Therefore, (a combination of) PGPR strains that can protect plants through different mechanisms are ideally suited to confer consistent, long-lasting protection of crops against various diseases and pests. Mixtures of PGPR strains have indeed been shown to enhance biological control in cucumber and radish against different pathogens compared to the effect of single treatments (Raupach and Kloepper 1998; De Boer et al. 2003).

References

Agrawal AA, Strauss SY, Stout MJ (1999) Costs of induced responses and tolerance to herbivory in male and female fitness components of wild radish. Evolution 53:1093–1104

Ahn I-P, Lee S-W, Suh S-C (2007) Rhizobacteria-induced priming in *Arabidopsis* is dependent on ethylene, jasmonic acid, and NPR1. Mol Plant Microbe Interact 20:759–768

Bakker PAHM, Pieterse CMJ, Van Loon LC (2007) Induced systemic resistance by fluorescent *Pseudomonas* spp. Phytopathology 97:239–243

Baldwin IT (1998) Jasmonate-induced responses are costly but benefit plants under attack in native populations. Proc Natl Acad Sci USA 95:8113–8118

Barto EK, Cipollini D (2005) Testing the optimal defense theory and the growth-differentiation balance hypothesis in *Arabidopsis thaliana*. Oecologia 146:169–178

Beckers GJM, Spoel SH (2006) Fine-tuning plant defence signalling: salicylate versus jasmonate. Plant Biol 8:1–10

Beckers GJM, Conrath U (2007) Priming for stress resistance: from the lab to the field. Curr Opin Plant Biol 10:425–431

Beckers GJM, Jaskiewicz M, Liu Y, Underwood WR, He SY, Zhang S, Conrath U (2009) Mitogen-activated proteinkinases 3 and 6 are required for full priming of stress responses in *Arabidopsis thaliana*. Plant Cell 21:944–953

Berger S, Sinha AK, Roitsch T (2007) Plant physiology meets phytopathology: plant primary metabolism and plant–pathogen interactions. J Exp Bot 58:4019–4026

Bloemberg GV, Lugtenberg BJJ (2001) Molecular basis of plant growth promotion and biocontrol by rhizobacteria. Curr Opin Plant Biol 4:343–350

Bostock RM (2005) Signal crosstalk and induced resistance: straddling the line between cost and benefit. Annu Rev Phytopathol 43:545–580

Bowling SA, Guo A, Cao H, Gordon AS, Klessig DF, Dong X (1994) A mutation in *Arabidopsis* that leads to constitutive expression of systemic acquired resistance. Plant Cell 6:1845–1857

Bruce TJA, Matthes MC, Napier JA, Pickett JA (2007) Stressful "memories" of plants: evidence and possible mechanisms. Plant Sci 173:603–608

Cameron RK, Paiva NC, Lamb CJ, Dixon RA (1999) Accumulation of salicylic acid and *PR* gene transcripts in relation to the systemic acquired resistance (SAR) response by *Pseudomonas syringae* pv. *tomato* in *Arabidopsis*. Physiol Mol Plant Pathol 55:121–130

Cao H, Bowling SA, Gordon AS, Dong X (1994) Characterization of an *Arabidopsis* mutant that is nonresponsive to inducers of systemic acquired resistance. Plant Cell 6:1583–1592

Cartieaux F, Contesto C, Gallou A, Desbrosses G, Kopka J, Taconnat L, Renou J-P, Touraine B (2008) Simultaneous interaction of *Arabidopsis thaliana* with *Bradyrhizobium* sp. strain ORS278 and *Pseudomonas syringae* pv. *tomato* DC3000 leads to complex transcriptome changes. Mol Plant Microbe Interact 21:244–259

Cipollini DF (2002) Does competition magnify the fitness costs of induced responses in *Arabidopsis thaliana*? A manipulative approach. Oecologia 131:514–520

Conrath U, Pieterse CMJ, Mauch-Mani B (2002) Priming in plant–pathogen interactions. Trends Plant Sci 7:210–216

Conrath U, Beckers GJM, Flors V, García-Agustín P, Jakab G, Mauch F, Newman M-A, Pieterse CMJ, Poinssot B, Pozo MJ, Pugin A, Schaffrath U, Ton J, Wendehenne W, Zimmerli L, Mauch-Mani B (2006) Priming: getting ready for battle. Mol Plant Microbe Interact 19:1062–1071

De Boer M, Bom P, Kindt F, Keurentjes JJB, Van der Sluis I, Van Loon LC, Bakker PAHM (2003) Control of Fusarium wilt of radish by combining *Pseudomonas putida* strains that have different disease-suppressive mechanisms. Phytopathology 93:626–632

De Bruijn I, De Kock MJD, Yang M, De Waard P, Van Beek TA, Raaijmakers JM (2007) Genome-based discovery, structure prediction and functional analysis of cyclic lipopeptide antibiotics in *Pseudomonas* species. Mol Microbiol 63:417–428

De Meyer G, Bigirimana J, Elad Y, Höfte M (1998) Induced systemic resistance in *Trichoderma harzianum* T39 biocontrol of *Botrytis cinerea*. Eur J Plant Pathol 104:279–286
De Vos M, Van Zaanen W, Koornneef A, Korzelius JP, Dicke M, Van Loon LC, Pieterse CMJ (2006) Herbivore-induced resistance against microbial pathogens in *Arabidopsis*. Plant Physiol 142:352–363
De Vos M, Van Oosten VR, Van Poecke RMP, Van Pelt JA, Pozo MJ, Muelle MJ, Buchala AJ, Métraux JP, Van Loon LC, Dicke M, Pieterse CMJ (2005) Signal signature and transcriptome changes of *Arabidopsis* during pathogen and insect attack. Mol Plant Microbe Interact 18:923–937
Delaney TP, Friedrich L, Ryals JA (1995) Arabidopsis signal transduction mutant defective in chemically and biologically induced disease resistance. Proc Natl Acad Sci USA 92: 6602–6606
Dietrich R, Ploss K, Heil M (2004) Constitutive and induced resistance to pathogens in *Arabidopsis thaliana* depends on nitrogen supply. Plant Cell Environ 27:896–906
Dietrich R, Ploss K, Heil M (2005) Growth responses and fitness costs after induction of pathogen resistance depend on environmental conditions. Plant Cell Environ 28:211–222
Dong J, Chen C, Chen Z (2003) Expression profiles of the *Arabidopsis* WRKY gene superfamily during plant defense response. Plant Mol Biol 51:21–37
Durrant WE, Dong X (2004) Systemic acquired resistance. Annu Rev Phytopathol 42:185–209
Ellis C, Turner JG (2001) The *Arabidopsis* mutant *cev1* has constitutively active jasmonate and ethylene signal pathways and enhanced resistance to pathogens. Plant Cell 13:1025–1033
Engelberth J, Alborn HT, Schmelz EA, Tumlinson JH (2004) Airborne signals prime plants against insect herbivore attack. Proc Natl Acad Sci USA 101:1781–1785
Fan WH, Dong XN (2002) In vivo interaction between NPR1 and transcription factor TGA2 leads to salicylic acid-mediated gene activation in *Arabidopsis*. Plant Cell 14:1377–1389
Frye CA, Innes RW (1998) An *Arabidopsis* mutant with enhanced resistance to powdery mildew. Plant Cell 10:947–956
Frye CA, Tang DZ, Innes RW (2001) Negative regulation of defense responses in plants by a conserved MAPKK kinase. Proc Natl Acad Sci USA 98:373–378
Glandorf DCM, Bakker PAHM, Van Loon LC (1997) Influence of the production of antibacterial and antifungal proteins by transgenic plants on the saprophytic soil microflora. Acta Bot Neerl 46:85–104
Goellner K, Conrath U (2008) Priming: it's all the world to induced disease resistance. Eur J Plant Pathol 121:233–242
Harman GE, Howell CR, Viterbo A, Chet I, Lorito M (2004) *Trichoderma* species-opportunistic, avirulent plant symbionts. Nat Rev Microbiol 2:43–56
Harrison MJ (2005) Signaling in the arbuscular mycorrhizal symbiosis. Annu Rev Microbiol 59:19–42
Hase S, Van Pelt JA, Van Loon LC, Pieterse CMJ (2003) Colonization of *Arabidopsis* roots by *Pseudomonas fluorescens* primes the plant to produce higher levels of ethylene upon pathogen infection. Physiol Mol Plant Pathol 62:219–226
Heidel AJ, Clarke JD, Antonovics J, Dong X (2004) Fitness costs of mutations affecting the systemic acquired resistance pathway in *Arabidopsis thaliana*. Genetics 168:2197–2206
Heil M (2002) Ecological costs of induced resistance. Curr Opin Plant Biol 5:345–350
Heil M, Baldwin IT (2002) Fitness costs of induced resistance: emerging experimental support for a slippery concept. Trends Plant Sci 7:61–67
Heil M, Hilper A, Kaiser W, Linsenmair KE (2000) Reduced growth and seed set following chemical induction of pathogen defence: does systemic acquired resistance (SAR) incur allocation costs? J Ecol 88:645–654
Herman MAB, Restrepo S, Smart CD (2007) Defense gene expression patterns of three SAR-induced tomato cultivars in the field. Physiol Mol Plant Pathol 71:192–200
Jones JDG, Dangl JL (2006) The plant immune system. Nature 444:323–329

Jung HW, Tschaplinski TJ, Wang L, Glazebrook J, Greenberg JT (2009) Priming in systemic plant immunity. Science 324:89–91

Kagaya Y, Hobo T, Murata M, Ban A, Hattori T (2002) Abscisic acid-induced transcription is mediated by phosphorylation of an abscisic acid response element binding factor, TRAB1. Plant Cell 14:3177–3189

Kloepper JW, Ryu C-M, Zhang SA (2004) Induced systemic resistance and promotion of plant growth by *Bacillus* spp. Phytopathology 94:1259–1266

Knoester M, Pieterse CMJ, Bol JF, Van Loon LC (1999) Systemic resistance in *Arabidopsis* induced by rhizobacteria requires ethylene-dependent signaling at the site of application. Mol Plant Microbe Interact 12:720–727

Kohler A, Schwindling S, Conrath U (2002) Benzothiadiazole-induced priming for potentiated responses to pathogen infection, wounding, and infiltration of water into leaves requires the NPR1/NIM1 gene in *Arabidopsis*. Plant Physiol 128:1046–1056

Koornneef A, Pieterse CMJ (2008) Cross talk in defense signaling. Plant Physiol 146:839–844

Koornneef A, Leon-Reyes A, Ritsema T, Verhage A, Den Otter FC, Van Loon LC, Pieterse CMJ (2008) Kinetics of salicylate-mediated suppression of jasmonate signaling reveal a role for redox modulation. Plant Physiol 147:1358–1368

Lawton K, Weymann K, Friedrich L, Vernooij B, Uknes S, Ryals J (1995) Systemic acquired resistance in *Arabidopsis* requires salicylic acid but not ethylene. Mol Plant Microbe Interact 8:863–870

Leeman M, Van Pelt JA, Den Ouden FM, Heinsbroek M, Bakker PAHM, Schippers B (1995) Induction of systemic resistance against fusarium wilt of radish by lipopolysaccharides of *Pseudomonas fluorescens*. Phytopathology 85:1021–1027

Leon-Reyes A, Spoel SH, De Lange ES, Abe H, Kobayashi M, Tsuda S, Millenaar FF, Welschen RAM, Ritsema T, Pieterse CMJ (2009) Ethylene modulates the role of NONEXPRESSOR OF PATHOGENESIS-RELATED GENES1 in cross talk between salicylate and jasmonate signaling. Plant Physiol 149:1797–1809

Liu J, Maldonado-Mendoza I, Lopez-Meyer M, Cheung F, Town CD, Harrison MJ (2007) Arbuscular mycorrhizal symbiosis is accompanied by local and systemic alterations in gene expression and an increase in disease resistance in the shoots. Plant J 50:529–544

Lorenzo O, Solano R (2005) Molecular players regulating the jasmonate signalling network. Curr Opin Plant Biol 8:532–540

Lorenzo O, Chico JM, Sanchez-Serrano JJ, Solano R (2004) *JASMONATE-INSENSITIVE1* encodes a MYC transcription factor essential to discriminate between different jasmonate-regulated defense responses in *Arabidopsis*. Plant Cell 16:1938–1950

Maleck K, Levine A, Eulgem T, Morgan A, Schmid J, Lawton KA, Dangl JL, Dietrich RA (2000) The transcriptome of *Arabidopsis thaliana* during systemic acquired resistance. Nat Genet 26:403–410

Mosher RA, Durrant WE, Wang D, Song J, Dong X (2006) A comprehensive structure–function analysis of *Arabidopsis* SNI1 defines essential regions and transcriptional repressor activity. Plant Cell 18:1750–1765

Mou Z, Fan W, Dong X (2003) Inducers of plant systemic acquired resistance regulate NPR1 function through redox changes. Cell 113:935–944

Nürnberger T, Brunner F, Kemmerling B, Piater L (2004) Innate immunity in plants and animals: striking similarities and obvious differences. Immunol Rev 198:249–266

Osbourn AE (1996) Preformed antimicrobial compounds and plant defense against fungal attack. Plant Cell 8:1821–1831

Pasquer F, Isidore E, Zarn J, Keller B (2005) Specific patterns of changes in wheat gene expression after treatment with three antifungal compounds. Plant Mol Biol 57:693–707

Pieterse CMJ, Dicke M (2007) Plant interactions with microbes and insects: from molecular mechanisms to ecology. Trends Plant Sci 12:564–569

Pieterse CMJ, Leon-Reyes A, Van der Ent S, Van Wees SCM (2009) Networking by small-molecule hormones in plant immunity. Nat Chem Biol 5:308–316

Pieterse CMJ, Van Wees SCM, Hoffland E, Van Pelt JA, Van Loon LC (1996) Systemic resistance in *Arabidopsis* induced by biocontrol bacteria is independent of salicylic acid accumulation and pathogenesis-related gene expression. Plant Cell 8:1225–1237

Pieterse CMJ, Van Wees SCM, Van Pelt JA, Knoester M, Laan R, Gerrits H, Weisbeek PJ, Van Loon LC (1998) A novel signaling pathway controlling induced systemic resistance in *Arabidopsis*. Plant Cell 10:1571–1580

Pozo MJ, Azcón-Aguilar C (2007) Unraveling mycorrhiza-induced resistance. Curr Opin Plant Biol 10:393–398

Pozo MJ, Van der Ent S, Van Loon LC, Pieterse CMJ (2008) Transcription factor MYC2 is involved in priming for enhanced defense during rhizobacteria-induced systemic resistance in *Arabidopsis thaliana*. New Phytol 180:511–523

Pré M, Atallah M, Champion A, De Vos M, Pieterse CMJ, Memelink J (2008) The AP2/ERF domain transcription factor ORA59 integrates jasmonic acid and ethylene signals in plant defense. Plant Physiol 147:1347–1357

Raupach GS, Kloepper JW (1998) Mixtures of plant growth-promoting rhizobacteria enhance biological control of multiple cucumber pathogens. Phytopathology 88:1158–1164

Reymond P, Farmer EE (1998) Jasmonate and salicylate as global signals for defense gene expression. Curr Opin Plant Biol 1:404–411

Rojo E, Solano R, Sanchez-Serrano JJ (2003) Interactions between signaling compounds involved in plant defense. J Plant Growth Regul 22:82–98

Roman G, Lubarsky B, Kieber JJ, Rothenberg M, Ecker JR (1995) Genetic analysis of ethylene signal transduction in *Arabidopsis thaliana*: five novel mutant loci integrated into a stress response pathway. Genetics 139:1393–1409

Schwessinger B, Zipfel C (2008) News from the frontline: recent insights into PAMP-triggered immunity in plants. Curr Opin Plant Biol 11:389–395

Segarra G, Van der Ent S, Trillas I, Pieterse CMJ (2009) MYB72, a node of convergence in induced systemic resistance triggered by a fungal and a bacterial beneficial microbe. Plant Biol 11:90–96

Shah J, Tsui F, Klessig DF (1997) Characterization of a salicylic acid-insensitive mutant (*sai1*) of *Arabidopsis thaliana*, identified in a selective screen utilizing the SA-inducible expression of the *tms2* gene. Mol Plant Microbe Interact 10:69–78

Shoresh M, Yedidia I, Chet I (2005) Involvement of jasmonic acid/ethylene signaling pathway in the systemic resistance induced in cucumber by *Trichoderma asperellum* T203. Phytopathology 95:76–84

Spaink HP (2000) Root nodulation and infection factors produced by rhizobial bacteria. Annu Rev Microbiol 54:257–288

Spoel SH, Johnson JS, Dong X (2007) Regulation of tradeoffs between plant defenses against pathogens with different lifestyles. Proc Natl Acad Sci USA 104:18842–18847

Spoel SH, Koornneef A, Claessens SMC, Korzelius JP, Van Pelt JA, Mueller MJ, Buchala AJ, Métraux J-P, Brown R, Kazan K, Van Loon LC, Dong X, Pieterse CMJ (2003) NPR1 modulates cross-talk between salicylate- and jasmonate-dependent defense pathways through a novel function in the cytosol. Plant Cell 15:760–770

Stein E, Molitor A, Kogel K-H, Waller F (2008) Systemic resistance in *Arabidopsis* conferred by the mycorrhizal fungus *Piriformospora indica* requires jasmonic acid signaling and the cytoplasmic function of NPR1. Plant Cell Physiol 49:1747–1751

Ton J, Van Pelt JA, Van Loon LC, Pieterse CMJ (2002) Differential effectiveness of salicylate-dependent and jasmonate/ethylene-dependent induced resistance in *Arabidopsis*. Mol Plant Microbe Interact 15:27–34

Ton J, D'Alessandro M, Jourdie V, Jakab G, Karlen D, Held M, Mauch-Mani B, Turlings TCJ (2007) Priming by airborne signals boosts direct and indirect resistance in maize. Plant J 49:16–26

Van Dam NM, Baldwin IT (2001) Competition mediates costs of jasmonate-induced defences, nitrogen acquisition and transgenerational plasticity in *Nicotiana attenuata*. Funct Ecol 15:406–415

Van der Ent S, Van Wees SCM, Pieterse CMJ (2009a) Jasmonate signals in plant interactions with beneficial microbes. Phytochemistry 70:1581–1588

Van der Ent S, Van Hulten MHA, Pozo MJ, Czechowski T, Udvardi MK, Pieterse CMJ, Ton J (2009b) Priming of plant innate immunity by rhizobacteria and beta-aminobutyric acid: differences and similarities in regulation. New Phytol 183:419–431

Van der Ent S, Verhagen BWM, Van Doorn R, Bakker D, Verlaan MG, Pel MJC, Joosten RG, Proveniers MCG, Van Loon LC, Ton J, Pieterse CMJ (2008) MYB72 is required in early signaling steps of rhizobacteria-induced systemic resistance in *Arabidopsis*. Plant Physiol 146:1293–1304

Van Hulten M, Pelser M, Van Loon LC, Pieterse CMJ, Ton J (2006) Costs and benefits of priming for defense in *Arabidopsis*. Proc Natl Acad Sci USA 103:5602–5607

Van Loon LC (1997) Induced resistance and the role of pathogenesis-related proteins. Eur J Plant Pathol 103:753–765

Van Loon LC (2000) Systemic induced resistance. In: Slusarenko AJ, Fraser RSS, Van Loon LC (eds) Mechanisms of resistance to plant diseases. Kluwer, Dordrecht, pp 521–574

Van Loon LC, Van Strien EA (1999) The families of pathogenesis-related proteins, their activities, and comparative analysis of PR-1 type proteins. Physiol Mol Plant Pathol 55:85–97

Van Loon LC, Bakker PAHM (2005) Induced systemic resistance as a mechanism of disease suppression by rhizobacteria. In: Siddiqui ZA (ed) PGPR: biocontrol and biofertilization. Springer, Dordrecht, pp 39–66

Van Loon LC, Bakker PAHM (2006) Root-associated bacteria inducing systemic resistance. In: Gnanamanickam SS (ed) Plant-associated bacteria. Springer, Dordrecht, pp 269–316

Van Loon LC, Bakker PAHM, Pieterse CMJ (1998) Systemic resistance induced by rhizosphere bacteria. Annu Rev Phytopathol 36:453–483

Van Loon LC, Rep M, Pieterse CMJ (2006) Significance of inducible defense-related proteins in infected plants. Annu Rev Phytopathol 44:135–162

Van Oosten VR, Bodenhausen N, Reymond P, Van Pelt JA, Van Loon LC, Dicke M, Pieterse CMJ (2008) Differential effectiveness of microbially induced resistance against herbivorous insects in *Arabidopsis*. Mol Plant Microbe Interact 21:919–930

Van Peer R, Schippers B (1992) Lipopolysaccharides of plant-growth promoting *Pseudomonas* sp. strain WCS417r induce resistance in carnation to fusarium wilt. Neth J Plant Pathol 98:129–139

Van Peer R, Niemann GJ, Schippers B (1991) Induced resistance and phytoalexin accumulation in biological control of fusarium wilt of carnation by *Pseudomonas* sp. strain WCS417r. Phytopathology 81:728–734

Van Wees SCM, Van der Ent S, Pieterse CMJ (2008) Plant immune responses triggered by beneficial microbes. Curr Opin Plant Biol 11:443–448

Van Wees SCM, Luijendijk M, Smoorenburg I, Van Loon LC, Pieterse CMJ (1999) Rhizobacteria-mediated induced systemic resistance (ISR) in *Arabidopsis* is not associated with a direct effect on expression of known defense-related genes but stimulates the expression of the jasmonate-inducible gene *Atvsp* upon challenge. Plant Mol Biol 41:537–549

Van Wees SCM, De Swart EAM, Van Pelt JA, Van Loon LC, Pieterse CMJ (2000) Enhancement of induced disease resistance by simultaneous activation of salicylate- and jasmonate-dependent defense pathways in *Arabidopsis thaliana*. Proc Natl Acad Sci USA 97:8711–8716

Van Wees SCM, Pieterse CMJ, Trijssenaar A, Van't Westende YAM, Hartog F, Van Loon LC (1997) Differential induction of systemic resistance in *Arabidopsis* by biocontrol bacteria. Mol Plant Microbe Interact 10:716–724

Verhagen BWM, Glazebrook J, Zhu T, Chang H-S, Van Loon LC, Pieterse CMJ (2004) The transcriptome of rhizobacteria-induced systemic resistance in *Arabidopsis*. Mol Plant Microbe Interact 17:895–908

Vinale F, Sivasithamparam K, Ghisalberti EL, Marra R, Woo SL, Lorito M (2008) *Trichoderma*–plant–pathogen interactions. Soil Biol Biochem 40:1–10

Waller F, Achatz B, Baltruschat H, Fodor J, Becker K, Fischer M, Heier T, Huckelhoven R, Neumann C, von Wettstein D, Franken P, Kogel K-H (2005) The endophytic fungus *Piriformospora indica* reprograms barley to salt-stress tolerance, disease resistance, and higher yield. Proc Natl Acad Sci USA 102:13386–13391

Walters D, Heil M (2007) Costs and trade-offs associated with induced resistance. Physiol Mol Plant Pathol 71:3–17

Walters DR (2009) Are plants in the field already induced? Implications for practical disease control. Crop Protect 28:459–465

Walters DR, Boyle C (2005) Induced resistance and allocation costs: what is the impact of pathogen challenge? Physiol Mol Plant Pathol 66:40–44

Walters DR, Paterson L, Walsh DJ, Havis ND (2009) Priming for plant defense in barley provides benefits only under high disease pressure. Physiol Mol Plant Pathol 73:95–100

Wang D, Amornsiripanitch N, Dong X (2006) A genomic approach to identify regulatory nodes in the transcriptional network of systemic acquired resistance in plants. PLoS Pathog 2:1042–1050

Weller DM, Raaijmakers JM, McSpadden Gardener BB, Thomashow LS (2002) Microbial populations responsible for specific soil suppressiveness to pathogens. Annu Rev Phytopathol 40:309–348

Zehnder GW, Murphy JF, Sikora EJ, Kloepper JW (2001) Application of rhizobacteria for induced resistance. Eur J Plant Pathol 107:39–50

Zeyen RJ, Carver TLW, Lyngkjær MF (2002) The formation and role of papillae. Carver APS, St Paul, Minnesota

Zhang H, Xie X, Kim M-S, Kornyeyev DA, Holaday S, Paré PW (2008) Soil bacteria augment *Arabidopsis* photosynthesis by decreasing glucose sensing and abscisic acid levels *in planta*. Plant J 56:264–273

Zimmerli L, Métraux J-P, Mauch-Mani B (2001) b-Aminobutyric acid-induced protection of *Arabidopsis* against the necrotrophic fungus *Botrytis cinerea*. Plant Physiol 126:517–523

Zimmerli L, Jakab G, Métraux J-P, Mauch-Mani B (2000) Potentiation of pathogen-specific defense mechanisms in *Arabidopsis* by b-aminobutyric acid. Proc Natl Acad Sci USA 97:12920–12925

Allelopathy and Exotic Plant Invasion

Amutha Sampath Kumar and Harsh P. Bais

Abstract Biologists have long searched for an explanation as to why some plant invaders become much more dominant in their naturalized range than in their native range, and, accordingly, several nonexclusive hypotheses have been proposed. Plants are unparalleled factories for the production of diverse biochemicals, and allelochemistry has recently re-emerged as a possible mechanism for the success of some invasive plants. The idea is that some invaders may succeed because they possess unique allelopathic biochemistry to which naïve natives have not adapted. Indeed, there are a number of studies that support this hypothesis. In this chapter, we revise and expand this biochemical hypothesis and discuss experimental and conceptual advances and limitations.

1 Introduction

Allelopathy "is the phytotoxicity of a compound or a group of compounds released from plant part by leaching, root exudation, volatilization, or residue decomposition to susceptible plants" (Inderjit et al. 2006). In terms of diversity, they encompass a wide range of compounds that vary in their chemical structure, concentration, and localization in the plants, tissues, and between species, which are all linked to the changes in the biotic and abiotic conditions (Inderjit and Duke 2003). Allelochemicals predominantly organic in its form can be highly selective in their influence or may include a broad target range. Allelopathy plays a pivotal role in understanding the mechanism of exotic plant invasion in naïve communities and has been studied in

A.S. Kumar and H.P. Bais (✉)
Department of Plant and Soil Sciences, University of Delaware, Newark, DE 19711, USA
e-mail: amutha@udel.edu

Delaware Biotechnology Institute, 15 Innovation Way, Newark, DE 19711, USA
e-mail: hbais@udel.edu

F. Baluška and V. Ninkovic (eds.), *Plant Communication from an Ecological Perspective*, Signaling and Communication in Plants,
DOI 10.1007/978-3-642-12162-3_4,

copious number of plants. This chapter enumerates how novel biochemicals drive exotic invasion around the world and explains how biochemicals along with the soil microorganisms have expanded our knowledge of success in some of the invasive plants.

2 Allelopathy and Exotic Plant Invasion

Exotic plant invasion has long been a poorly understood ecological phenomenon until several authors (Williams 1954; Elton 1958; Crawley 1987, 1997; Mack et al. 2000; Maron and Vilà 2001; Levine et al. 2002) proposed the "natural enemies hypothesis". Convincingly, this hypothesis explains the importance of natural enemies hypothesis with several successful plant invasions (Wolfe 2002; Siemann and Rogers 2003; Reinhart et al. 2003; Callaway et al. 2004a, b; DeWalt et al. 2004; Jakobs et al. 2004). However, sequential evidence supporting minimal effects of natural enemies on invaders (Lesica and Hanna 2004), similar consumer based effects on the natives and invaders (Beckstead and Parker 2003; Maron and Vilà 2001; Reinhart and Callaway 2004), and lack of relative population ecology studies on the invaders in the invaded and the native range led to the dawn of the "novel weapons hypothesis (NWH)". The NWH "holds that some exotics transform from native weaklings to invasive bullies by biosynthesizing biochemicals that are allelopathic to plants or soil microbes in the invaded communities, but relatively ineffective against natural neighbors that had adapted over time" (Callaway and Ridenour 2004). Though a parallel concept "evolution of increased competitive ability (EICA)" was also argued regarding the shedding off of costly traits used for survival in the native range and more resource allocation for competitive advantage in the new invaded range, the NWH addressed the concept of invasion of exotic species and also the evolution of higher competitive ability.

Invasion of plants in the invaded communities is through the release of novel phytochemicals called allelopathic compounds, which reduce the fitness of the neighboring plants or are phytotoxic to the non-coevolved neighbors (Callaway and Aschehoug 2000). Allelopathy can be either direct wherein the invading plants release the alleochemicals that encourage the spread of the invader in the invaded range or indirect which encompasses the soil microorganisms that mediate changes affecting the ecosystem. Root-mediated allelopathy is often due to the enhanced overall production of the root exudates when plants encounter stress conditions such as UV exposure, drought, and temperature (Pramanik et al. 2000; Inderjit and Weston 2003). The plant root exudates represent a fairly major input of plant-based chemicals in the rhizosphere and therefore the large source of allelochemicals (Bertin et al. 2003).

A well documented example of root-mediated allelopathy in an invasive plant species is the exudation of racemic catechin by the spotted knapweed *Centaurea maculosa* (Callaway et al. 1999; Callaway and Aschehoug 2000) in the invaded North American range. One of the well characterized compound released from the roots through its rhizosphere is (±)-catechin (Bais et al. 2003; Ridenour et al. 2008; Blair et al. 2005; Weir et al. 2006; Perry et al. 2005, 2007). Research indicates that the invasiveness of *C. maculosa* is attributed to the stronger allelopathic effects on the native North American species than on the related European species which is one of the unquestionable aspects of the "NWH." A parallel argument is that *C. maculosa* plants from the invaded range are evolved to be better competitors than the con-specifics from the European range which is paraphrased as "EICA" (Inderjit et al. 2006; Thorpe et al. 2009). However He et al. (2009) argue that apart from the competitive ability and evolution of the increased size that contributes to the invasive success of *C. maculosa* in North America, it is the composition of the plant community in every invaded region, i.e., biogeographic differences that are the deciding factors contributing to the competitive effect of the *C. maculosa*.

Several authors have documented the phytotoxicity of (±)-catechin (Simões et al. 2008); however, controversies surround the phytotoxicity at very high levels (Blair et al. 2005; Furubayashi et al. 2007). Blair et al. (2005, 2006) and Duke et al. (2009) maintained that catechin is an antioxidant and may not participate in the oxidative stress. However, Kaushik et al. (2010) emphasize the decisive conditions that have to be rigidly followed to evaluate the phytotoxicity studies involving catechin. The authors have explained that catechin is indeed a ROS activator in *Arabidopsis* and there was an up regulation of genes corresponding to cell death (*acd2* and *cad1*) and oxidative stress signaling pathways (Bais et al. 2003). In addition, the authors have also proved the induction of the pro-oxidant activity of catechin due to divalent transition metal ions in the growth medium.

Another known invasive weed that requires in depth study is *Phragmites australis*, which invades marsh and wetland ecosystems in the North America, especially in the eastern United States. Attempts to characterize the potent allelochemical responsible for displacing native *Spartina alterniflora* in North America revealed the phytotoxin gallic acid (3,4,5-trihydroxybenzoic acid) (Rudrappa et al. 2007, 2009). Acute rhizotoxicity of gallic acid was tested in *A. thaliana*, where gallic acid triggered elevated levels of reactive oxygen species (ROS) in the roots, leading to the ROS-mediated microtubule disruption, which was deduced as the primary method of destruction to eliminate the native species in the invaded range (Fig. 1). The sequential analysis of the putative allelochemical gallic acid on several plant species revealed that it is indeed a potent broad-spectrum rhizotoxin. In a concomitant report Bains et al. (2009) have sustained the phytotoxic nature of gallic acid in the invasive *P. australis*.

A similar example of a wetland invasive plant that is considered to be a putative source of allelochemicals is the narrow leaved cattail *Typha augustifolia* which is a native European plant species that occurs as monocultures in the wetland ecosystems in the US dominating the broad leaved cattail *T. latifolia* (Jarchow and Cook 2009).The root-mediated allelopathic effects in *T. augustifolia* were studied by growing it along with the native North American bulrush (*Bolboschoenus fluviatilis*) which is common in the wetland communities of North America. In the presence of activated carbon (that adsorbs the exuded phytochemical) the bulrush showed favorable growth responses (increase in the biomass, increase in the leaf length)

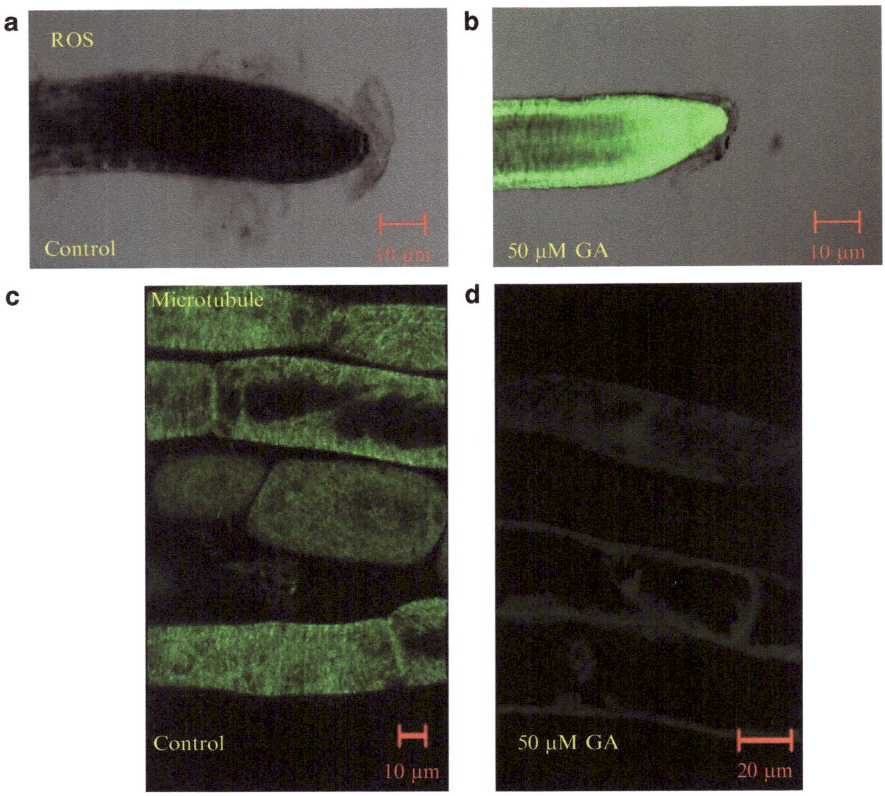

Fig. 1 Gallic acid (GA) a secreted phenolic acid from *Phragmites australis* triggers oxidative stress response (**a–b**) in *Arabidopsis thaliana* roots which causes microtubule disruption (**c–d**) leading to rhizotoxicity

as compared to those grown in the absence of activated carbon. The authors did not track down the key ingredient involved in allelopathy but explained a variation in the soluble phenolics in the roots of *T. augustifolia* and *T. latifolia* which in turn affects the exudation of the allelopathic compounds in the wetlands. In the absence of the carbon treatment, bulrush out-competed *T. augustifolia* (total biomass was 87% lower) indicating resource competition. It must not be forgotten that in wetland ecosystem target phytochemicals have to reach sufficiently higher levels to become allelopathic. The authors also speculate that mycorrhizal degradation of the exuded phytochemical might have occurred in the natural ecosystem.

Volatile root leachates have been described as the primary weapon in the spread of the bitou bush (*Chrysanthemoides monilifera* spp. *Rotundata*) in the New South Wales coastline of Australia. This bush has been reported to suppress the germination of native plants including dominant *Acacia longifolia* var. *sophorae*. It was identified by Ens et al. (2009) that the invasive bitou bush produces volatile allelochemicals especially sesquiterpenes which were found to be in high concentrations in the bitou bush dominated soils and exhibited phytotoxicity on the native plants by diminishing seedling growth and establishment. Terpenes in general play a significant role in the ecosystem function and composition. They play a vital role in the plant defense against vertebrates, invertebrates, and microbes. One major terpene identified in the bitou bush exudates was 5-hydroxycalamenene which constituted the bulk of the fraction analyzed and was a phytotoxin when tested in the native sedge *I. nodosa*. It has been hypothesized that the native plant species *Acacia* release alkanes, which are sources of carbon for microbes, and that the absence of any of these alkanes in the bitou bush-dominated soil may restructure the entire soil community, thereby nurturing the spread of the exotic bitou bush in the Australian coastlines.

3 Allelopathy and Microbial Accomplices in Invasion

Soil microorganisms regulate multiple processes in the ecosystem (Wolfe and Klironomos 2005) and plants widen their understanding with the soil micro biota through the supply of resources that include leaf litter inputs, root exudates, and deposition of organic compounds (Pivato et al. 2007). From the ecological perspective, soil biota have profound positive and negative effects on plants. The microbes can either have a mutualistic relationship as in a root-fungal association or can be responsible for driving vital nutrient cycles. In exchange, plants involve themselves in a dynamic negative or positive feedback. The positive feedback constitutes accumulation of the soil microbes such as mycorrhizal fungi and nitrogen fixers in their roots that promote survival. On the contrary, plants encourage the growth and accumulation of pathogenic microbes that inhibit survival (negative feedback) (Callaway et al. 2004a, b).

The relevance of the microbes in exotic species invasion gained attention when biologists insisted on conducting studies and impacts of the invaders in the home range, i.e., biogeographical approach (Callaway et al. 2004a, b; Inderjit et al. 2008) and evidence indicates that they have powerful effects on the invasions (Reinhart and Callaway 2004). One of the causal reasons for variations in the allelopathic effects is mainly due to the differences in the soil in different locations. (Inderjit et al. 1996). Shifts in the soil pH, salinity, carbon and nitrogen content, and moisture all contribute to the significant modification in the below ground communities (Ehrenfeld 2003).

Invasive species in the home land may have accumulated resident microbial pathogens (negative feedback) and these species in the invaded range demonstrate positive feedbacks. The influence of the microbes and the extent of their involvement in allelopathy might differ, i.e., the released phytochemical might be toxic to the plants at higher concentrations or can be reduced in its toxicity by the microbes. In the second case, the released compound can be innocuous but the product after degradation can be toxic (Inderjit and Dakshini 1999). The persistence of most of the allelochemicals is dependent on factors such as retention, transport, and transformation in the soil before the allelochemicals can be actually phytotoxic to the plants (Inderjit 2005). This in turn is dependent on the soil microflora, the type of the allelochemical, and the substrate conditions. With an invader in the new range, communication between aboveground and below ground, flora and fauna can be disrupted transforming pre-established communities (Kisa et al. 2007). Underground microbial input towards exotic plant invasion has been well studied in few invading plant species. However, with regard to most of the invaders, the studies are in a very preliminary stage of investigation. In most of the invading systems there is a community effect, i.e., the causal organism was not identified. In the subsequent pages the impact of microbes and their involvement in promoting allelopathy and invasion are discussed.

The involvement of the allelochemicals in structuring the soil communities has only been recently investigated. Bais et al. (2003) and recently Tharayil and Triebwasser, (2010) showed that the spotted knapweed (*C. maculosa*) exuded (±)-catechin. In another *Centaurea* species, *C. diffusa* (Eurasian diffuse knapweed), which invades many parts of western North America, 8-hydroxyquinoline was identified as the potent allelochemical exuded from its roots (Vivanco et al. 2004; Tharayil et al. 2008). It is speculated that this allelochemical may cause a shift in the soil microbial community (Callaway et al. 2004a, b) thus helping the spread of this noxious invader. For both these *Centaurea* species the concentration of their respective allelochemicals (±)-catechin and 8-hydroxyquinoline is twice as high in the rhizosphere of the invasive plants than in the native plants.

Garlic mustard (*Alliaria petiolata*) is another exotic plant from Europe (garlic mustard) which belongs to the family Brassicaceae known for the production of glucosinolates (Wolfe and Klironomos 2005). The hydrolysis of these glucosinolates yielded allyl isothiocyanate, benzyl isothiocyanate, and glucotropaeolin, which were proven to have allelopathic effects on certain plants. These compounds enter the soil system through litter decomposition and root exudation, which in turn alters the soil microbial community (Vaughn and Berhow 1999). Garlic mustard has turned out to be an invader by inhibiting Arbuscular Mycorrhizal Fungi (AMF) population probably through phytochemical inhibition (Callaway et al. 2008; Stinson et al. 2006).

A justification to the logical extension of the NWH in the invasion of the exotic reed *P. australis* has been put forth by Bains et al. (2009). Their findings have shown that elevated levels of gallotannins in the rhizosphere of the native *Phragmites* are available for attack by the enzyme tannase produced by the soil bacteria in the invaded range, which releases gallic acid, thereby aggravating the invasiveness of the exotic *Phragmites*. Moreover, the native plant communities also produce higher tannase, resulting in them getting targeted by release of gallic acid from exotic *Phragmites* (Fig. 2). This suggests the potential linking of the native plant and microbe interaction that encourages invasive traits of the exotic *Phragmites*. However, the soil microbiota in the home/invaded range needs to be analyzed for their positive/negative effect.

A prototype for negative feedback effect of the soil pathogens to suppress the native population was observed in the invasion of *Chromolaena odorata*

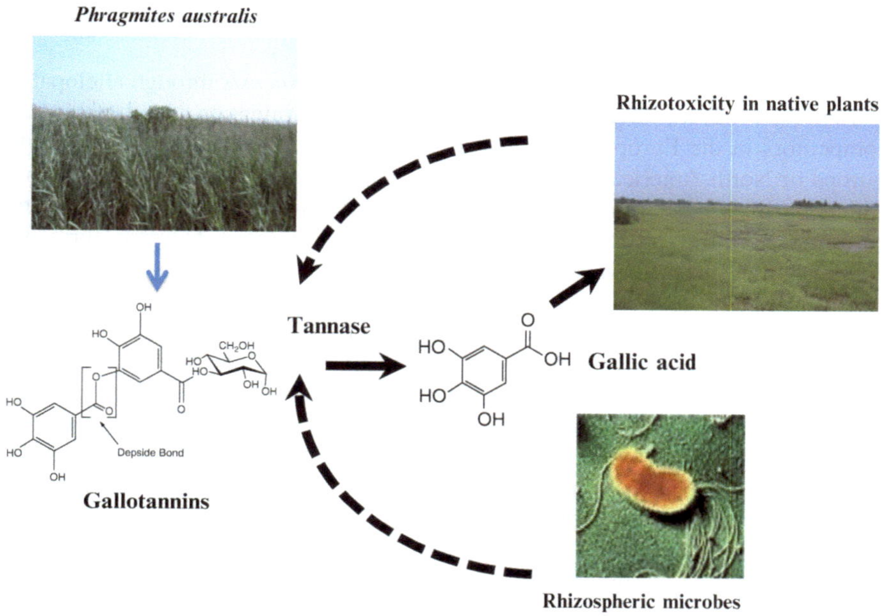

Fig. 2 A schematic to show the gallotannin degradation in rhizosphere triggered by tannase donors. The major difference between exotic and native *Phragmites australis* is at the level of gallotannin production that can be acted upon by native plant and microbial tannase

(Mangla et al. 2008). Through experimentation, *Chromolaena* was observed to suppress the native populations in the Western Ghats region in India by recruiting *Fusarium* species which fuels the spread of the invading *Chromolaena*. Direct allelopathic experiments using activated carbon revealed that the native species were unaffected. When compared to the soils without *Chromolaena*, soils with *Chromolaena* had very high abundance of the phytopathogen *Fusarium*. Addition of the leachate to the soil increased the spore count of *Fusarium* and the attack of the leachate on the native plants proceeded via a dose dependent strategy, i.e., after a particular level of the inoculum (*Fusarium*) was reached, the invader suppressed the native species. Even in the absence of the leachate, in the autoclaved soil, there was no growth inhibition symptoms observed with the native plants. Nevertheless, in the non-autoclaved soil, native plant seed germination was severely affected thus supporting the hypothesis that the leachate was indeed culturing the growth of the *Fusarium* to limit the growth of the native species, thus imparting a negative feedback on the native plants species in the invaded range. It must be borne in mind that *Fusarium* species attack young plants and seedlings. Since the experiments were performed with the well-grown plants, the effect of the *Fusarium* on the invader could not be evaluated. One likely explanation is that the invader spreads via below ground vegetative propagation which might completely rule out the involvement of the seed to seedling phase which is vulnerable to the *Fusarium* attack.

Invasion success in the perennial herb *Solidago canadensis* through allelopathy and soil microbes was investigated by comparing the germination and growth of the competitors in the European range with the *Solidago* populations that belong to Europe or North America. Seven European grassland species were tested, among which five plant species in the invaded range revealed a reduction in the growth and performance in the soil conditioned with *S. canadensis*, which naturally formed a dominant stand in the invaded European range (Abilasha et al. 2008). Activated carbon was added to the soil in these experiments to minimize the potential effects of the putative allelochemical (see Callaway and Aschehoug 2000). Though the active compounds were not identified in their study they included the polyacetylene derivative and three diterpene lactone derivatives.

Though these compounds are being considered allelopathic from the ecological standpoint, there are several factors that might influence their production, such as humidity and composition of the microbial community (Blair et al. 2006). These compounds support the theory of the novel weapons hypothesis, as a majority of the European competitors tested were negatively affected when they were exposed to compounds produced by the invading *Solidago*. The cross design was employed to study various factors that influence the growth of the European competitors. Findings show that even when activated carbon was used to remove the exuded phytochemical from the soil, the soil type, the proximity of the invasive/native plants, and the soil microbial communities were found to be detrimental factors.

3.1 Nitrogen Fixers

Invasive non-native plants can either enter the new range with their own symbionts or can enter into associations with the resident symbionts in the invaded range (Ehrenfeld 2003). Whether all these associations take place in the presence of the natives in the invaded range is regardless (Weir et al. 2004). Though it has been reported that a threshold density of the nitrogen fixing bacteria is virtually necessary for the development of nodules in the nitrogen fixing invading legumes (Parker 2001), the actinorhizal nitrogen fixer fire tree (*Myrica faya*) has invaded the volcanic soils of Hawaii (Vitousek and Walker 1989). Apparently, its success in Hawaii was due to the successful symbiosis with the nitrogen fixer actinomycete *Frankia*. Flourishing invasions of *Myrica* altered not only the nitrogen cycle in the Hawaiian ecosystems but also native plant communities. Concurrently, another invasive nonnative nitrogen fixer was negatively inhibited, for example, *Bromus diandrus*, which accumulated higher shoot and root biomass when grown in the artificially treated soil that killed native nitrogen fixers (Maron and Connors 1996).

3.2 Mutualists

Arbuscular mycorrhizal (AM) fungi unlike the ericoid mycorrhizas and ectomycorrhizas lack host specificity and hence they neither limit nor facilitate invasion (Richardson et al. 2000). An interesting observation with AM fungi was found when combinations of various AM fungal species were tested with native and non-native species (Klironomos 2003). Even within the native range, the associations ranged from highly parasitic to mutualistic associations and this entirely depended on the AM fungal species taken into study. Incorporation of AM fungi in studies with the invaders may reveal trends that correspond to the success of a particular invader, i.e., a mutualist may encourage an invader while a non-native parasite may repel an invader (Reinhart and Callaway 2004).

A classical example of the modification of the soil mutualists by the invading exotic species was illustrated by Kourtev et al. (2002, 2003). Invasion of two exotic plant species *Berberis thunbergii* (Japanese barberry) and *Microstegium vimineum* (Japanese stilt grass) and their effect on the soil biota in North America showed a vast variation in the microbial community in the invaded range than in the native range. In the barberry soil, there was an increase in the bacterial population and relative decrease in the fungal abundance, whereas in the stilt grass soils, colonization of the AMF was more pronounced. Whether allelochemicals were involved in the modification was not confirmed.

The invasive species *C. maculosa* in North America did not show significant increase in the growth and biomass when grown in the sterilized North American soil compared to an astounding increase in the biomass when grown in the sterilized European soil. This observation clearly justifies that *C. maculosa* in Europe was affected by the soil micro biota in their home range when compared to the invaded range and that it survives and spreads in the invaded range due to lack of detrimental micro biota. However when the *C. maculosa* was grown in the American soil without sterilization, the growth was exuberant confirming that in the new invaded range there may be a relationship between *C. maculosa* and the beneficial microbes that foster its rapid growth and spread. The invasive species has apparently modified the soil community for its own advantage. Mutualistic microbes such as the AMF not only infect specific hosts but also a broad range of hosts making it available to the invading plants in the new region.

AMF colonization was higher in the invaded serpentine soils of *C. solstitialis* (yellow starthistle) and *Aegilops triuncialis* (barb goatgrass) (Batten et al. 2006). Sepentine soils are highly infertile with low calcium, magnesium ratio, low nitrogen, and high heavy metal content, and in addition have a low water holding capacity. Through the phospholipid fatty acid analysis (PLFA), it was shown that invaded communities had higher levels of the biomarkers for sulfate reducing bacteria and AMF compared to the native soil communities. There has been a variation in the microbial communities in the areas invaded by these two species as compared to the original native soil communities.

4 Future Directions

Whilst investigating the mechanism of exotic plant invasion, discrepancies relating to the phytotoxicity of allelochemicals arise. Failure in the reproduction of the results published in one species elsewhere in the world occurs. It must be recalled that the time and place of the soil sampling, the duration between various soil sampling, experimentation and storage conditions of the soil during the transit from the site of collection to the laboratory, techniques used for the bioassay, and environmental conditions all contribute to a completely different set of results (Inderjit et al. 2006). The studies performed have to be repeatable, and stringent working conditions need to be established to avoid contradictory theories on the invasion of the same exotic species. Several mechanisms of invasion, soil biota feedbacks have been presented in the past. However only in a very few invasive species have the direct and indirect mechanisms been explained. Lack of identification of key microbial ingredient exists with most successful invasions. Assimilation of a microbial profile of the invasive plant in question both in the invaded and the home range would contribute towards designing templates for formulating bioherbicides. More reliable conclusions regarding invasive plants especially those altering the underground microbial communities need to be made for a better understanding for which an explicit approach of plant-soil biota feedback experiments need to be performed. With the soil microbiota in question, the rates of degradation of the allelochemical, the population density of the microbe, and the rates of microbial degradation all come in to the study and have to be evaluated for a more realistic conclusion. Besides the extension and supporting theories of the NWH, the concept of EICA needs to be addressed in plant invasion ecology (Callaway and Ridenour 2004). The plant nutrient trade-offs for defense and growth in the new environment would allow us to understand the need for a plant to choose either one in the newly invaded range. Successful failure of some of the invasions with significance in agro-forestry (Scheffer 2003) will also improve our understanding of the belowground flora and fauna which might mitigate our lack of knowledge in invasions.

Acknowledgements H.P.B. acknowledges the support from University of Delaware and University of Delaware Research Foundation (UDRF).

References

Abilasha D, Quintana N, Vivanco J, Joshi J (2008) Do allelopathic compounds in invasive *Solidago canadensis* s.l. restrain the native European flora? J Ecol 96:993–1001

Bains G, Amutha S, Rudrappa T, Alff E, Hanson TE, Bais HP (2009) Native plant and microbial contributions to a negative plant–plant interaction. Plant Physiol 151:2145–2151

Bais HP, Vepachedu R, Gilroy S, Callaway RM, Vivanco JM (2003) Allelopathy and exotic plant invasion: from molecules and genes to species interactions. Science 301:1377–1379

Batten KM, Scow KM, Davies KF, Harrison SP (2006) Two invasive plants alter soil microbial community composition in serpentine grasslands. Biol Invasions 8:217–230

Beckstead J, Parker IM (2003) Invasiveness of *Ammophila arenaria*: release from soil-borne pathogens? Ecology 84:2824–2831

Bertin C, Yang X, Weston LA (2003) The role of root exudates and allelochemicals in the rhizosphere. Plant Soil 256:67–83

Blair AC, Hanson BD, Bunk GR, Marrs RA, Westra P, Nissen SJ, Hufbauer RA (2005) New techniques and findings in the study of a candidate allelochemical implicated in invasion success. Ecol Lett 8:1039–1047

Blair AC, Nissen SJ, Brunk GR, Hufbauer RA (2006) A lack of evidence for an ecological role of the putative allelochemical (±)-catechin in spotted knapweed invasion success. J Chem Ecol 32:2327–2331

Callaway RM, Aschehoug ET (2000) Invasive plants versus their new and old neighbors: a mechanism for exotic invasion. Science 290:521–523

Callaway RM, Ridenour WM (2004) Novel weapons: invasive success and the evolution of increased competitive ability. Front Ecol Environ 2:436–443

Callaway RM, DeLuca T, Ridenour WM (1999) Herbivores used for biological control may increase the competitive ability of the noxious weed *Centaurea maculosa*. Ecology 80:1196–1201

Callaway R, Thelen GC, Barth S, Ramsey GJE (2004a) Soil fungi alter interactions between the invader *Centaurea maculosa* and North American natives. Ecology 85:1062–1071

Callaway RM, Thelen G, Rodriguez A, Holben WE (2004b) Soil biota and exotic plant invasion. Nature 427:731–733

Callaway RM, Cipollini D, Kathryn B, Thelen GC, Hallett SG, Prati D, Stinson K, Klironomos J (2008) Novel weapons:Invasive plant suppresses fungal mutualists in America but not in its native Europe. Ecology 89:1043–1055

Crawley MJ (1987) What makes a community invasible? In: Gray AJ, Crawley MJ, Edwards PJ (eds) Colonization, succession and stability. Blackwell, London

Crawley MJ (1997) Plant ecology. Blackwell, London

DeWalt SJ, Denslow JS, Ickes K (2004) Natural-enemy release facilitates habitat expansion of the invasive tropical shrub *Clidemia hirta*. Ecology 85:471–483

Duke SO, Blair AC, Dayan FE, Johnson RD, Meepagala KM, Cook D, Bajsa J (2009) Is (−)-Catechin a novel weapon of spotted knapwed (*Centaurea stoebe*)? J Chem Ecol 35:141–153

Ehrenfeld JG (2003) Effects of exotic plant invasions on soil nutrient cycling processes. Ecosystems 6:503–523

Elton CS (1958) The ecology of invasions. Methuen, London

Ens EJ, Bremner JB, French K, Korth J (2009) Identification of volatile compounds released by roots of an invasive plant, bitou bush (*Chrysanthemoides monilifera* spp. *rotundata*), and their inhibition of native seedling growth. Biol Invasions 11:275–287

Furubayashi A, Hiradate S, Fujii Y (2007) Role of catechol structure in the adsorption and transformation reactions of L-DOPA in soils. J Chem Ecol 33:239–250

He W, Feng Y, Ridenour WM, Thelen G, Pollock JL, Diaconu A, Callaway RM (2009) Novel weapons and invasion:biogeographic differences in the competitive effects of *Centaurea maculosa* and its root exudates (±)-catechin. Oecologia 15:803–815

Inderjit (2005) Soil microorganisms: an important determinant of allelopathic acivity. Plant Soil 274:227–236

Inderjit, Dakshini KMM (1999) Allelopathic potential of well water from *Pluchea lanceolata*-infested cultivated fields. J Chem Ecol 22:1132–1131

Inderjit, Duke SO (2003) Ecophysiological aspects of allelopathy. Planta 217:529–539

Inderjit, Weston LA (2003) Root exudation: an overview. In: Root ecology. de Kroon H, Visser EJW (eds) Springer, Heidelberg

Inderjit, Kaur S, Dakshini KMM (1996) Determination of allelopathic potential of a weed *Pluchea lanceolata* through a multi-faceted approach. Can J Bot 74:1445–1450

Inderjit, Callaway RM, Vivanco JM (2006) Can plant biochemistry contribute to understanding of invasion ecology? Trends Plant Sci 11:574–580

Inderjit, Seastedt TR, Callaway RM, Pollock JL, Kaur J (2008) Allelopathy and plant invasions: traditional, congeneric, and bio-geographical approaches. Biol Invasions 10:875–890

Jakobs G, Weber E, Edwards PJ (2004) Introduced plants of the invasive *Solidago gigantea* (*Asteraceae*) are larger and grow denser than conspecifics in the native range. Divers Distrib 10:11–19

Jarchow ME, Cook BJ (2009) Allelopathy as a mechanism for the invasion of *Typha augustifolia*. Plant Ecol 204:113–124

Kaushik S, Bais HP, Biedrzycki ML, Venkatachalam L (2010) Catechin is a phytototoxin and a pro-oxidant secreted from the roots of *Centaurea stoebe*. Plant Sig Behav (In Press)

Kisa M, Sanon A, Thioulouse J, Assigbetse K, Sylla S, Spichiger R, Dieng L, Berthelin J, Prin Y, Galiana A, Lepage M, Duponnois R (2007) Arbuscular mycorrhizal symbiosis counterbalance the negative influence of the exotic tree species *Eucalyptus camaldulensis* on the structure and functioning of soil microbial communities in a Sahelian soil. FEMS Microbiol Ecol 62:32–44

Klironomos JN (2003) Variation in plant response to native and exotic arbuscular mycorrhizal fungi. Ecology 84:2292–2301

Kourtev PS, Ehrenfeld JG, Haggblom M (2002) Exotic plant species alter the microbial community structure and function in the soil. Ecology 83:3152–3166

Kourtev PS, Ehrenfeld JG, Haggblom M (2003) Experimental analysis of the effect of exotic and native plant species on the structure and function of soil microbial communities. Soil Biol Biochem 35:895–905

Lesica P, Hanna D (2004) Indirect effects of biological control on plant diversity vary across sites in Montana grasslands. Conserv Biol 18:444–454

Levine JM, Vilá M, D'Antonio CM et al (2002) Mechanisms underlying the impacts of exotic plant invasions. Proc R Soc Lond B Biol Sci 270:775–781

Mack RN, Simberloff D, Lonsdale WM, Evans H, Clout M, Bazzaz FA (2000) Biotic invasions: causes, epidemiology, global consequences, and control. Ecol Appl 10:689–710

Mangla S, Inderjit, Callaway RM (2008) Exotic invasive plant accumulates native soil pathogens which inhibit native plants. J Ecol 96:58–67

Maron JL, Connors PG (1996) A native nitrogen-fixing shrub facilitates weed invasion. Oecologia 105:302–312

Maron JL, Vilà M (2001) Do herbivores affect plant invasion? Evidence for the natural enemies and biotic resistance hypotheses. Oikos 95:361–373

Parker MA (2001) Mutualism as a constraint on invasion success for legumes and rhizobia. Divers Distrib 7:125–136

Perry LG, Thelen GC, Ridenour WM, Weir TL, Callaway RM, Paschke MW, Vivanco JM (2005) Dual role for an allelochemical:(±)-catechin from *Centaurea maculosa* root exudates regulates conspecific seedling establishment. J Ecol 93:1125–1136

Perry LG, Thelen GC, Ridenour WM, Callaway RM, Paschke MW, Vivanco JM (2007) Concentrations of the allelochemical (±)-catechin in *Centaurea maculosa* soils. J Chem Ecol 33:2337–2344

Pivato B, Mazurier S, Lemanceau P, Siblot S, Berta G, Mougel C, van Tuinen D (2007) *Medicago* species affect the community of arbuscular mycorrhizal fungi associated with roots. New Phytol 176:197–210

Pramanik MHR, Nagal M, Asao M, Matsui Y (2000) Effect of temperature and photoperiod on phytotoxic root exudates of cucumber (*Cucumis sativus*) in hydroponic culture. J Chem Ecol 26:1953–1967

Reinhart KO, Callaway RM (2004) Soil biota facilitate exotic *Acer* invasions in Europe and North America. Ecol Appl 14:1737–1745

Reinhart KO, Packer A, Van der Putten WH, Clay K (2003) Plant–soil biota interactions and spatial distribution of black cherry in its native and invasive ranges. Ecol Lett 6:1046–1050

Richardson DM, Allsopp N, D'Antonio CM, Milton SJ, Rajmanek M (2000) Plant invasions – the role of mutualisms. Biol Rev 75:65–93

Ridenour WM, Vivanco JM, Feng Y, Horiuchi J, Callaway RM (2008) No evidence for tradeoffs: *Centaurea* plants from America are better competitors and defenders than plants from the native range. Ecol Monogr 78:369–386

Rudrappa T, Bonsall J, Gallagher JL, Seliskar DM, Bais HP (2007) Root-secreted allelochemical in the noxious weed *Phragmites australis* deploys a reactive oxygen species response and microtubule assembly disruption to execute rhizotoxicity. J Chem Ecol 33:1898–1918

Rudrappa T, Choi YS, Levia DF, Legates DR, Lee KH, Bais HP (2009) *Phragmites australis* root secreted phytotoxin undergoes photo-degradation to execute severe phytotoxicity. Plant Signal Behav 4:506–513

Scheffer RP (2003) The nature of disease in plants. Cambridge University Press, Cambridge

Siemann E, Rogers WE (2003) Increased competitive ability of an invasive tree limited by an invasive beetle. Ecol Appl 13:1503–1507

Simões K, Du J, Kretzshmar FS, Broeckling CD, Stermiz FS, Vivanco JM, Braga MR (2008) Phytotoxic catechin exuded by seeds of the tropical weed *Sesbania virgata*. J Chem Ecol 34:681–687

Stinson KA, Campbell SA, Powell JR, Wolfe BE, Callaway RM, Thelen GC, Hallett SG, Prati D, Klironomos JN (2006) Invasive plant suppresses the growth of native tree seedlings by disrupting belowground mutualisms. PloS Biol 4:727–731

Tharayil N, Bhowmik PC, Xing B (2008) Bioavailability of allelochemicals as affected by companion compounds in soil matrices. J Agric Food Chem 56:3706–3713

Thorpe AS, Thelen GC, Diaconu A, Callaway RM (2009) Root exudates is allelopathic in invaded community but not in native community: field evidence for the novel weapons hypothesis. J Ecol 97:641–645

Tharayil N, Triebwasser DJ (2010) Elucidation of diurnal pattern of catechin exudation by *Centaurea stoebe*. J Chem Ecol 36:200–204

Vaughn SF, Berhow MA (1999) Allelochemicals isolated from tissues of the invasive weed garlic mustard (*Alliaria petiolata*). J Chem Ecol 25:2495–2504

Vitousek PM, Walker LR (1989) Biological invasion by *Myrica faya* in Hawai'i: plant demography, nitrogen fixation, ecosystem effects. Ecol Monogr 59:247–265

Vivanco JM, Bais HP, Stermitz FR, Thelen GC, Callaway RM (2004) Biogeographical variation in community response to root allelochemistry:novel weapons and exotic invasion. Ecol Lett 7:285–292

Weir TL, Park SW, Vivanco JM (2004) Biochemical and physiological mechanisms mediated by allelochemicals. Curr Opin Plant Biol 7:472–479

Weir TL, Bais HP, Stull VJ, Callaway RM, Thelen GC, Ridenour WM, Bhamidi S, Stermitz FR, Vivanco JM (2006) Oxalate contributes to the resistance of *Gaillardia grandiflora* and *Lupinus sericeus* to a phytotoxin produced by *Centaurea maculosa*. Planta 223:785–795

Williams JR (1954) The biological control of weeds. In: Report of the Sixth Commonwealth Entomological Congress. London, UK

Wolfe LM (2002) Why alien invaders succeed: support for the escape-from-enemy hypothesis. Am Nat 160:705–711

Wolfe BE, Klironomos JN (2005) Breaking new ground: soil communities and exotic plant invasion. Bioscience 55:477–488

Volatile Interaction Between Undamaged Plants: A Short Cut to Coexistence

Velemir Ninkovic

Abstract Coexistence with other plants is one of the most important factors affecting the growth of plant individuals and the distribution of species. Most research in this area has focused on competition between plants, but recently an understanding has emerged that coexistence can take other forms. Plant released volatile organic compounds (VOCs) can be phytotoxic for receiving plants and are commonly employed in plant competition. VOCs can also be used as a source of information by the plant. The information available to the plant through the interpretation of plant volatile signals can help it adapt to surrounding conditions, such as competition, via morphological and physiological changes. These changes can result in alterations of leaf temperature and biomass allocation patterns. The plant can then react to compete when necessary or avoid unfavourable competition. Studies indicate that plants can distinguish between related and non-related plants by means of VOCs. Therefore, VOCs may also play an important role in the kin selection. Volatile signal interpretation and its effects on the plant and subsequent trophic levels have been given the name *allelobiosis*.

1 Introduction

During the cyclical development of an individual from propagation unit to propagation unit, plants are exposed to a broad range of stress elements. During their evolution, plants have developed different ecological means to adapt their development, physiology, and life history to these challenges. Studies have begun to focus on the ability of plants to make adaptive decisions about the challenges they face

V. Ninkovic
Department of Ecology, Swedish University of Agricultural Science, Box 7044, 750 07 Uppsala, Sweden
e-mail: Velemir.Ninkovic@ekol.slu.se

F. Baluška and V. Ninkovic (eds.), *Plant Communication from an Ecological Perspective*, Signaling and Communication in Plants,
DOI 10.1007/978-3-642-12162-3_5,

based on cues and signals they perceive from their environment (Silvertown and Gordon 1989; Sultan 2007). Competition between plants has been recognised as one of the most important factors affecting the growth of individuals and the distribution of species, and plants utilise a wide spectrum of behaviours to compete with or resist competition from their neighbours (Trewavas 2009).

For many decades, it has been known that chemical messages play an important role in these behavioural mechanisms. The process was given the name allelopathy by Molisch (1937). There are several hundred allelochemicals released from plants which are known to interfere with physiological processes in the receiving plant (Einhellig 1995), and several studies have clearly demonstrated the mode of action of allelochemicals (Batish et al. 2006; Field et al. 2006; Filella et al. 2009; Hierro and Callaway 2003).

Recent research has also shown that this type of chemical interaction may represent not only direct competition but also a source of information that can be used to adapt to upcoming events. The term 'allelobiosis' has been introduced to describe this process and its effects on plants and at higher trophic levels (Ninkovic et al. 2006; Pettersson et al. 2003).

In this chapter, the ecological importance of volatile chemical communication between undamaged plants is summarised and discussed, highlighting the effects of allelobiosis on plant growth (Fig. 1).

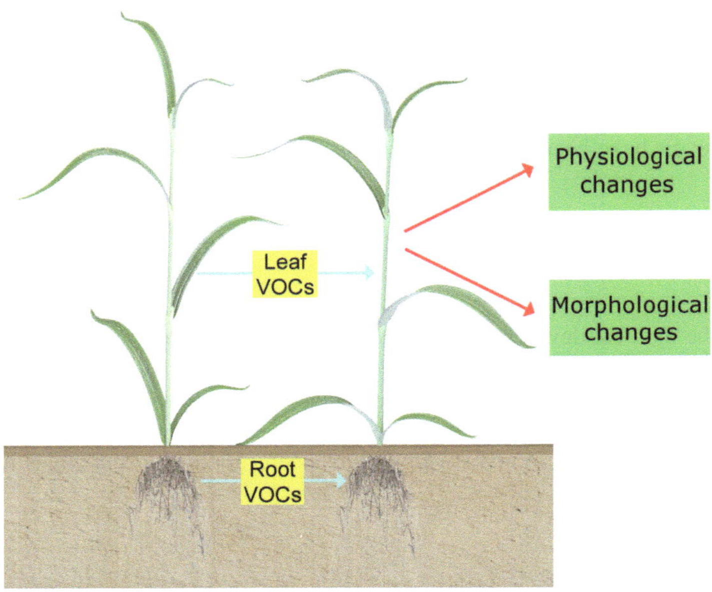

Fig. 1 Allelobiosis. Volatile chemical interactions between undamaged plants induce morphological and physiological changes in the responding plant

2 Role of Plant Volatiles

Plants release volatile organic compounds (VOCs) into the surrounding environment from organs such as leaves, flowers and roots. More than 1700 volatile compounds have been identified (Chapter 'Allelopathy and Exotic Plant Invasion'; Knudsen et al. 2006), serving various ecological roles. Emitted airborne VOCs aid plant reproduction by attracting pollinators (Reinhard et al. 2004) and seed dispersers (Luft et al. 2003; Steiger et al. 2008). They are part of the plant inducible defence system against herbivores and pathogens (e.g. Shiojiri et al. 2006) and may repel herbivores directly (De Moraes et al. 2001; Kessler and Baldwin 2001) or induce indirect defence against insects (Arimura et al. 2000; Ton et al. 2007) or pathogens (Kishimoto et al. 2006) as well as protect plants via tritrophic interactions by attracting herbivore natural enemies (e.g. Arimura et al. 2000; Ninkovic et al. 2001). In addition, emission of VOCs can contribute in other stress acclimatisations such as thermotolerance (Sharkey et al. 2001) and removal of reactive oxygen species (Loreto and Velikova 2001).

The production of VOCs is complex and constantly altered by interactions of plants with biotic and abiotic factors in continuously changing environments (Pichersky et al. 2006). Biotic stress such as intraspecific and interspecific competition between plants have both positive and negative effects on the amount and composition of volatile emissions, depending on the species a seedling coexists with (Ormeño et al. 2007). Herbivory or pathogen attack modify the plant's volatile profile by induction of semiochemicals and odour blends associated with leaf damage and plant volatile induced defence (Dicke 2009; Chapters 'The Chemistry of Plant Signalling', 'Within-Plant Signalling by Volatiles Triggers Systemic Defences' 'Plant Volatiles: Useful Signals to Monitor Crop Health Status in Greenhouses'). Production of VOCs may also be sensitive to different abiotic factors such as severe water stress (Delfine et al. 2005; Filella et al. 2009; Ormeño et al. 2007), temperature and light intensity (Wassner and Ravetta 2005; Yuan et al. 2009), different nutrition levels in the soil (Ormeño et al. 2008), ultraviolet radiation (Zavala and Ravetta 2002) and ozone (Yuan et al. 2009). Almost all changes in abiotic factors, except air humidity, cause small but significant changes in the relative ratios among the different compounds in odour blends after an herbivore attack (Gouinguene and Turlings 2002).

The concentration and composition of plant-generated VOCs can carry information about plants, their physiological status and the stress that they have been exposed to. Therefore, VOCs can play an important role in chemical interaction between plants and insects (Chapters 'Volatile Chemical Interaction Between Undamaged Plants: Effects at Higher Trophic Levels', 'Within-Plant Signalling by Volatiles Triggers Systemic Defences', 'Volatile Interactions Between Undamaged Plants: Effects and Potential for Breeding Resistance to Aphids', 'Photosensory Cues in Plant–Plant Interactions: Regulation and Functional Significance of Shade Avoidance Responses', 'Exploiting Plant Signals in Sustainable Agriculture', and 'Plant

Volatiles: Useful Signals to Monitor Crop Health Status in Greenhouses') and between plant individuals.

2.1 The Role of Plant Volatiles in Allelopathy

The groundbreaking empirical studies of volatile involvement in plant–plant interactions were done by Muller and colleagues. These studies were based on observations of the special patterns of annual grassland species and of colonies of *Salvia leucophylla* Greene and *Artemisia californica* (Less.) (Muller et al. 1964). Both species are shrubs that form chaparral vegetation adjacent to areas of annual grassland. Within and surrounding shrub thickets, graminaceous species are excluded and the thickets are frequently surrounded by areas of bare soil without plants. The areas without plants beyond the lateral spread of the shrub system suggested that a volatile mediated interaction was involved. Studies completely relying on laboratory assays have demonstrated that volatile monoterpenoids can have phytotoxic effects on recipient plants (Abrahim et al. 2000; Muller 1965) and alter soil microflora (Badri and Vivanco 2009).

Laboratory studies have indicated that VOCs released from leaves of one plant can induce inhibitory responses in neighbouring plants, reducing their germination and root elongation (Alonso-Amelot et al. 2006; Viles and Reese 1996). Many of these studies used concentrations of active compounds much higher (e.g. >1000-fold) than those observed under field conditions (Barney et al. 2005). It was suggested that VOCs with inhibitory effects on annual grass released from above-ground plant parts can be absorbed by soil (Muller and Delmoral 1966). Recent results from greenhouse studies have shown that volatiles released from above-ground parts of *Artemisia vulgaris* L. have little direct activity and increased VOCs concentration in the soil within and bordering the *A. vulgaris* stand negatively affects performance of native grasses (Barney et al. 2009). Inhibitory effects of soil-absorbed monoterpenoids on seed germination and growth can be a result of alterations in the root phospholipid and sterol composition (Zunino and Zygadlo 2005) and a reduction of both nuclear and organelle DNA synthesis in the root apical meristem (Nishida et al. 2005).

Ethylene is an important volatile plant hormone that modulates diverse growth responses to a wide variety of environmental cues. Whether ethylene can affect neighbouring plants was tested using genetically transformed plants (Inderjit et al. 2009). Root growth of wild type *Nicotiana attenuata* Torr. ex S. Watson used as receivers was negatively correlated with the number of wild type tobacco emitting plants and headspace ethylene concentrations. This effect was not observed when wild type plants were exposed to volatiles from genetically transformed plants with already known reduced ability of ethylene production. Furthermore, transgenic plants with low ethylene production were more sensitive to the presence of ethylene from their neighbours than were wild type plants.

The main opinion is that floral VOCs are involved in the attraction of pollinators, but a recent study reported that these volatiles can be involved in plant–plant interactions as well (Horiuchi et al. 2007). Such VOCs from snapdragon flowers, *Antirrhinum majus* L., inhibit *Arabidopsis* root growth by more than 50% relative to untreated control plants. Methyl benzoate was found to be primarily responsible for the inhibition of root growth, and treatment with this component induced expression of cytokinin, auxin and other plant-hormone-related genes and genes related to seed germination processes in *Arabidopsis*.

Root-emitted volatile compounds in the rhizosphere can also be involved in plant–plant signalling. However, their effects are most likely limited to immediate neighbours and competitors because of their relatively low transport rates (van der Putten 2001). For instance, specific volatile compounds, mostly sesqui- and diterpenes, released by roots of the invasive Bitou bush into the soil had inhibitory effects on native plant seedling growth (Ens et al. 2009). High concentrations of VOCs in the soil can be a result of fast root growth and great root biomass production of the Bitou bush, which can confer superior allelopathic efficiency and high competitive ability to this species. Volatiles released by roots into the soil also have phytotoxic effects on seed germination and growth of competing neighbouring plants.

Inhibitory effects of allelochemicals, including VOCs released from invasive/allelopathic plants, have been suggested to contribute to the success of some invasive exotic plants in establishing monocultures. This phenomenon may be amplified by the absence of adequate mechanisms among native plants to protect themselves (Barney et al. 2009). It has often been suggested that exudation/emission of compounds into the surrounding environment by native plant species have only minor effects on surrounding plants (e.g. Chapter 'Allelopathy and exotic plant invasion'). This is a result of adaptation by associated plant species to the presence of neighbouring plants that produce specific chemical compounds (Grøndahl and Ehlers 2008) and the fact that local plant communities can co-evolve in the presence of 'chemical neighbours' (Callaway et al. 2005; Ehlers and Thompson 2004). However, few studies have focused on how individual plants or plant species adapt to the presence of other plant individuals or species.

2.2 VOCs as Cues in Plant–Plant Interaction

VOCs released from neighbouring plants may have informative value for responding/listening plants (Ninkovic et al. 2006). These volatile signals may initiate responses in receiving plants that represent adaptations to present or future competitive conditions. Useful information must be relevant for future competition with neighbouring plants since plant responses, especially those related to development of new organs and resource allocation, require time (see review Novoplansky 2009). In this way, plants can modify their development in response to environmental conditions to maximise growth.

It has been suggested that plasticity of biomass allocation is consistent with optimal partitioning theory, e.g. biomass allocation in response to limiting resources (Chapin et al. 1987). An increased allocation to root biomass is a well known response to nitrogen limitation (Glimskär and Ericsson 1999; Ingestad and Agren 1991) and may be a general response across plant species (Reynolds and Dantonio 1996). Trade-off between allocation to shoots and roots may be one of the plant's primary responses to competition with other plants (e.g. Grime 2001). Plants increase their root biomass in the presence of neighbouring roots before physical contact, via a mechanism that involves allelochemicals (e.g. Grøndahl and Ehlers 2008) even when nutrients are not limiting (reviewed by Schenk 2006). These results demonstrate that root-growth patterns of plants grown with different types of neighbours vary between species and suggest that factors additional to resource depletion could be involved in their development.

Laboratory studies show that the pattern of biomass allocation of barley seedlings was changed in response to volatiles from undamaged neighbouring plants with which interaction between roots was prevented (Ninkovic 2003). Plants exposed to volatiles from another cultivar allocated more biomass to roots and less to leaves than plants exposed to volatiles from the same cultivar, or to air alone. Given that the total dry biomass of exposed plants did not significantly differ between treatments, the observed changes in biomass allocation between different organs may represent a response to competitive cues generated by another plant. Plants can discriminate kin in competitive interactions, and the root interactions may provide the cue for kin recognition (Dudley and File 2007). The annual plant *Cakile edentula* increased allocation to roots when groups of non-kin shared a common pot, but not when groups of siblings shared a pot (Dudley and File 2007). Volatile chemical interaction between plants as reported by Ninkovic (2003) can be consistent with kin selection because greater allocation to roots is claimed to increase below-ground competitive ability, even though the barley cultivars used have been developed by plant breeding. However, the mechanisms of individual recognition are less than intuitive in plants, and have received relatively little attention (Callaway and Mahall 2007).

Shade avoidance is another example of plant modification of growth in response to neighbours. These growth adjustments help plants to reach the light and outgrow their competitors by enhanced elongation of stems and petioles. Action of phytocrome photoreceptors that sense the red:far-red (R:FR) ratio in light reflected by neighbours initiate and contribute to shade avoidance responses (see Chapter 'Photosensory Cues in Plant–Plant Interactions: Regulation and Functional Significance of Shade Avoidance Responses'). Furthermore, the volatile hormone ethylene has been associated with shade avoidance, both as a direct volatile neighbour detection signal (through atmospheric accumulation) that induces strong elongation in seedlings and as a downstream target for photoreceptor signalling (Pierik et al. 2004, 2007).

Early signal detection followed by appropriate response to the actual presence of potentially competing plants should have great advantage for the individual plant. It is therefore suggested that plants have developed selective perception systems to respond to significant information contributing to the maximisation of their competitive responsiveness towards worthy targets (Novoplansky 2009). Since VOCs can carry information about neighbouring plants, the perception system for volatile cues may allow plants to adapt their establishment strategy depending on the presence of competitors during early growth phases and avoid wasteful competition against self and kin, or the waging of hopeless battles against overwhelmingly strong competitors.

3 Plant Responses to Allelobiosis, Methyl Salicylate and Methyl Jasmonate

Plants exposed to different types of stress release stress related volatile chemicals, such as methyl salicylate (MeSA) and methyl jasmonate (MeJA) into the environment (Shulaev et al. 1997). Both MeSA and MeJA may have important roles in plant responses to biotic and abiotic stresses, acting as signals that trigger the oxidative pathways needed to activate gene expression in plant stress responses. In laboratory experiments barley plants exposed to MeSA, MeJA and to allelobiotic interactions via volatiles from another barley cultivar exhibited changes in leaf temperature measured by thermal imaging. Plants exposed toMeSA orMeJA had significantly higher leaf temperatures than unexposed barley plants, indicating stress responses in the exposed plants, whereas plants exposed to volatiles from another barley cultivar showed the opposite reaction (Fig. 2). The thermal signal measured by an infra-red camera generally depends on differences in transpiration rate, where high temperature reflects closed stomata and low temperature reflects open stomata (see review Chaerle et al. 2007). Stress generated by pathogens can

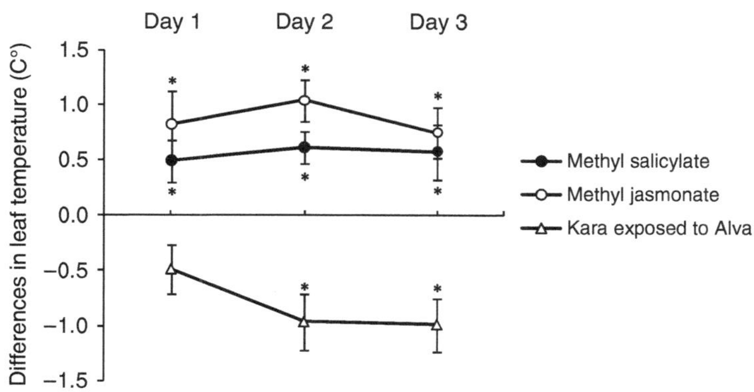

Fig. 2 Differences in leaf temperature measured by IR camera on plants of the barley cultivar Kara exposed to MeSA, MeJA and volatiles from the cultivar Alva compared to unexposed Kara. Plants were constantly exposed and the number of days denotes the time from start of the exposure. The exposure to MeSA and MeJA was performed as described in Glinwood et al. (2007). *Asterisks* denote statistical significance at $p < 0.05$ (Tukey test, $n = 10$)

influence stomatal opening and closing by the release of specific compounds inducing plant resistance responses, or by interfering with water transport (Chaerle et al. 2004; Jones 2004; Melotto et al. 2006; Prats et al. 2006). Stomatal closure as a consequence of accumulation of salicylic acid causes increase in leaf temperature (Chaerle et al. 1999).

Different effects of VOCs on leaf temperature indicate that plant responses are dependent on the nature of the volatile cue. Exposure of barley to MeSA induces *BCI-4* (barley chemical inducible gene-4) whereas MeJA induces *AOS2* (allene oxide synthase) (Glinwood et al. 2007). *BCI-4* is a putative Ca2+ -binding EF-hand protein that is upregulated by external application of benzo-(1,2,3) thiadiazole-7-carbothioic acid S-methyl ester (BTH), and belongs to a group of chemically induced barley genes that correlate with resistance to powdery mildew (Beßer et al. 2000). *AOS2* is known as a stress related gene in barley (Maucher et al. 2000). These two genes were not upregulated in plants treated with volatiles from undamaged plants of another barley cultivar.

Observed decreases in leaf temperature following exposure to volatiles from an undamaged barley cultivar suggest increased transpiration rates in exposed barley plants. This change of transpiration rate generates an increased need for water that may be compensated by greater allocation of available biomass to roots in exposed plants (Ninkovic 2003). It has been shown that MeJA can reduce root growth (Nakagawa and Kawaguchi 2006; Staswick et al. 1992), suggesting that stress related VOCs negatively affect growth rate and may therefore reduce competitive ability of treated plants. It seems that the allelobiotic response is different from the effect caused by the stress related substances MeSA and MeJA. It is still unknown if this allelobiotically induced response can help plants reduce the negative effects of competition with other plants. Pettersson et al. (1999) showed that significant reduction in leaf temperature occurred only when certain cultivars were exposed to volatiles from certain other cultivars. This principle can be ubiquitously important for plant's decision-making system, especially in competitive settings where the success of each genotype is inherently dependent on its response to competition from neighbouring genotypes/species.

The results of the studies presented here suggest that plant VOCs carry information about the physiological status of neighbouring plants and that they may help the plant to adapt its growth strategy depending on the presence of competitors. Thus, plants may have developed selective perception systems to detect volatile cues, and respond by activating different responding mechanisms depending on which volatile cues are emitted by neighbouring plants.

4 Conclusions

Coexistence with other plants is an important factor affecting the growth of plant individuals and the distribution of species. VOCs may be a valuable source of information in competition and avoidance of competition, and the information available to plants through the interpretation of such signals may help them adapt to surrounding conditions by means of physiological as well as morphological changes. Recent research has confirmed that chemical interaction may represent not only an aggressive tool but also a source of information used by plants to adapt to anticipated future events. The complex composition of plant generated VOCs is dynamic and offers a tool for responding plants and immediate information on actual status of neighbouring plants. This perception system may enable plants to adapt their growth strategy depending on the presence of competitors during early growth phases and to avoid wasteful competition against self and kin or overwhelmingly powerful competitors. Thus the term 'allelobiosis' has been introduced to stress the role of VOCs in intra- as well as multi-trophic ecological mechanisms.

The information carried by volatile substances can have a multitude of effects on plants that are exposed to them, ranging from phytotoxicity to beneficial adaptations to competition. Therefore, the ecological importance of such exchange may be more important than currently recognised. Volatile communication between plants is potentially very finely tuned, and even small changes in the composition of the volatile blend may have great importance for neighbouring plants. From an ecological point of view, the importance of understanding the effects of different substances and blends in the volatile profile cannot be overestimated, and more research is needed in this field.

Acknowledgements This work was financially supported by the Swedish Foundation for Strategic Environmental Research (Mistra) through the PlantComMistra programme and by the Swedish Research Council for Environment, Agricultural Sciences and Spatial Planning (FORMAS). Torgny Näsholm, Robert Glinwood, Jan Pettersson, Iris Dahlin, Martin Kellner and Dimitrije Markovic are thanked for reviewing the manuscript.

References

Abrahim D, Braguini WL, Kelmer-Bracht AM, Ishii-Iwamoto EL (2000) Effects of four monoterpenes on germination, primary root growth, and mitochondrial respiration of maize. J Chem Ecol 26:611–624

Alonso-Amelot ME, Usubillaga A, Avila-Nunez JL, Oliveros-Bastidas A, Avendano M (2006) Effects of *Minthostachys mollis* essential oil and volatiles on seedlings of lettuce, tomato, cucumber and *Bidens pilosa*. Allelopathy J 18:267–275

Arimura G-i, Ozawa R, Shimoda T, Nishioka T, Boland W, Takabayashi J (2000) Herbivory-induced volatiles elicit defence genes in lima bean leaves. Nature 406:512–515

Badri DV, Vivanco JM (2009) Regulation and function of root exudates. Plant Cell Environ 32:666–681

Barney J, Sparks J, Greenberg J, Whitlow T, Guenther A (2009) Biogenic volatile organic compounds from an invasive species: impacts on plant–plant interactions. Plant Ecol 203:195–205

Barney JN, Hay AG, Weston LA (2005) Isolation and characterization of allelopathic volatiles from mugwort (*Artemisia vulgaris*). J Chem Ecol 31:247–265

Batish DR, Singh HP, Setia N, Kaur S, Kohli RK (2006) 2-Benzoxazolinone (BOA) induced oxidative stress, lipid peroxidation and changes in some antioxidant enzyme activities in mung bean (*Phaseolus aureus*). Plant Physiol Biochem 44:819–827

Beßer K, Jarosch B, Langen G, Kogel K-H (2000) Expression analysis of genes induced in barley after chemical activation reveals distinct disease resistance pathways. Mol Plant Pathol 1:277–286
Callaway RM, Mahall BE (2007) Plant ecology: family roots. Nature 448:145–147
Callaway RM, Ridenour WM, Laboski T, Weir T, Vivanco JM (2005) Natural selection for resistance to the allelopathic effects of invasive plants. J Ecol 93:576–583
Chaerle L, Hagenbeek D, De Bruyne E, Valcke R, Van Der Straeten D (2004) Thermal and chlorophyll-fluorescence imaging distinguish plant–pathogen interactions at an early stage. Plant Cell Physiol 45:887–896
Chaerle L, Leinonen I, Jones HG, Van Der Straeten D (2007) Monitoring and screening plant populations with combined thermal and chlorophyll fluorescence imaging. J Exp Bot 58:773–784
Chaerle L, Van Caeneghem W, Messens E, Lambers H, Van Montagu M, Van Der Straeten D (1999) Presymptomatic visualization of plant–virus interactions by thermography. Nat Biotechnol 17:813–816
Chapin FS, Bloom AJ, Field CB, Waring RH (1987) Plant responses to multiple environmental factors. BioScience 37:49–57
De Moraes CM, Mescher MC, Tumlinson JH (2001) Caterpillar-induced nocturnal plant volatiles repel conspecific females. Nature 410:577–580
Delfine S, Loreto F, Pinelli P, Tognetti R, Alvino A (2005) Isoprenoids content and photosynthetic limitations in rosemary and spearmint plants under water stress. Agric Ecosyst Environ 106:243–252
Dicke M (2009) Behavioural and community ecology of plants that cry for help. Plant Cell Environ 32:654–665
Dudley SA, File AL (2007) Kin recognition in an annual plant. Biol Lett 3:435–438
Ehlers BK, Thompson J (2004) Do co-occurring plant species adapt to one another? The response of *Bromus erectus* to the presence of different *Thymus vulgaris* chemotypes. Oecologia 141:511–518
Einhellig FA (1995) Mechanism of action of allelochemicals in allelopathy. In: Inderjit A, Dakshini KMM, Einhellig FA (eds) Allelopathy: organisms, processes, and applications. American Chemical Society, Washington DC, pp 96–116
Ens E, Bremner J, French K, Korth J (2009) Identification of volatile compounds released by roots of an invasive plant, bitou bush (*Chrysanthemoides monilifera* spp. *rotundata*), and their inhibition of native seedling growth. Biol Invasions 11:275–287
Field B, Jordan F, Osbourn A (2006) First encounters – deployment of defence-related natural products by plants. New Phytol 172:193–207
Filella I, Peñuelas J, Seco R (2009) Short-chained oxygenated VOC emissions in *Pinus halepensis* in response to changes in water availability. Acta Physiol Plant 31:311–318
Glimskär A, Ericsson TOM (1999) Relative nitrogen limitation at steady-state nutrition as a determinant of plasticity in five grassland plant species. Ann Bot 84:413–420
Glinwood R, Gradin T, Karpinska B, Ahmed E, Jonsson L, Ninkovic V (2007) Aphid acceptance of barley exposed to volatile phytochemicals differs between plants exposed in daylight and darkness. Plant Signal Behav 2:321–326
Gouinguene SP, Turlings TCJ (2002) The effects of abiotic factors on induced volatile emissions in corn plants. Plant Physiol 129:1296–1307
Grime JP (2001) Plant strategies, vegetation processes, and ecosystem properties. Wiley, Chichester
Grøndahl E, Ehlers BK (2008) Local adaptation to biotic factors: reciprocal transplants of four species associated with aromatic *Thymus pulegioides* and *T. serpyllum*. J Ecol 96:981–992
Hierro JL, Callaway RM (2003) Allelopathy and exotic plant invasion. Plant Soil 256:29–39
Horiuchi JI, Badri DV, Kimball BA, Negre F, Dudareva N, Paschke MW, Vivanco JM (2007) The floral volatile, methyl benzoate, from snapdragon (*Antirrhinum majus*) triggers phytotoxic effects in *Arabidopsis thaliana*. Planta 226:1–10

Inderjit, von Dahl C, Baldwin I (2009) Use of silenced plants in allelopathy bioassays: a novel approach. Planta 229:569–575

Ingestad T, Agren GI (1991) The influence of plant nutrition on biomass allocation. Ecol Appl 1:168–174

Jones HG (2004) Irrigation scheduling: advantages and pitfalls of plant-based methods. J Exp Bot 55:2427–2436

Kessler A, Baldwin IT (2001) Defensive function of herbivore-induced plant volatile emissions in nature. Science 291:2141–2144

Kishimoto K, Matsui K, Ozawa R, Takabayashi J (2006) Analysis of defensive responses activated by volatile allo-ocimene treatment in *Arabidopsis thaliana*. Phytochemistry 67:1520–1529

Knudsen JT, Eriksson R, Gershenzon J, Stahl B (2006) Diversity and distribution of floral scent. Bot Rev 72:1–120

Loreto F, Velikova V (2001) Isoprene produced by leaves protects the photosynthetic apparatus against ozone damage, quenches ozone products, and reduces lipid peroxidation of cellular membranes. Plant Physiol 127:1781–1787

Luft S, Curio E, Tacud B (2003) The use of olfaction in the foraging behaviour of the golden-mantled flying fox, *Pteropus pumilus*, and the greater musky fruit bat, *Ptenochirus jagori* (Megachiroptera: Pteropodidae). Naturwissenschaften 90:84–87

Maucher H, Hause B, Feussner I, Ziegler J, Wasternack C (2000) Allene oxide synthases of barley (*Hordeum vulgare* cv. Salome): tissue specific regulation in seedling development. Plant J 21:199–213

Melotto M, Underwood W, Koczan J, Nomura K, He SY (2006) Plant stomata function in innate immunity against bacterial invasion. Cell 126:969–980

Molisch H (1937) Der Einfluss einer Pflanze auf die andere-Allelopathie. Fischer, Jena

Muller CH (1965) Inhibitory terpenes volatilized from *Salvia shrubs*. Bull Torrey Bot Club 92:38–45

Muller CH, Delmoral R (1966) Soil toxicity induced by trepenes from *Salvia leucophylla*. Bull Torrey Bot Club 93:130

Muller CH, Haines BL, Muller WH (1964) Volatile growth inhibitors produced by aromatic shrubs. Science 143:471

Nakagawa T, Kawaguchi M (2006) Shoot-applied MeJA suppresses root nodulation in *Lotus japonicus*. Plant Cell Physiol 47:176–180

Ninkovic V (2003) Volatile communication between barley plants affects biomass allocation. J Exp Bot 54:1931–1939

Ninkovic V, Al Abassi S, Pettersson J (2001) The influence of aphid-induced plant volatiles on ladybird beetle searching behavior. Biol Control 21:191–195

Ninkovic V, Glinwood R, Pettersson J (2006) Communication between undamaged plants by volatiles: the role of allelobiosis. In: Baluska F, Mancuso S, Volkmann D (eds) Communication in plants: neuronal aspects of plant life. Springer, Berlin, pp 421–434

Nishida N, Tamotsu S, Nagata N, Saito C, Sakai A (2005) Allelopathic effects of volatile monoterpenoids produced by *Salvia leucophylla*: inhibition of cell proliferation and DNA synthesis in the root apical meristem of *Brassica campestris* seedlings. J Chem Ecol 31:1187–1203

Novoplansky A (2009) Picking battles wisely: plant behaviour under competition. Plant Cell Environ 32:726–741

Ormeño E, Baldy V, Ballini C, Fernandez C (2008) Production and diversity of volatile terpenes from plants on calcareous and siliceous soils: effect of soil nutrients. J Chem Ecol 34:1219–1229

Ormeño E, Fernandez C, Mévy J-P (2007) Plant coexistence alters terpene emission and content of Mediterranean species. Phytochemistry 68:840–852

Pettersson J, Ninkovic V, Ahmed E (1999) Volatiles from different barley cultivars affect aphid acceptance of neighbouring plants. Acta Agric Scand B Soil Plant Sci 49:152–157

Pettersson J, Ninkovic V, Glinwood R (2003) Plant activation of barley by intercropped conspecifics and weeds: allelobiosis. British Crop Protection Council, pp 1135–1144

Pichersky E, Noel JP, Dudareva N (2006) Biosynthesis of plant volatiles: nature's diversity and ingenuity. Science 311:808–811

Pierik R, Sasidharan R, Voesenek L (2007) Growth control by ethylene: adjusting phenotypes to the environment. J Plant Growth Regul 26:188–200

Pierik R, Whitelam GC, Voesenek LACJ, Hd K, Visser EJW (2004) Canopy studies on ethylene-insensitive tobacco identify ethylene as a novel element in blue light and plant–plant signalling. Plant J 38:310–319

Prats E, Gay AP, Mur LAJ, Thomas BJ, Carver TLW (2006) Stomatal lock-open, a consequence of epidermal cell death, follows transient suppression of stomatal opening in barley attacked by *Blumeria graminis*. J Exp Bot 57:2211–2226

Putten WH van der (2001) Interactions of plants, soil pathogens and their antagonists in natural ecosystems. In: Jeger MJ, Spence NJ (eds) Biotic interactions in plant-pathogen associations. CABI, Wallingford, pp 285–305

Reinhard J, Srinivasan MV, Zhang S (2004) Olfaction: scent-triggered navigation in honeybees. Nature 427:411–411

Reynolds HL, Dantonio C (1996) The ecological significance of plasticity in root weight ratio in response to nitrogen: opinion. Plant Soil 185:75–97

Schenk HJ (2006) Root competition: beyond resource depletion. J Ecol 94:725–739

Sharkey TD, Chen X, Yeh S (2001) Isoprene increases thermotolerance of fosmidomycin-fed leaves. Plant Physiol 125:2001–2006

Shiojiri K et al (2006) Changing green leaf volatile biosynthesis in plants: an approach for improving plant resistance against both herbivores and pathogens. Proc Natl Acad Sci USA 103:16672–16676

Shulaev V, Silverman P, Raskin I (1997) Airborne signalling by methyl salicylate in plant pathogen resistance. Nature 385:718–721

Silvertown J, Gordon DM (1989) A framework for plant behavior. Annu Rev Ecol Syst 20:349–366

Staswick PE, Su W, Howell SH (1992) Methyl jasmonate inhibition of root growth and induction of a leaf protein are decreased in an *Arabidopsis thaliana* mutant. Proc Natl Acad Sci USA 89:6837–6840

Steiger SS, Fidler AE, Valcu M, Kempenaers B (2008) Avian olfactory receptor gene repertoires: evidence for a well-developed sense of smell in birds? Proc R Soc B Biol Sci 275:2309–2317

Sultan SE (2007) Development in context: the timely emergence of eco-devo. Trends Ecol Evol 22:575–582

Ton J, D'Alessandro M, Jourdie V, Jakab G, Karlen D, Held M, Mauch-Mani B, Turlings TCJ (2007) Priming by airborne signals boosts direct and indirect resistance in maize. Plant J 49:16–26

Trewavas A (2009) What is plant behaviour? Plant Cell Environ 32:606–616

Viles AL, Reese RN (1996) Allelopathic potential of *Echinacea angustifolia* D.C. Environ Exp Bot 36:39–43

Wassner DF, Ravetta DA (2005) Temperature effects on leaf properties, resin content, and composition in *Grindelia chiloensis* (Asteraceae). Ind Crops Prod 21:155–163

Yuan JS, Himanen SJ, Holopainen JK, Chen F, Stewart CN Jr (2009) Smelling global climate change: mitigation of function for plant volatile organic compounds. Trends Ecol Evol 24:323–331

Zavala JA, Ravetta DA (2002) The effect of solar UV-B radiation on terpenes and biomass production in *Grindelia chiloensis* (Asteraceae), a woody perennial of Patagonia, Argentina. Plant Ecol 161:185–191

Zunino MP, Zygadlo JA (2005) Changes in the composition of phospholipid fatty acids and sterols of maize root in response to monoterpenes. J Chem Ecol 31:1269–1283

Volatile Chemical Interaction Between Undamaged Plants: Effects at Higher Trophic Levels

Robert Glinwood

Abstract This chapter discusses whether plant chemical communication is a mechanism by which plant genetic diversity can affect the natural enemies of herbivores. Plant genetic diversity influences natural enemies, and these insects use volatile chemical cues to locate suitable habitats. However, the importance of chemical communication for these interactions has not been considered. In this chapter, the latest research on chemical communication between undamaged plants is reviewed. Evidence for the fact that volatile chemical communication between weeds and barley, and between different barley genotypes, can influence host plant and habitat selection by aphid natural enemies, polyphagous ladybirds and parasitic wasps, is presented. The results suggest that aphid natural enemies may recognise the effects of plant–plant volatile interaction and volatile mixing as cues denoting favourable habitats. This represents a new aspect of the ecology of plant communication that may be exploited for sustainable plant protection.

1 Introduction

Plants and insects have to overcome many challenges to survive and reproduce and need to respond to information in their environment. Chemical information is abundant in the habitats in which plants and insects live, and while we have known for a long time that insects use chemical cues from other insects and plants, it took longer to establish that plants also use volatile chemical signals in ways that appear to be adaptive. It is now generally accepted that plants respond to defence-inducing volatile signals released after attack by herbivores (Chapter 'Within-Plant Signalling by Volatiles Triggers Systemic Defences'), and their informational role is gradually

R. Glinwood
Department of Ecology, Swedish University of Agricultural Sciences, 750 07 Uppsala, Sweden
e-mail: robert.glinwood@ekol.slu.se

F. Baluška and V. Ninkovic (eds.), *Plant Communication from an Ecological Perspective*, Signaling and Communication in Plants,
DOI 10.1007/978-3-642-12162-3_6,

being revealed (Heil and Bueno 2007; Halitschke et al. 2008; Chapter 'Within-Plant Signalling by Volatiles Triggers Systemic Defences'). However, plants also respond to chemicals released by undamaged plants (e.g. Runyon et al. 2006).

In allelopathy, for example, plant substances that escape into the environment may affect the growth and development of neighbours (Rice 1984). Although allelopathy is an important issue in agricultural science, affecting many aspects of plant coexistence and competition (Weston and Duke 2003), investigation of its effects at higher trophic levels such as herbivores and their natural enemies has started only recently (Ninkovic et al. 2006). The natural enemies of herbivores respond to habitat cues denoting prey abundance and quality. Specialised natural enemies, such as parasitic wasps, use cues indicating the presence of their hosts (Vet and Dicke 1992), whereas polyphagous predators respond to more general indicators of habitat quality (Pettersson et al. 2005). The vegetational composition of habitats is therefore of great relevance to natural enemies.

There is growing interest in the effects of plant diversity on higher trophic levels. In many cases, increased plant genetic diversity has been found to influence herbivore natural enemies in a positive way (Root 1973; Russell 1989; Cook et al. 2007). Given that these insects are often well developed in their use of chemical cues, and that they must be able to recognise plant genetic diversity in order to respond, it is surprising that the role of chemical interactions is only now starting to receive attention. In this chapter I ask the question 'is plant chemical communication a mechanism by which plant genetic diversity can affect herbivore natural enemies?' I review evidence from investigations on plant volatile interactions affecting aphids on barley, which has opened an exciting new perspective on the ecology of plant communication.

2 Increased Plant Genetic Diversity Affects Higher Trophic Levels

2.1 Between Species Diversity

A concept emerging from the study of ecosystem function is that multitrophic interactions can be directly affected by plant biodiversity (Hooper et al. 2005). The greatest focus has been on increased diversity of species, which has been found to support higher diversity and abundance of both herbivorous and predatory arthropod species (Siemann et al. 1998; Haddad et al. 2001). Much of the classic work on plant diversity in relation to agricultural systems relates to mixed species intercropping, and has led to the development of a number of theories explaining how complex plant habitats suppress pest insects and benefit their natural enemies (Root 1973; Russell 1989; Andow 1991). These theories, while not universally proven, have been supported by experiments in agricultural (Andow 1991) and natural ecosystems (e.g. Koricheva et al. 2000; Haddad et al. 2001; Otway et al. 2005).

Despite extensive research on the effects of plant diversity at higher trophic levels, understanding of the mechanistic background to these responses is still limited. Plants sharing a common environment can interact chemically with each other, and can contribute to the combined chemical profile of the habitat. However, whereas chemical mechanisms have been mentioned among theories seeking to explain the effects of mixed cropping on herbivores and natural enemies (e.g. Uvah and Coaker 1984), the role of direct chemical interaction between plants has not been widely considered.

2.2 Within-Species Genotype Diversity

Emphasis on the importance of plant species diversity for trophic interactions may have hidden the fact that diversity between different genotypes of the same plant species has its own ecological significance. Due to the hierarchical structure of trophic food webs, it has been pointed out that plant traits specifically influencing natural enemies should be relatively fewer than those affecting their herbivore prey. It has therefore been assumed that variation resulting from combining genotypes of a single plant species should have only a weak affect at the third trophic level (Vinson 1999; Johnson and Agrawal 2005). This view appears to be changing, and evidence for effects of within species genotype diversity on higher trophic levels is starting to accumulate. For example, Johnson (2008) found that plant genotype diversity can directly affect the abundance of insect predators in a natural ecosystem. However, the strongest effects of plant genotype diversity may be found in managed crop systems rather than natural ones because, against the relatively homogenous genetic background typical of most agricultural monocultures, genetic diversity within the plant species may be more apparent. This also opens the possibility to exploit insect responses for sustainable crop protection.

In fact, mixing different genotypes of the same species of crop plant has been shown to reduce the incidence of damage-causing organisms that use the plants as hosts (Power 1991; Mundt 2002; Ninkovic et al. 2002; Cadet et al. 2007). Again, it is not known to what extent chemical signalling is involved in these interactions. However, the role of chemical signalling as a mechanism contributing to the impact of plant genotype interactions at higher trophic levels has been more extensively studied in a model system consisting of barley, aphids and their natural enemies, and this work will be described next.

3 Volatile Interactions Between Undamaged Plants Affect Higher Trophic Levels

It has become apparent that chemical interaction between undamaged plants is a complex process that affects trophic interactions beyond the plant level (Glinwood et al. 2003; Ninkovic et al. 2006), and may even have an informational value for the receiving plant. This has been shown in interactions between barley and weeds and between different genotypes of barley itself. Barley exposed to volatiles from undamaged plants of a different cultivar adjusts its biomass allocation to favour roots rather than shoots, a response that may indicate increased readiness for competition with a neighbouring plant (Ninkovic 2003). Barley also responds to volatiles from weeds and other barley genotypes by becoming less acceptable as a host plant to aphid herbivores (Pettersson et al. 1999; Ninkovic et al. 2002; Glinwood et al. 2004; Glinwood et al. 2007). These interactions are reviewed in Chapter 'Volatile Interactions Between Undamaged Plants: Effects and Potential for Breeding Resistance to Aphids'.

These findings have led to the introduction of the term, allelobiosis to describe the wider trophic effects of plant interaction via chemicals (Pettersson et al. 2003; Ninkovic et al. 2006). The three key aspects of the definition of allelobiosis are (1) the chemical interaction occurs between undamaged plants, (2) the interaction may be beneficial for the receiving plant and (3) the responses of the receiving plant affect organisms at other trophic levels. Aspect 1 separates allelobiosis from a large body of research on interplant signalling, which focuses on signals released by infected/infested plants, while aspect 3 separates allelobiosis from the plant-focused approach of allelopathy.

Volatile interaction between undamaged plants can also affect the third trophic level. There are two main mechanisms by which this may occur, outlined in Fig. 1. First by allelobiosis (Fig. 1a), in which direct volatile interaction between neighbouring plants induces responses in the receiving plant that affect its interactions with natural enemies, via either volatile emission or quality of the herbivore prey. Secondly, if there is variation in the volatile profiles of the different plant genotypes, they can combine to give an increased volatile chemical diversity that natural enemies could use as an indicator of habitat quality (Fig. 1b). The evidence for these different mechanisms in the barley model system will be discussed next, focusing separately on interactions between different plant species and between genotypes of the same species.

3.1 Allelobiosis Between Different Plant Species

The seven spot ladybird *Coccinella septempunctata* is a polyphagous predator that uses aphids as a high quality food source (Hodek and Honek 1996). It is commonly found in barley fields where it is an important member of the complex of natural enemies that regulate aphid populations. Experiments with adult *C. septempunctata*, barley and two common weeds, *Elytrigia repens* and *Cirsium arvense* have demonstrated how both allelobiosis and plant volatile diversity may support the habitat preferences of adult *C. septempunctata* in barley.

Surveys in a barley field showed that adult ladybirds occurred in significantly greater numbers in patches containing either *E. repens* or *C. arvense* than in weedless patches (Ninkovic and Pettersson 2003). The plant biomass was similar

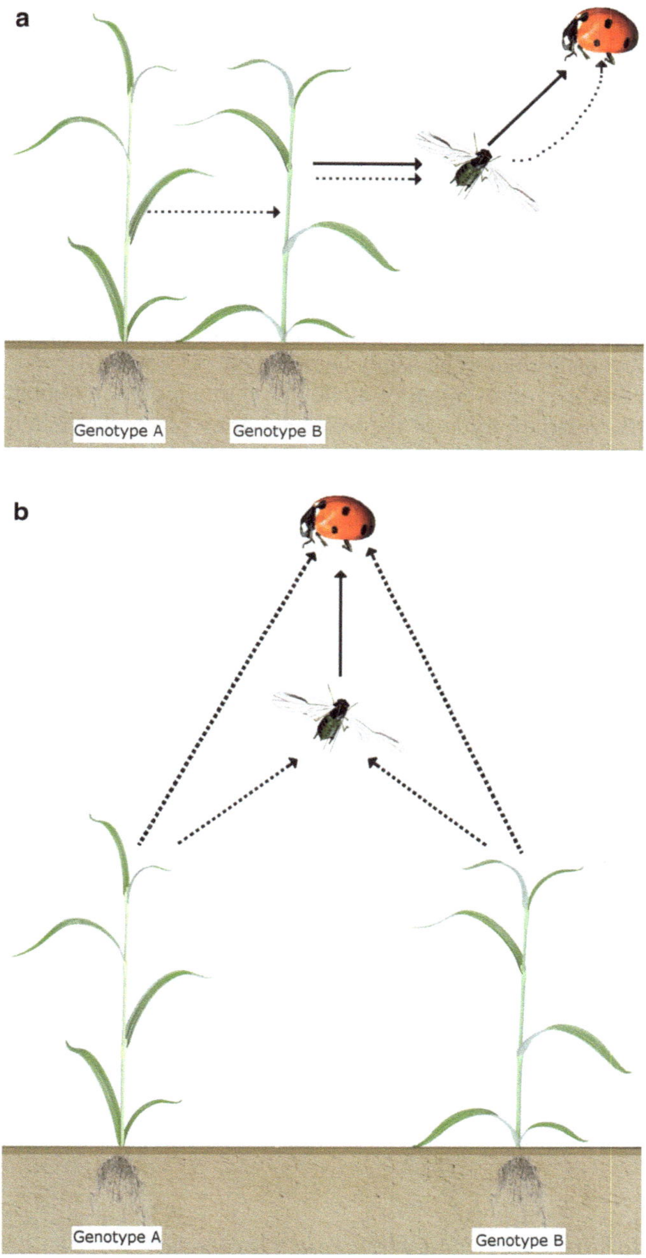

Fig. 1 **a** and **b**. Two mechanisms by which mixing different plant genotypes may affect the natural enemies of herbivores via volatile chemicals (*dotted lines*) and host quality (*bold lines*). In (**a**) volatiles from one plant genotype induce responses in a second genotype that affect plant volatile emission and herbivore host quality. In (**b**) mixed odours of the different plant genotypes affect natural enemy orientation

in the different patches, the weeds were not flowering and there were no aphids in the crop. This suggests that the ladybird distribution represented a habitat preference influenced by the presence of the weeds. In a series of laboratory experiments, the importance of plant volatile cues for the observed habitat preference was investigated (Ninkovic and Pettersson 2003). Volatiles of neither weed were directly attractive to ladybirds, however mixed odours of either weed together with barley were more attractive than the odour of barley alone. When the *E. repens* plants were removed from the system, barley plants that had been exposed to volatiles from the weed were not more attractive than unexposed barley, suggesting that volatile mixing was responsible for the previously observed attraction. However, barley plants that had been exposed to volatiles from *C. arvense* continued to be attractive to ladybirds even after the weed had been removed.

This suggests the involvement of allelobiosis, in which weed volatiles induced changes in the volatile profile of the exposed barley plants that made them more attractive to ladybirds. Interestingly, barley exposed to volatiles from this weed were found to be less acceptable to aphids both on contact and via their odour (Glinwood et al. 2004; Chapter 'Volatile Interaction Between Undamaged Plants: A Short Cut to Coexistence'), indicating that effects of allelobiosis occur over at least two trophic levels.

Aphids are attacked by small parasitic wasps in the subfamily Aphidiinae, and these too are considered important members of the aphid natural enemy complex. Allelobiosis between weeds and barley also affects olfactory orientation by these insects. While odour of the thistle *Cirsium vulgare* was repellent to female *Aphidius ervi*, barley that had been exposed to volatiles from *C. vulgare* became significantly more attractive than unexposed plants (Fig. 2). Presumably the thistle itself represents a non-host plant for a parasitoid that has been reared on a cereal–cereal aphid system, while attraction to exposed barley suggests induced changes that modify the volatile profile.

The responses of ladybirds in the experiments described above are consistent with theories on the positive effects of plant diversity on natural enemies (Root 1973; Bach 1980; Russell 1989) and also support previous studies on the importance of plant composition for the ladybird habitat preference (Leather et al. 1999; Elliott et al. 2002). They add a new perspective to the theoretical debate, namely that chemical interaction between plants, either directly by allelobiosis or via odour mixing, may comprise a mechanism by which insects detect plant diversity.

3.2 Allelobiosis Between Different Genotypes of the Same Plant Species

Several studies have shown that allelobiosis between different cultivars of barley affects both plant biomass allocation and interaction with aphid herbivores

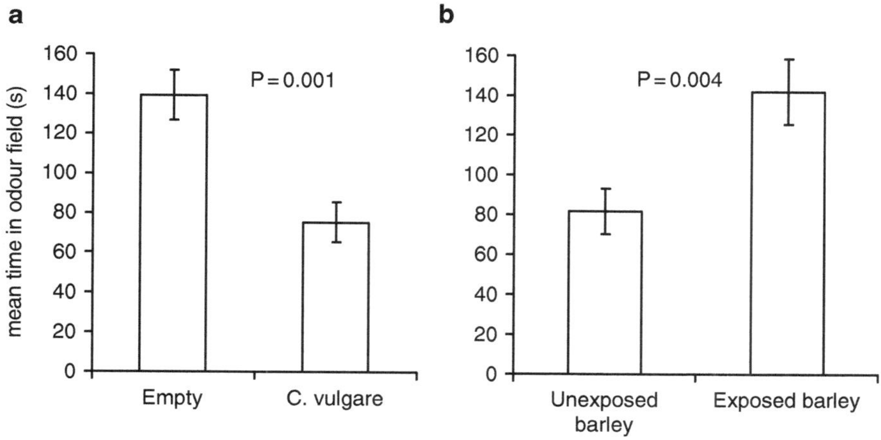

Fig. 2 Olfactory preference of the aphid parasitoid *Aphidius ervi* in a two-way olfactometer given a choice between (**a**) odour from the thistle *Cirsium vulgare* and clean air or (**b**) odour of barley previously exposed to volatiles from *C. vulgare* and unexposed barley. Values of P from Wilcoxon tests

(Ninkovic 2003; Ninkovic et al. 2002; Glinwood et al. 2007; Ninkovic and Åhman 2009; Chapter 'Volatile Interactions Between Undamaged plants: Effects and Potential for Breeding Resistance to Aphids'). As is the case for weed–barley interactions, volatile exposures between barley cultivars also affect aphid natural enemies. The effects appear to follow the same pattern as found for aphids, with only specific cultivar combinations influencing natural enemy foraging behaviour. In a field experiment, significantly more ladybird adults were recorded in stands of a mixture of two particular barley cultivars than in stands of either cultivar alone (Ninkovic et al. unpublished). Laboratory experiments showed that, although neither cultivar was more attractive, volatile exposure of one cultivar to the other caused the odour of exposed plants to become more attractive to ladybirds. When the odours of the cultivars were combined, this same cultivar combination was also more attractive than the odour of either cultivar alone. This is the first evidence that direct chemical interaction and odour mixing in genotypes of the same plant species can affect olfactory orientation by a polyphagous predator.

These findings are backed up by results with further barley cultivar combinations, in which attraction of ladybirds persisted in exposed plants for up to 7 days after the end of the exposure period, suggesting long term changes in plant status that are reflected by the profile of volatile emission (Glinwood et al. 2009). Interestingly there was a close agreement between those cultivar combinations that gave both negative effects on aphid plant acceptance and positive effects on ladybird attraction, suggesting involvement of a common mechanism. Aphid parasitoids of the species *Aphidius colemani* were also attracted to the odour of exposed plants from these combinations.

The adaptive significance of these insect responses is still not apparent, however aphid natural enemies are known to be sensitive to the quality of their aphid hosts, and should forage optimally to locate and exploit the best quality hosts. Barley plants that became attractive to aphid natural enemies after volatile exposure also appear to be less preferred as hosts by aphids. However, experiments suggest that aphids developing on plants exposed to volatiles from a different cultivar represent higher quality hosts for aphid natural enemies. For example, ladybirds consumed more aphids that had developed on exposed plants than on unexposed plants, and female parasitoids attacked and laid more eggs in aphids that had developed on exposed plants (Glinwood et al. 2009). Parasitoid egg development did not appear to be affected, suggesting that natural enemies may have handled aphid prey items more efficiently, perhaps due to decreased defensive behaviour by the aphids.

In addition to the effects of allelobiosis, odour mixing from different barley cultivars affects ladybird olfactory orientation. The combined odours of certain pair wise combinations of barley cultivars were more attractive to ladybirds than an equivalent biomass of either individual cultivar alone (Fig. 3). Interestingly the cultivar combinations that gave the clearest effects via mixed odours were those in which one cultivar became more attractive after exposure to volatiles from the other. The more specialised aphid parasitoids, however, did not show a clear pattern of attraction to mixed odours.

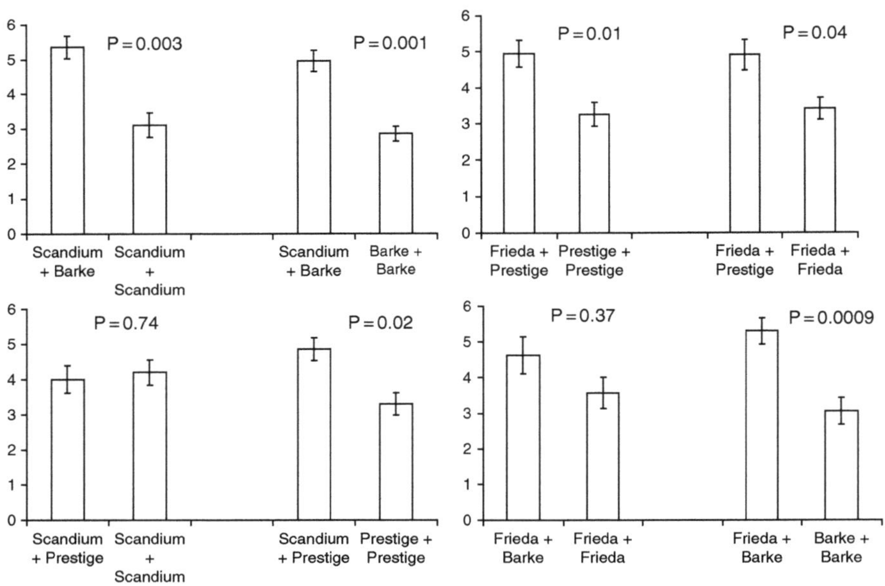

Fig. 3 Olfactory preference of the seven spot ladybird *Coccinella septempunctata* in a two-way olfactometer, given a choice between mixed odours of two different barley cultivars and odour of an equal biomass of each cultivar alone. Mean no. observations in odour field. Values of P from Wilcoxon tests (Glinwood et al. (2009), with kind permission of Springer Science and Business Media)

4 Does Plant Genetic Diversity Affect Natural Enemies Via Volatile Interaction and Diversity?

The findings on chemical interaction in barley reviewed here give a new perspective on one possible route linking plant and natural enemy levels. For a polyphagous predator like *C. septempunctata*, high quality habitats are characterised by the availability of a range of different food sources (Elliott et al. 2002; Pettersson et al. 2008). So, while aphids are a required food for this species, other small insects and pollen are also important and increased plant diversity should thus be favourable. However in order to respond to plant diversity, insects must be able to detect it. The work on barley has shown that chemical interactions between undamaged plants can make them more attractive to aphid natural enemies. It is still not clear what changes are induced in barley plants exposed to allelobiosis. It is possible that changes in the odour profile of exposed plants result from physiological events connected with growth and biomass allocation (Ninkovic 2003) or are coupled to a general stress condition induced by the presence of a close neighbour (Chapter 'Volatile Interaction Between Undamaged Plants: A Short Cut to Coexistence').

In either case, these volatile cues may enable natural enemies to locate habitats or patches with a favourable degree of plant diversity. They may then benefit from more diverse and abundant food in the case of ladybirds, or be able to forage more efficiently due to the condition of their prey in the case of both ladybirds and parasitoids. *C. septempunctata* also responded positively to the combined odours of weeds and barley and certain mixtures of barley cultivars. These responses may represent a further mechanism by which ladybirds can detect a favourable level of increased plant genetic diversity. Aphid parasitoids did not respond to the odour mixtures as strongly as ladybirds, which is expected considering their higher degree of specialisation.

These hypotheses seem more appropriate for the interactions between weeds and barley than for those between different barley genotypes, since variation in plant characteristics is presumably much lower within species than between species. However, if ladybirds can detect indicators of plant genetic diversity through volatile cues, the behavioural responses mediating this may be triggered by relatively small increases in odour diversity against a homogenous background, as is the case in agricultural monocrops such as cereals. Enhancement of natural enemies in mixed cropping has been explained by the provisioning of alternative resources (Root 1973) but it is unlikely that cultivars of the same plant species would fulfil this role for a generalist predator such as *C. septempunctata*, so whether these responses are adaptive for ladybirds foraging in barley is still unclear and needs to be investigated. Nevertheless, the work discussed here opens the way for further investigation on how chemical cues communicate plant genetic diversity to higher trophic levels.

5 Conclusions

We are now beginning to understand the wider aspects of plant volatile chemical signalling. The ecological role of damage-induced cues is becoming clear (Halitschke et al. 2008; Chapter 'Within-Plant Signalling by Volatiles Triggers Systemic Defences') and explanations for the adaptive significance of induced volatile emission are emerging (Heil and Ton 2008). Our understanding of the broader implications for chemical signalling between undamaged plants is, however, still in its infancy. The current chapter has proposed a mechanism contributing to the effects of plant genetic diversity at the natural enemy level, while the work reviewed in Chapters 'Volatile Interaction Between Undamaged Plants: A Short Cut to Coexistence' and 'Volatile Interactions Between Undamaged Plants: Effects and Potential for Breeding Resistance to Aphids' suggests these interactions can be exploited for insect pest management. Although most of the knowledge is limited to a model system with barley, there is increasing evidence from agricultural systems that mixing different genotypes of the same plant species can affect organisms that use the plants as hosts.

Research on these questions should continue along two closely coordinated lines:

1. Investigation of the ecosystem role of chemical signalling between undamaged plant individuals within habitats and between plant habitats and insects. Chemical interaction and chemical diversity may be previously unconsidered mechanisms linking genetic diversity at the plant level with behaviour and population dynamics at the primary and secondary consumer levels. Since even polyphagous insects have a specific set of nutritional requirements, the strongest effects are expected to occur in combinations of plant genotypes with particular characteristics, rather than with continuous increases in plant diversity *per se*. This is reflected in the research on barley in which effects at higher trophic levels occur only in particular genotype combinations.
2. Exploitation of these interactions by using mixed cropping to manipulate pest herbivores and their natural enemies in agricultural systems (Exploiting Plant Signals in Sustainable Agriculture). Although intercropping with different plant species is long established, it may be possible to achieve similar results by mixing genotypes of the same plant species, giving a more practical approach to sustainable plant protection.

We are beginning to view plant chemical communication as a natural facet of the behaviour of plants under normal conditions and not only as a response to attack. Key questions to be answered include whether this occurs commonly in natural or managed ecosystems and whether it represents a genuine exchange of information between plants. Further study is likely to reveal fascinating new aspects of the ecology of plant communication.

References

Andow DA (1991) Vegetational diversity and arthropod population responses. Annu Rev Entomol 36:561–586
Bach CE (1980) Effect of plant diversity and time of colonization on an herbivore–plant interaction. Oecologia 44:319–326
Cadet P, Berry SD, Leslie GW, Spaull VW (2007) Management of nematodes and a stalk borer by increasing within-field sugarcane cultivar diversity. Plant Pathol 56:526–535
Cook SM, Khan ZR, Pickett JA (2007) The use of push–pull strategies in integrated pest management. Annu Rev Entomol 52:375–400
Elliott NC, Kieckhefer RW, Michels GJ, Giles KL (2002) Predator abundance in alfalfa fields in relation to aphids, within-field vegetation, and landscape matrix. Environm Entomol 31:253–260
Glinwood RT, Pettersson J, Ninkovic V, Ahmed E, Birkett M, Pickett JA (2003) Change in acceptability of barley plants to aphids after exposure to allelochemicals from couch-grass *(Elytrigia repens)*. J Chem Ecol 29:259–272
Glinwood R, Ninkovic V, Ahmed E, Pettersson J (2004) Barley exposed to aerial allelopathy from thistles *(Cirsium* spp.) becomes less acceptable to aphids. Ecol Entomol 29:188–195
Glinwood RT, Gradin T, Karpinska B, Ahmed E, Jonsson LMV, Ninkovic V (2007) Aphid acceptance of barley exposed to volatile phytochemicals differs between plants exposed in daylight and darkness. Plant Signal Behav 2:205–210
Glinwood R, Ahmed E, Qvarfordt E, Ninkovic V, Pettersson J (2009) Airborne interactions between undamaged plants of different cultivars affect insect herbivores and natural enemies. Arthropod–Plant Interact 3:219–224
Haddad NM, Tilman D, Haarstad J, Ritchie M, Knops JMH (2001) Contrasting effects of plant richness and composition on insect communities: a field experiment. Am Nat 158:17–35
Halitschke R, Stenberg JA, Kessler D, Kessler A, Baldwin IT (2008) Shared signals – ‘alarm calls’ from plants increase apparency to herbivores and their enemies in nature. Ecol Lett 11:24–34
Heil M, Bueno JCS (2007) Within-plant signaling by volatiles leads to induction and priming of an indirect plant defense in nature. Proc Natl Acad Sci USA 104:5467–5472
Heil M, Ton J (2008) Long-distance signalling in plant defence. Trends Plant Sci 13:264–272
Hodek I, Honek A (1996) Ecology of coccinellidae. Kulwer, Dordrecht
Hooper DU, Chapin FS III, Ewel JJ, Hector A, Inchausti P, Lavorel S, Lawton JH, Lodge DM, Loreau M, Naeem S, Schmid B, Setälä H, Symstad AJ, Vandermeer J, Wardle DA (2005) Effects of biodiversity on ecosystem functioning: a consensus of current knowledge. Ecol Monogr 75:3–35
Johnson MTA, Agrawal AA (2005) Plant genotype and environment interact to shape a diverse arthropod community in evening primrose (*Oenothera biennis*). Ecology 86:875–885
Johnson MTA (2008) Bottom-up effects of plant genotype on aphids, ants, and predators. Ecology 89:145–154
Koricheva J, Mulder CPH, Schmid B, Joshi J, Huss-Danell K (2000) Numerical responses of different trophic groups of invertebrates to manipulations of plant diversity in grasslands. Oecologia 125:271–282
Leather SR, Cooker RCA, Fellowes MDE, Rombe R (1999) Distribution and abundance of ladybirds (Coleoptera: Coccinellidae) in non-crop habitats. Eur J Entomol 96:23–27
Mundt CC (2002) Use of multiline cultivars and cultivar mixtures for disease management. Annu Rev Phytopathol 40:381–410
Ninkovic V (2003) Volatile communication between barley plants affects biomass allocation. J Exp Bot 54:1931–1939
Ninkovic V, Åhman I (2009) Aphid acceptance of *Hordeum* genotypes is affected by plant volatile exposure and is correlated with aphid growth. Euphytica 169:177–185

Ninkovic V, Pettersson J (2003) Searching behaviour of sevenspotted ladybird, *Coccinella septempunctata* – effects of plant–plant odour interaction. Oikos 100:65–70

Ninkovic V, Olsson U, Pettersson J (2002) Mixing barley cultivars affects aphid host plant acceptance in field experiments. Entomol Exp Appl 102:177–182

Ninkovic V, Glinwood R, Pettersson J (2006) Communication between undamaged plants by volatiles: the role of allelobiosis. In: Baluška F, Mancuso S, Volkmann D (eds) Communication in plants: neuronal aspects of plant life. Springer-Verlag, Berlin, pp 421–434

Otway SJ, Hector A, Lawton JH (2005) Resource dilution effects on specialist insect herbivores in a grassland biodiversity experiment. J Anim Ecol 74:234–240

Pettersson J, Ninkovic V, Ahmed E (1999) Volatiles from different barley cultivars affect aphid acceptance of neighbouring plants. Acta Agric Scand Sect B 49:152–157

Pettersson J, Ninkovic V, Glinwood R (2003) Plant activation of barley by intercropped conspecifics and weeds: allelobiosis. BCPC Crop Sci Technol 2:1135–1144

Pettersson J, Ninkovic V, Glinwood R, Birkett MA, Pickett JA (2005) Foraging in a complex environment – semiochemicals support searching behaviour of the seven spot ladybird. Eur J Entomol 102:365–370

Pettersson J, Ninkovic V, Glinwood R, Al Abassi S, Birkett M, Pickett J, Wadhams L (2008) Chemical stimuli supporting foraging behaviour of *Coccinella septempunctata* L (Coleoptera: Coccinellidae): volatiles and allelobiosis – a minireview. Appl Entomol Zool 43:315–321

Power AG (1991) Virus spread and vector dynamics in genetically diverse plant populations. Ecology 72:232–241

Rice EL (1984) Allelopathy, 2nd edn. Academic, New York

Root RB (1973) Organization of plant-arthropod association in simple and diverse habitats: The fauna of collards (*Brassica oleraceae*). Ecol Mon 43:95–124

Runyon JB, Mescher MC, Moraes CMD (2006) Volatile chemical cues guide host location and host selection by parasitic plants. Science 313:964–1967

Russell EP (1989) Enemies hypothesis: a review of the effects of vegetational diversity on predatory insects and parasitoids. Environ Entomol 18:590–599

Siemann E, Tilman D, Haarstad J, Ritchie M (1998) Experimental tests of the dependence of arthropod diversity on plant diversity. Am Nat 152:738–750

Uvah III, Coaker TH (1984) Effect of mixed cropping on some insect pests of carrots and onions. Entomol Exp Appl 36:159–167

Vet LEM, Dicke M (1992) Ecology of infochemical use by natural enemies in a tritophic context. Annu Rev Entomol 37:141–172

Vinson SB (1999) Parasitoid manipulation as a plant defense strategy. Ann Ent Soc Am 92:812–828

Weston LA, Duke SO (2003) Weed and crop allelopathy. Crit Rev Plant Sci 22:367–389

Within-Plant Signalling by Volatiles Triggers Systemic Defences

Martin Heil

Abstract Plants use both internal and external signals to mount their systemic responses to local enemy attack. Resistance to herbivores being induced by airborne cures (VOCs, volatile organic compounds) has originally been discovered in the context of 'communication' among independent individual plants. Because the phenomenon of plants 'helping' their non-related neighbours to survive apparently contradicts evolutionary theory, this phenomenon remained controversial for many years. Only recently, several groups reported that VOCs released from damaged organs can also trigger a systemic resistance in as yet intact organs of a plant. This mechanism may represent the evolutionary origin of resistance induction by VOCs, as it opens the potential for fitness benefits that are achieved by the emitter itself. The observation comes, however, with its own problems: Why do plants use VOCs, whose movements are out of their control, instead of relying exclusively on internal vascular signals? Until now, the phenomenon of within-plant signalling by VOCs has been described for only four plant species: sagebrush, lima bean, poplar and blueberry. Generalisations are, thus, difficult to make at the present stage. Likely benefits of airborne as compared to vascular signalling comprise the speed of information transfer, the independence from the vascular system (VOCs can reach organs that lack direct vascular connections with the attacked one or that insert on spatially close, yet anatomically independent branches) and the option of priming: VOCs can prime intact plant tissues, thereby preparing them for a likely attack without the need for immediately investing in full resistance expression. Future research is required to understand how common and how important within-plant signalling by volatile compounds is within the plant kingdom.

M. Heil
Departamento de Ingeniería Genética. CINVESTAV – Irapuato, Km. 9.6 Libramiento Norte, CP 36821 Irapuato, Guanajuato, México
e-mail: mheil@ira.cinvestav.mx

F. Baluška and V. Ninkovic (eds.), *Plant Communication from an Ecological Perspective*, Signaling and Communication in Plants,
DOI 10.1007/978-3-642-12162-3_7,

1 Introduction

Plants are sessile organisms that cannot move to reach favourable conditions. A considerable aspect of their strategy to survive in an ever-changing environment is therefore represented by phenotypic plasticity, that is, the ability of a given genotype to express different phenotypes in response to changing environmental conditions (Sultan 2000; Agrawal 2001). Phenotypic plasticity applies to most aspects of plant life and becomes particularly obvious when plants alter their transcriptome after enemy attack. Plants respond to attack by pathogens or herbivores with extensive changes in gene expression patterns, which lead to induced resistance or tolerance: various traits are then expressed de novo or at much higher intensities to reduce or prevent further damage or reduce the consequences of damage for plant fitness (Walling 2000; Kessler and Baldwin 2002; Durrant and Dong 2004). The resulting alterations affect the metabolic, chemical and morphological phenotype and allow plants to more successfully cope with their enemies (Karban and Baldwin 1997).

As both pathogens and herbivores are mobile, such plant responses are usually expressed systemically, in as yet undamaged organs. In addition to the immediate resistance induction after attack, plants can also mount an adequate level of resistance using cues that are indicative of future attack, such as elicitors released during insect egg deposition (Meiners and Hilker 2000; Hilker and Meiners 2006) and volatiles that are released from damaged neighbours or damaged parts of the same plant (Heil and Ton 2008; Heil and Karban 2010). Since the earliest descriptions of a plant-wide expression of a phenotypic resistance that is induced by local virus infection (Gilpatrick and Weintraub 1952; Ross 1961) or insect feeding (Green and Ryan 1972), researchers have spent entire careers on the search for the genetic basis of this phenomenon, for the identity of the phenotypically active resistance traits and for the mobile signals that underlie the (systemic) resistance expression.

Early research on the mobile signals focused on compounds that move through the vascular system and identified the two plant hormones, jasmonic acid (JA) and salicylic acid (SA) and their derivatives (such as precursors of JA and the methylated forms of both hormones, methyl salicylate, MeSA and methyl jasmonate, MeJA) as the most common mobile players associated with systemic resistance expression (Shulaev et al. 1997; Durrant and Dong 2004; Schilmiller and Howe 2005; Park et al. 2007; Heil and Ton 2008; Chini et al. 2009). Recent studies have, however, revealed that long-distance signalling can also be mediated by volatile compounds in a plant's headspace (Karban et al. 2006; Frost et al. 2007; Heil and Silva Bueno 2007b; Rodriguez-Saona et al. 2009). Since such volatiles are released from the plant surface and move through the air, they can affect systemic parts of the same plant as well as neighbouring plants, then mediating a phenomenon of 'plant–plant communication' (Karban 2008).

Resistance expression in intact plants that is caused by cues from other plants being currently under attack was advertised as 'talking trees' or 'plant communication' and has been intensively reviewed and discussed over the recent years (Dicke and Bruin 2001b; Baldwin et al. 2006; Heil and Ton 2008; Karban 2008; Heil and Karban 2010). The focus of this chapter is, by contrast, a within-signalling that is mediated by volatiles. I discuss the benefits and putative shortcomings of systemic within-plant signalling as compared to vascular signalling and I will present arguments making it likely that this phenomenon, which so far has been described for four plant species, is a rather common one.

2 Talking Trees

The first reports that plant resistance expression can be triggered by airborne cues date back to the early eighties. In 1983, David F. Rhoades reported increased resistance in undamaged Sitka willow trees growing close to herbivore-infested conspecific plants (Rhoades 1983) and Ian T. Baldwin and Jack C. Schultz found increased levels of chemical anti-herbivore defence in undamaged poplar and sugar maple saplings when these shared the air with damaged plants (Baldwin and Schultz 1983). These seminal studies did, however, not identify the chemical cues that triggered the responses in the receiving plants and were criticised for a lack of independent repetitions and for not ruling out alternative explanations such as unexplored sources of mortality (Fowler and Lawton 1985; Dicke and Bruin 2001a; Preston et al. 2001).

Over the next decade, the question as to whether plants 'really' communicate by means of airborne cues was discussed intensively and more experiments were presented. Several studies failed to find any evidence for an information transfer from damaged to undamaged plants by volatiles (Myers and Williams 1984; Lin et al. 1990; Preston et al. 1999), whereas others demonstrated the phenomenon in the laboratory (Farmer and Ryan 1990; Bruin et al. 1992; Shulaev et al. 1997) and under natural conditions (Dolch and Tscharntke 2000; Karban et al. 2000; Karban and Maron 2002). Significant breakthroughs in the effort to understand the underlying mechanisms were reached with the discovery of MeJA and MeSA as volatile cues that trigger defence expression in tomato and tobacco, respectively (Farmer and Ryan 1990; Shulaev et al. 1997) and with the demonstration that the expression of defence-related genes in lima bean, *Phaseolus lunatus*, was induced by several green leaf volatiles (GLVs, small C_6 volatile compounds such as alcohols and aldehydes that are formed from pre-existing lipid precursors by pre-existing enzymes in response to plant tissue disruption) (Arimura et al. 2000). By now, a functioning 'plant communication' that is mediated by airborne cues released from neighbouring plants has been demonstrated for at least 14 plant species from nine families, covering both monocots and dicots (Heil 2009a; Heil and Karban 2010).

After the initial identifications of MeSA and MeJA as airborne signals, further studies identified several volatile organic compounds that triggered genotypic and/ or phenotypic resistance expression in undamaged plants. In bean and cotton, (*Z*)-jasmone can trigger defensive responses via airborne transport, although this herbivore-induced volatile activates different sets of genes than MeJA (Birkett et al. 2000, 2009; Bruce et al. 2008). Compounds that have been reported to prime or induce gene activity or phenotypic defences in intact maize plants comprise (*Z*)-3-hexen-1-ol, (*Z*)-3-hexenal and (*Z*)-3-hexenyl acetate (Engelberth et al. 2004; Farag et al. 2005; Ruther and Kleier 2005). In Lima bean, (*Z*)-3-hexenyl acetate was found to induce extrafloral nectar (EFN) secretion, an indirect defence mechanism (Kost and Heil 2006; Heil et al. 2008). Fewer data exist on plant communication in the context of pathogen resistance. MeSA released from virus-infected tobacco plants was, however, found to induce resistance in as yet uninfected plants (Shulaev et al. 1997) and MeSA and nonanal that were released from *Pseudomonas syringae*-infected lima beans made neighbouring plants more resistant to future infection with the bacterial pathogen (Yi et al. 2009).

In several cases, VOCs did not fully induce defence expression but rather caused priming: a sensitization of the plant's defence arsenal that prepares the plant to respond more rapidly and/or effectively to subsequent attack (Conrath et al. 2006; van Hulten et al. 2006; Bruce et al. 2007; Goellner and Conrath 2008). Hence, primed plants show no enhanced defence activity at the phenotypic level, but they respond much faster or stronger to wounding or infection than un-primed plants. For example, intact maize or lima bean plants that had been exposed to VOCs released from herbivore-damaged conspecific plants responded to herbivore damage with a faster expression of defence genes, a stronger induction of endogenous JA synthesis, or higher rates of EFN secretion, than plants that were damaged without prior exposition to VOCs (Engelberth et al. 2004; Heil and Kost 2006; Ton et al. 2007). Similarly, nonanal primed the expression of pathogenesis-related (PR) genes in lima bean (Yi et al. 2009). Such responses can be highly specific as, for example, nonanal primed PR-gene expression but not EFN secretion in lima bean. The significant benefit of being primed by VOCs that are released from an attacked neighbour is that a plant becomes prepared for an attack by enemies that eventually leave the damaged plant, whilst it does not invest high amounts of resources before an active resistance is actually needed (Heil and Ton 2008).

Interestingly, most scientists who investigated plant–plant communication focused on alterations in defence expression by the receiver. For two systems (sagebrush and lima bean), even positive effects on fitness-relevant traits of the VOC-receiving plant under field conditions could be found (Karban and Maron 2002; Kost and Heil 2006; Heil and Silva Bueno 2007b). By contrast, I am not aware of a single report showing that an induction of defence expression in the surrounding plants can benefit the emitter: a situation making the evolutionary explanation of this phenomenon problematic. Four non-exclusive hypotheses have recently been presented and comprise the role of volatiles (1) in direct defence, (2) as traits that synergistically interact with other defences, (3) as cues among kin and (4) as within-plant signals (Heil and Karban 2010). Future empirical studies are, however, required to elucidate which of these non-exclusive factors are important for the evolutionary onset and maintenance of VOC-mediated plant–plant communication.

3 VOCs as Within-Plant Signals

Although we lack knowledge on the fitness consequences of information transfer among independent neighbouring plants, the phenomenon undoubtedly exists, is mediated by volatile cues and can affect the receivers under natural conditions. By contrast, mobile signals that underlie the systemic resistance expression within a single plant have been searched for – and found – mainly in the vascular system. Damaging a single leaf usually elicits the strongest responses in orthostichous leaves (Orians 2005), and the spatio-temporal patterns in the resistance expression to herbivores in poplar, *Populus trichocarpa x deltoides* (Davis et al. 1991), tomato, *Lycopersicon esculentum* (Orians et al. 2000), wild tobacco, *Nicotiana attenuata* (Schittko and Baldwin 2003) and clover, *Trifolium repens* (Gómez and Stuefer 2006) matched the respective patterns of assimilate transport. Furthermore, girdling of the petioles of pathogen-infected cucumber leaves blocked systemic resistance expression against pathogens, indicating an involvement of phloem transport (Guedes et al. 1980). For these reasons, research on long-distance signals usually focused on chemical or – to a lesser extent – electrical (Stankovic and Davies 1996, 1998; Brenner et al. 2006; Maffei et al. 2007) signals that travel through the vascular system (Heil and Ton 2008).

In some cases, however, phenotypic resistance was expressed quickly and strongly in distal leaves that lacked a direct vascular connexion to the attacked leaf (Mutikainen et al. 1996; Kiefer and Slusarenko 2003), an observation suggesting additional routes that allow the induction of systemic resistance in source leaves or leaves that are non-orthostichous to the damaged ones (Kiefer and Slusarenko 2003; Van Bel and Gaupels 2004). What transportation ways could exist if the signal does not move through the vascular system? Considering the above-mentioned phenomenon of 'talking trees' it might not appear too surprising that volatile cues were found to act in within-plant signalling as well. This idea was originally presented by Farmer (2001) and Orians (2005), but first empirical confirmations were reported not earlier than 2006. Richard Karban and co-workers found that sagebrush fails to express systemic resistance to herbivores when airflow between a damaged and the systemic, as yet undamaged branches is prevented (Karban et al. 2006). Similarly, EFN secretion by intact lima bean leaves was primed by VOCs released from damaged neighbouring leaves, and external rather than vascular signals were found to underlie the systemic response of lima bean to local damage (Heil and Silva Bueno 2007b). In following studies, a VOC-mediated priming of defence expression in systemic parts of a locally damaged plant was also found for poplar, *Populus deltoides x nigra*, and blueberry, *Vaccinium corymbosum* (Frost et al. 2007; Rodriguez-Saona et al. 2009).

3.1 *Shortcomings of Signalling by VOCs*

Plants use both internal and external signals to mount their systemic resistance to local attack by pathogens or herbivores. What are the shortcomings and benefits of this dual system? A within-plant signalling 'worn on the outside' beyond doubts suffers from significant shortcomings with respect to (1) the capacity of the emitting plant to control the spatiotemporal distribution of the signals and (2) the access of other organisms to the information. Volatile organic compounds move freely through the air, their spatiotemporal distribution is controlled by air movements and other abiotic factors rather than by the plant itself, and VOCs rapidly dilute, likely losing their activity at short distances from the emitting organ. I am not aware of a study that systematically investigated distances over which herbivore-induced VOCs can affect resistance in other plants, or parts of plants. All positive reports on plant–plant communication used, however, rather short distances: less than a meter in the alder system (Dolch and Tscharntke 2000) and even less than 50 cm for tobacco and sagebrush (Karban et al. 2003, 2006) and for lima bean (Heil and Silva Bueno 2007b).

Second, VOCs 'blow in the wind' (Preston et al. 2001) and can therefore in principle be used as a source of information by every organism that is capable of perceiving them. The perception of herbivore-induced VOCs can benefit the emitting plant when it serves to attract carnivorous or parasitoid arthropods and thereby leads to an indirect defence (Dicke and Sabelis 1988; Turlings et al. 1990; Dicke 1994; Tumlinson et al. 1999; van Loon et al. 2000; Kessler and Baldwin 2001) via the enhancement of tritrophic interactions (see Dicke et al. 2003; Turlings and Wäckers 2004; Heil 2008 for reviews). However, these cues can also be used by herbivorous insects for their own purposes. In fact, the localisation of suitable host plants by herbivores that are attracted to host-released odours likely represents a common phenomenon. For example, small potato plants which are normally not attractive for the Colorado potato beetle (*Leptinotarsa decemlineata*) became attractive after being mechanically damaged and the intensity and duration of the effect depended on the damaging regime (Bolter et al. 1997). Damaged *Adenostyles alliariae* host plants were more attractive to the leaf beetle, *Oreina caraliae*, than undamaged ones (Kalberer et al. 2001), female plant bugs (*Lygus rugulipennis*) responded positively to volatiles released from *Vicia faba* plants when these were damaged or exposed to oviposition by conspecifics (Frati et al. 2009), herbivore-induced monoterpenes released by poplar helped beetles (*Chrysomela populi*) to localise their host plant (Brilli et al. 2009) and the same function was attributed to odours of neotropical ginger (Zingiberales) (Garcia-Robledo and Horvitz 2009) and mulberry (*Morus alba*) plants (Mooney et al. 2009), which became more attractive to specialist herbivores when being damaged by conspecific animals.

Interestingly, this response seems to be dose-dependent, thus allowing herbivores to avoid plants that are too heavily damaged and therefore being unlikely to serve as a suitable food source (Heil 2004). From the point of view of the plant, the attraction of specialised herbivores can cause significant 'ecological costs' of a systemic signalling that is 'worn on the outside' and thus prone to eavesdropping by other plants or by animals (Heil and Karban 2010). We would expect, therefore, the existence of significant benefits of airborne within-plant signals that outweigh the potential ecological costs that result from enemy attraction.

3.2 Benefits of VOCs as Plant Signals

Plants possess a highly evolved and efficient vascular system but still take the above discussed risks that come with external signalling. What are the benefits of within-plant signalling by volatile cues? First, VOCs have several direct, positive effects on the emitting plant, because they can repel herbivores (Bernasconi et al. 1998; de Moraes et al. 2001; Kessler and Baldwin 2001; Heil 2004; Bruinsma et al. 2008), have antimicrobial effects (Nakamura and Hatanaka 2002; Dilantha Fernando et al. 2005; Matsui 2006; Shiojiri et al. 2006) or protect plants from abiotic stress (Loreto and Velikova 2001; Behnke et al. 2007). Second, as discussed above, VOCs move independently of vascular connections and can, thus, reach also no-orthostichous leaves and leaves that are localised downstream in assimilate transport. This characteristic can be particularly important in anatomically complex plants. Most herbivores and pathogens are mobile and can easily spread to spatially neighbouring leaves. In shrubs, trees and lianae, leaves that are spatially close to the attacked leaf might, however, insert on a different branch or shoot, then being separated anatomically by meters from the source of actual danger (Heil and Silva Bueno 2007a). By moving through the air, VOCs likely spread in patterns that are similar to the ones in which spores and insects distribute (Fig. 1). Third, green-leaf volatiles

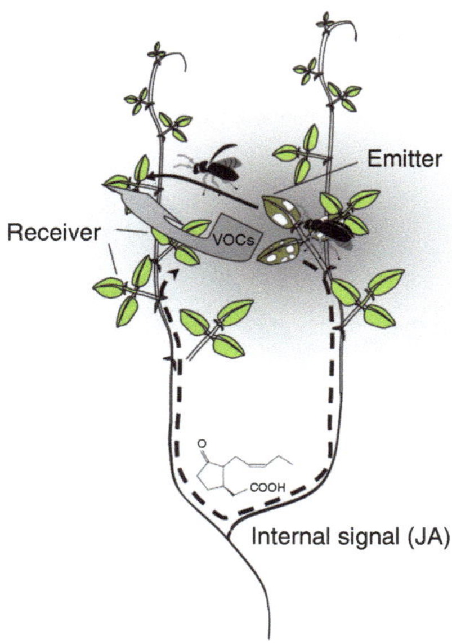

Fig. 1 Within-plant signalling by VOCs circumvents long-distance internal transport of signals. Volatile organic compounds (VOCs, *shaded area* and *grey arrow*) are released from beetle-damaged leaves and spread through the aerial space around this emitter leaf. Thus, they can rapidly reach and induce the spatially neighbouring receiver, which can easily be reached by the mobile, herbivorous beetle and thus are in high danger of immediate attack. By contrast, internal signals such as jasmonic acid (JA) would have to move through a much longer way (*dashed arrow*) in order to reach the leaf in danger. Volatile-mediated signalling thus allow the induction of those parts of a locally damaged plants that are at high risk: the leaves being spatially, but not necessarily anatomically localised close to the site of attack (redrawn from Heil and Silva Bueno 2007a)

are synthesised during cell disruption when membrane-bound lipids become exposed to pre-existing enzymes and are, thus, rapidly and inevitably released immediately upon tissue damage (Turlings et al. 1998). Such cues represent reliable indicators of the 'damaged self' (Heil 2009b) and are, thus, suitable signals to quickly 'warn' the yet undamaged parts of the same plant.

Finally, VOCs at lower concentrations usually prime resistance rather than fully inducing it (Engelberth et al. 2004; Heil and Kost 2006; Ton et al. 2007; Frost et al. 2008). Induced resistance has likely evolved because constitutive resistance expression is too costly (Heil and Baldwin 2002; Cipollini et al. 2003; Walters and Heil 2007). A full systemic resistance expression after every local enemy attack comes, therefore, with the risk of investing in a defence that is not needed when local resistance expression sufficed to prevent pathogens from spreading or when herbivores eventually leave the plant for other reasons. Priming enables plants to prepare themselves for future attack without investing in costly phenotypic resistance expression, and self-priming indeed has been reported for lima bean, poplar and blueberry (Frost et al. 2007; Heil and Silva Bueno 2007b; Rodriguez-Saona et al. 2009). It appears, thus, likely that VOCs serve as a first, rapid signal in a two-component system for the plant-wide regulation of future resistance (Fig. 2): VOCs are released quickly from the damaged site and prepare systemic organs for the likely spreading of pathogens and herbivores. Full (and costly) resistance expression then requires a confirmatory second trigger, which can be the arrival of the vascular signal or of the plant enemy itself (Heil and Ton 2008). Combining rapid

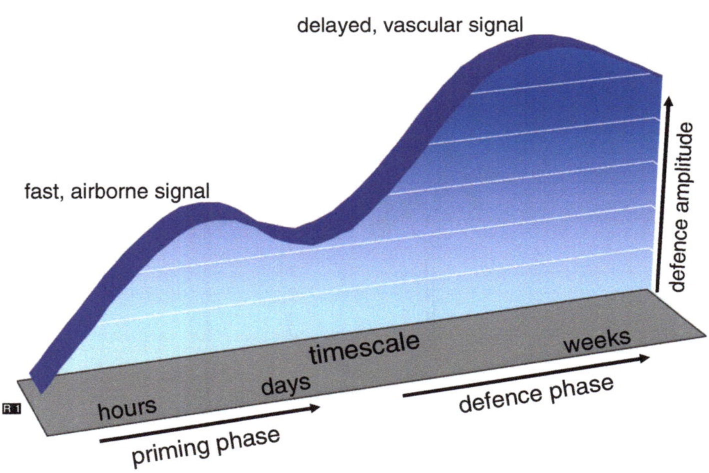

Fig. 2 Within-plant allows preparing systemic tissue by priming. Airborne signals can trigger full induced defence when they reach close leaves (being at high risk) at relatively high concentrations, while leaves inserting at longer distances are only primed for an enhanced defence induction once. Rapid airborne signals can, thus, prime systemic tissues (minutes – hours: priming phase) at low costs because they cause only relatively little phenotypic defence expression. The costly, full defence expression then requires confirmation by the delayed arrival of the vascular signal or by the plant enemy itself (days – weeks, defence phase). (redrawn from Heil and Ton 2008)

volatile and slower vascular signals allows plants to prepare systemic organs for future problems without investing limited resources in a resistance that perhaps is never needed.

3.3 Airborne Cues as Within-Plant Signals: Exotic Cases or Common Strategy?

Within-plant signalling via airborne cues by now has been reported for four plant species: wild tobacco, lima bean, poplar, and blueberry (Karban et al. 2006; Frost et al. 2007; Heil and Silva Bueno 2007b; Rodriguez-Saona et al. 2009). Do these findings represent the first cases of a common strategy that have been discovered, or rather some interesting but somehow 'exotic' cases of restricted general relevance? One study tried to exclude airborne signalling within the same plant (Park et al. 2007). Park and colleagues studied the role of the volatile compound, MeSA, in the systemic resistance expression of tobacco and reported its vascular transport without finding clear hints towards a significant effect of airborne MeSA. However, I am not aware of any further studies that actively controlled for the possibility of airborne signalling. The lack of reports can, thus, as probably result from a lack of studies that have looked for the phenomenon or from a lack of its general occurrence.

Several lines of argument favour the first interpretation. First, several studies found patterns of resistance expression at the phenotypic level that were inconsistent with an exclusively vascular transport of the signals (Mutikainen et al. 1996; Farmer 2001; Kiefer and Slusarenko 2003; Van Bel and Gaupels 2004; Orians 2005). Second, resistance induction in plants can be caused by GLVs and by the volatile derivatives of the central hormones, MeSA and MeJA. Resistance induction by these compounds in intact plants has been reported for a variety of species that do not share any close phylogenetic ancestors (see Sect. 2 and Heil and Karban 2010). Because all plants that have been investigated so far are capable to release VOCs after enemy attack and as so many plants respond to these compounds once they are present in their headspace, it appears reasonable to assume that the majority of plants will also respond to the same cues when they are released from their own leaves.

Within-plant signalling via airborne cues likely represents a general phenomenon that has been overseen due to the lack of adequate controls in the traditional studies on systemically induced resistance. We can expect, however, that the phenomenon is more important in large and anatomically complex plants because here the plant-internal distance between two spatially neighbouring leaves (see Fig. 1) can easily amount to several meters or even tens of meters (imagine two spatially neighbouring leaves that insert at the tips of two neighbouring branches: an internal signal would have to move from the damaged leaf down to the trunk and then up the neighbouring branch in order to reach the leaf in danger). Intriguingly, those four plant species for which the phenomenon has been reported by now represent a liana (lima bean), two shrubs (sagebrush and blueberry) and a tree species (poplar). Future studies should screen plants representing different taxonomical groups, life history strategies and anatomical types, in order to investigate whether within-plant signalling by volatiles represents a common or a rare phenomenon and whether its importance relative to vascular signalling depends on plant anatomical features.

References

Agrawal AA (2001) Phenotypic plasticity in the interactions and evolution of species. Science 294:321–326

Arimura G-I, Ozawa R, Shimoda T, Nishioka T, Boland W, Takabayashi J (2000) Herbivory-induced volatiles elicit defence genes in lima bean leaves. Nature 406:512–515

Baldwin IT, Halitschke R, Paschold A, von Dahl CC, Preston CA (2006) Volatile signaling in plant–plant interactions: "Talking trees" in the genomics era. Science 311:812–815

Baldwin IT, Schultz JC (1983) Rapid changes in tree leaf chemistry induced by damage: evidence for communication between plants. Science 221:277–279

Behnke K, Ehlting B, Teuber M, Bauerfeind M, Louis S, Hasch R, Polle A, Bohlmann J, Schnitzler JP (2007) Transgenic, non-isoprene emitting poplars don't like it hot. Plant J 51:485–499

Bernasconi ML, Turlings TCJ, Ambrosetti L, Bassetti P, Dorn S (1998) Herbivore-induced emissions of maize volatiles repel the corn leaf aphid, *Rhopalosiphum maidis*. Entomol Exp Appl 87:133–142

Birkett MA, Blassioli-Moraes MC, Borges M, Bruce TJA, Gordon-Weels R, Goulart-Santana AE, Smart LE, Woodcock CM, Pickett JA (2009) *cis*-Jasmone as an activator of plant defence: understanding underlying meachnisms and towards field deployment. In: 25th Annual Meeting of the International Society of Chemical Ecology, Neuchatel, p 107

Birkett MA, Campbell CAM, Chamberlain K, Guerrieri E, Hick AJ, Martin JL, Matthes M, Napier JA, Pettersson J, Pickett JA, Poppy GM, Pow EM, Pye BJ, Smart LE, Wadhams GH, Wadhams LJ, Woodcock CM (2000) New roles for cis-jasmone as an insect semiochemical and in plant defense. Proc Natl Acad Sci USA 97:9329–9334

Bolter CJ, Dicke M, vanLoon JJA, Visser JH, Posthumus MA (1997) Attraction of Colorado potato beetle to herbivore-damaged plants during herbivory and after its termination. J Chem Ecol 23:1003–1023

Brenner ED, Stahlberg R, Mancuso S, Vivanco J, Baluska F, Van Volkenburgh E (2006) Plant neurobiology: an integrated view of plant signaling. Trends Plant Sci 11:413–419

Brilli F, Ciccioli P, Frattoni M, Prestininzi M, Spanedda AF, Loreto F (2009) Constitutive and herbivore-induced monoterpenes emitted by *Populus x euroamericana* leaves are key volatiles that orient *Chrysomela populi* beetles. Plant Cell Environ 32:542–552

Bruce TJA, Matthes MC, Chamberlain K, Woodcock CM, Mohib A, Webster B, Smart LE, Birkett MA, Pickett JA, Napier JA (2008) cis-Jasmone induces *Arabidopsis* genes that affect the chemical ecology of multitrophic interactions with aphids and their parasitoids. Proc Natl Acad Sci USA 105:4553–4558

Bruce TJA, Matthes MC, Napier JA, Pickett JA (2007) Stressful "memories" of plants: evidence and possible mechanisms. Plant Sci 173:603–608

Bruin J, Dicke M, Sabelis MW (1992) Plants are better protected against spider-mites after exposure to volatiles from infested conspecifics. Experientia 48:525–529

Bruinsma M, Ildema H, van Loon JJA, Dicke M (2008) Differential effects of jasmonic acid treatment of *Brassica nigra* on the attraction of pollinators, parasitoids, and butterflies. Entomol Exp Appl 128:109–116

Chini A, Boter M, Solano R (2009) Plant oxylipins: COI1/JAZs/MYC2 as the core jasmonic acid-signalling module. FEBS J 276:4682–4692
Cipollini DF, Purrington CB, Bergelson J (2003) Costs of induced responses in plants. Basic Appl Ecol 4:79–89
Conrath U, Beckers GJM, Flors V, Garcia-Agustin P, Jakab G, Mauch F, Newman MA, Pieterse CMJ, Poinssot B, Pozo MJ, Pugin A, Schaffrath U, Ton J, Wendehenne D, Zimmerli L, Mauch-Mani B (2006) Priming: getting ready for battle. Mol Plant Microbe Interact 19:1062–1071
Davis MA, Gordon MP, Smit BA (1991) Assimilate movement dictates remote sites of wound-induced gene expression in poplar leaves. Proc Natl Acad Sci USA 88:2393–2396
de Moraes CM, Mescher MC, Tumlinson JH (2001) Caterpillar-induced nocturnal plant volatiles repel conspecific females. Nature 410:577–580
Dicke M (1994) Local and systemic production of volatile herbivore-induced terpenoids: their role in plant-carnivore mutualism. J Plant Physiol 143:465–472
Dicke M, Bruin J (2001a) Chemical information transfer between damaged and undamaged plants. Biochem Syst Ecol 29:979–980
Dicke M, Bruin J (2001b) Chemical information transfer between plants: back to the future. Biochem Syst Ecol 29:981–994
Dicke M, Sabelis MW (1988) How plants obtain predatory mites as bodyguards. Neth J Zool 38:148–165
Dicke M, van Poecke RMP, de Boer JG (2003) Inducible indirect defence of plants: from mechanisms to ecological functions. Basic Appl Ecol 4:27–42
Dilantha Fernando WG, Ramarathnam R, Krishnamoorthy AS, Savchuk SC (2005) Identification and use of potential bacterial organic antifungal volatiles in biocontrol. Soil Biol Biochem 37:955–964
Dolch R, Tscharntke T (2000) Defoliation of alders (*Alnus glutinosa*) affects herbivory by leaf beetles on undamaged neighbours. Oecologia 125:504–511
Durrant WE, Dong X (2004) Systemic acquired resistance. Annu Rev Phytopathol 42:185–209
Engelberth J, Alborn HT, Schmelz EA, Tumlinson JH (2004) Airborne signals prime plants against insect herbivore attack. Proc Natl Acad Sci USA 101:1781–1785
Farag MA, Fokar M, Zhang HA, Allen RD, Paré PW (2005) (Z)-3-Hexenol induces defense genes and downstream metabolites in maize. Planta 220:900–909
Farmer EE (2001) Surface-to-air signals. Nature 411:854–856
Farmer EE, Ryan CA (1990) Interplant communication: airborne methyl jasmonate induces synthesis of proteinase inhibitors in plant leaves. Proc Natl Acad Sci USA 87:7713–7716
Fowler SV, Lawton JH (1985) Rapidly induced defenses and talking trees: the devil's advocate position. Am Nat 126:181–195
Frati F, Chamberlain K, Birkett M, Dufour S, Mayon P, Woodcock C, Wadhams L, Pickett J, Salerno G, Conti E, Bin F (2009) *Vicia faba–Lygus rugulipennis* interactions: induced plant volatiles and sex pheromone enhancement. J Chem Ecol 35:201–208
Frost C, Appel H, Carlson J, De Moraes C, Mescher M, Schultz J (2007) Within-plant signalling by volatiles overcomes vascular constraints on systemic signalling and primes responses against herbivores. Ecol Lett 10:490–498
Frost C, Mescher MC, Carlson J, De Moraes CM (2008) Plant defense priming against herbivores: getting ready for a different battle. Plant Physiol 146:818–824
Garcia-Robledo C, Horvitz CC (2009) Host plant scents attract rolled-leaf beetles to Neotropical gingers in a Central American tropical rain forest. Entomol Exp Appl 131:115–120
Gilpatrick JD, Weintraub M (1952) An unusual type of protection with the carnation mosaic virus. Science 115:701–702
Goellner K, Conrath U (2008) Priming: it's all the world to induced disease resistance. Eur J Plant Pathol 121:233–242
Gómez S, Stuefer JF (2006) Members only: induced systemic resistance to herbivory in a clonal plant network. Oecologia 147:461–468

Green TR, Ryan CA (1972) Wound-induced proteinase inhibitor in plant leaves: a possible defense mechanism against insects. Science 175:776–777

Guedes MEM, Richmond S, Kuc J (1980) Induced systemic resistance to anthracnose in cucumber as influenced by the location of the inducer inoculation with *Colletotrichum lagenarium* and the onset of flowering and fruiting. Physiol Plant Pathol 17:229–233

Heil M (2004) Direct defense or ecological costs? Responses of herbivorous beetles to volatiles released by wild Lima bean (*Phaseolus lunatus*). J Chem Ecol 30:1289–1295

Heil M (2008) Indirect defence via tritrophic interactions. New Phytol 178:41–61

Heil M (2009a) Airborne induction and priming of defences. In: Baluška F (ed) Plant – environment interactions. Springer, Berlin, pp 137–152

Heil M (2009b) Damaged-self recognition in plant herbivore defence. Trends Plant Sci 14:356–363

Heil M, Baldwin IT (2002) Fitness costs of induced resistance: emerging experimental support for a slippery concept. Trends Plant Sci 7:61–67

Heil M, Karban R (2010) Explaining the evolution of plant communication by airborne signals. Trends Ecol Evol 25:137–144

Heil M, Kost C (2006) Priming of indirect defences. Ecol Lett 9:813–817

Heil M, Silva Bueno JC (2007a) Herbivore-induced volatiles as rapid signals in systemic plant responses. Plant Signal Behav 2:191–193

Heil M, Silva Bueno JC (2007b) Within-plant signaling by volatiles leads to induction and priming of an indirect plant defense in nature. Proc Natl Acad Sci USA 104:5467–5472

Heil M, Ton J (2008) Long-distance signalling in plant defence. Trends Plant Sci 13:264–272

Heil M, Lion U, Boland W (2008) Defence-inducing volatiles: in search for the active motif. J Chem Ecol 34:601–604

Hilker M, Meiners T (2006) Early herbivore alert: insect eggs induce plant defense. J Chem Ecol 32:1379–1397

Kalberer NM, Turlings TCJ, Rahier M (2001) Attraction of a leaf beetle (*Oreina cacaliae*) to damaged host plants. J Chem Ecol 27:647–661

Karban R (2008) Plant behaviour and communication. Ecol Lett 11:727–739

Karban R, Baldwin I, Baxter K, Laue G, Felton G (2000) Communication between plants: induced resistance in wild tobacco plants following clipping of neighboring sagebrush. Oecologia 125:66–71

Karban R, Baldwin IT (1997) Induced responses to herbivory. University of Chicago Press, Chicago

Karban R, Maron J (2002) The fitness consequences of interspecific eavesdropping between plants. Ecology 83:1209–1213

Karban R, Maron J, Felton GW, Ervin G, Eichenseer H (2003) Herbivore damage to sagebrush induces resistance in wild tobacco: evidence for eavesdropping between plants. Oikos 100:325–332

Karban R, Shiojiri K, Huntzinger M, McCall AC (2006) Damage-induced resistance in sagebrush: volatiles are key to intra- and interplant communication. Ecology 87:922–930

Kessler A, Baldwin IT (2001) Defensive function of herbivore-induced plant volatile emissions in nature. Science 291:2141–2144

Kessler A, Baldwin IT (2002) Plant responses to insect herbivory: the emerging molecular analysis. Annu Rev Plant Biol 53:299–328

Kiefer IW, Slusarenko AJ (2003) The pattern of systemic acquired resistance induction within the *Arabidopsis* rosette in relation to the pattern of translocation. Plant Physiol 132:840–847

Kost C, Heil M (2006) Herbivore-induced plant volatiles induce an indirect defence in neighbouring plants. J Ecol 94:619–628

Lin H, Kogan M, Fischer D (1990) Induced resistance in soybean to the Mexican bean beetle (Coleoptera: Coccinellidae): comparisons of inducing factors. Environ Entomol 119:1852–1857

Loreto F, Velikova V (2001) Isoprene produced by leaves protects the photosynthetic apparatus against ozone damage, quenches ozone products, and reduces lipid peroxidation of cellular membranes. Plant Physiol 127:1781–1787

Maffei ME, Mithöfer A, Boland W (2007) Before gene expression: early events in plant–insect interaction. Trends Plant Sci 12:310–316
Matsui K (2006) Green leaf volatiles: hydroperoxide lyase pathway of oxylipin metabolism. Curr Opin Plant Biol 9:274–280
Meiners T, Hilker M (2000) Induction of plant synomones by oviposition of a phytophagous insect. J Chem Ecol 26:221–232
Mooney AC, Robertson HM, Wanner KW (2009) Neonate silkworm (*Bombyx mori*) larvae are attracted to mulberry (*Morus alba*) leaves with conspecific feeding damage. J Chem Ecol 35:552–559
Mutikainen P, Walls M, Ovaska J (1996) Herbivore-induced resistance in *Betula pendula*: the role of plant vascular architecture. Oecologia 108:723–727
Myers JH, Williams KS (1984) Does tent caterpillar attack reduce the food quality of red alder foliage? Oecologia 62:74–79
Nakamura S, Hatanaka A (2002) Green-leaf-derived C6-aroma compounds with potent antibacterial action that act on both gram-negative and gram-positive bacteria. J Agric Food Chem 50:7639–7644
Orians C (2005) Herbivores, vascular pathways, and systemic induction: facts and artifacts. J Chem Ecol 31:2231–2242
Orians CM, Pomerleau J, Rico R (2000) Vascular architecture generates fine scale variation in the systemic induction of proteinase inhibitors in tomato. J Chem Ecol 26:471–485
Park SW, Kaimoyo E, Kumar D, Mosher S, Klessig DF (2007) Methyl salicylate is a critical mobile signal for plant systemic acquired resistance. Science 318:113–116
Preston CA, Laue G, Baldwin IT (2001) Methyl jasmonate is blowing in the wind, but can it act as a plant–plant airborne signal? Biochem Syst Ecol 29:1007–1023
Preston CA, Lewandowski C, Enyedi AJ, Baldwin IT (1999) Tobacco mosaic virus inoculation inhibits wound-induced jasmonic acid-mediated responses within but not between plants. Planta 209:87–95
Rhoades DF (1983) Responses of alder and willow to attack by tent caterpillars and webworms: evidence for pheromonal sensitivity of willows. In: Hedin PA (ed) Plant resistance to insects. American Chemical Society, Washington, pp 55–68
Rodriguez-Saona CR, Rodriguez-Saona LE, Frost CJ (2009) Herbivore-induced volatiles in the perennial shrub, *Vaccinium corymbosum*, and their role in inter-branch signaling. J Chem Ecol 35:163–175
Ross AF (1961) Systemic acquired resistance induced by localized virus infection in plants. Virology 14:340–358
Ruther J, Kleier S (2005) Plant–plant signaling: ethylene synergizes volatile emission in *Zea mays* induced by exposure to (Z)-3-Hexen-1-ol. J Chem Ecol 31:2217–2222
Schilmiller AL, Howe GA (2005) Systemic signaling in the wound response. Curr Opin Plant Biol 8:369–377
Schittko U, Baldwin IT (2003) Constraints to herbivore-induced systemic responses: bidirectional signaling along orthostichies in *Nicotiana attenuata*. J Chem Ecol 29:763–770
Shiojiri K, Kishimoto K, Ozawa R, Kugimiya S, Urashimo S, Arimura G, Horiuchi J, Nishioka T, Matsui K, Takabayashi J (2006) Changing green leaf volatile biosynthesis in plants: an approach for improving plant resistance against both herbivores and pathogens. Proc Natl Acad Sci USA 103:16672–16676
Shulaev V, Silverman P, Raskin I (1997) Airborne signalling by methyl salicylate in plant pathogen resistance. Nature 385:718–721
Stankovic B, Davies E (1996) Both action potentials and variation potentials induce proteinase inhibitor gene expression in tomato. FEBS Lett 390:275–279
Stankovic B, Davies E (1998) The wound response in tomato involves rapid growth and electrical responses, systemically up-regulated transcription of proteinase inhibitor and calmodulin and down-regulated translation. Plant Cell Physiol 39:268–274

Sultan SE (2000) Phenotypic plasticity for plant development, function and life history. Trends Plant Sci 5:537–542
Ton J, D'Allesandro M, Jourdie V, Jakab G, Karlen D, Held M, Mauch-Mani B, Turlings TCJ (2007) Priming by airborne signals boosts direct and indirect resistance in maize. Plant J 49:16–26
Tumlinson JH, Paré PW, Turlings TCJ (1999) Plant production of volatile semiochemicals in response to insect-derived elicitors. In: Chadwick DJ, Goode JA (eds) Insect–plant interactions and induced plant defence. Wiley, Chichester, pp 95–109
Turlings TCJ, Lengwiler UB, Bernasconi ML, Wechsler D (1998) Timing of induced volatile emissions in maize seedlings. Planta 207:146–152
Turlings TCJ, Tumlinson JH, Lewis WJ (1990) Exploitation of herbivore-induced plant odors by host-seeking parasitic wasps. Science 250:1251–1253
Turlings TCJ, Wäckers FL (2004) Recruitment of predators and parasitoids by herbivore-injured plants. In: Cardés RT, Millar JG (eds) Advances in insect chemical ecology. Cambridge University Press, Cambridge, pp 21–75
Van Bel AJE, Gaupels F (2004) Pathogen-induced resistance and alarm signals in the phloem. Mol Plant Pathol 5:495–504
van Hulten M, Pelser M, van Loon LC, Pieterse CMJ, Ton J (2006) Costs and benefits of priming for defense in *Arabidopsis*. Proc Natl Acad Sci USA 103:5602–5607
van Loon JJA, de Boer JG, Dicke M (2000) Parasitoid-plant mutualism: parasitoid attack of herbivore resistance increases plant reproducion. Entomol Exp Appl 97:219–227
Walling LL (2000) The myriad plant responses to herbivores. J Plant Growth Reg 19:195–216
Walters D, Heil M (2007) Costs and trade-offs associated with induced resistance. Physiol Mol Plant Pathol 71:3–17
Yi H-S, Heil M, Ballhorn D, Ryu C-M (2009) Airborne induction and priming of plant resistance to a bacterial pathogen. Plant Physiol 151(4):2152–2161

Volatile Interactions Between Undamaged Plants: Effects and Potential for Breeding Resistance to Aphids

Inger Åhman and Velemir Ninkovic

Abstract Various theories about why vegetational diversity may affect herbivore abundance have been put forward over the years. In this chapter, we discuss one possible mechanism that has not been widely studied, involving plant volatiles and using an aphid (*Rhopalosiphum padi* L.) and one of its hosts (*Hordeum vulgare* L., barley) as test organisms. Volatiles from neighbouring plants of a different species or even from plants of the same species have been found to alter aphid acceptance of the receiving plant. Similar effects have been found earlier with volatiles from damaged plants, but here the volatile-emitting plants are apparently undamaged. In the majority of plant combinations tested, host acceptance is decreased but only when certain emitters are combined with certain receivers. Exposure of barley to volatiles from the common weeds *Chenopodium album* L., *Cirsium* spp. and *Solanum nigrum* L. resulted in reduced host acceptance by the aphid, but exposing barley to volatiles from many other weed species had no effect. The same was true for intra-specific interactions; only when certain barley genotypes were exposed to volatiles from specific barley genotypes did the aphids respond differently. Such induced effects correlated with aphid growth rates in a set of barley genotypes representing a wide range of host suitability to the aphid. Pedigree information suggested that the ability to become induced is heritable in barley, something that might be exploited in breeding. More crop/pest combinations should be investigated for these effects, and favourable interactions should be exploited in new cropping systems as plant mixtures or in monocultures with chemical elicitors applied according to forecasts of pest attacks.

I. Åhman (✉)
Faculty of Landscape Planning, Horticulture and Agricultural Science, Swedish University of Agricultural Sciences, P.O. Box 104, SE-230 53 Alnarp, Sweden
e-mail: Inger.Ahman@ltj.slu.se

V. Ninkovic
Department of Ecology, Swedish University of Agricultural Sciences, P.O. Box 7044, SE-750 07 Uppsala, Sweden
e-mail: Velemir.Ninkovic@ekol.slu.se

F. Baluška and V. Ninkovic (eds.), *Plant Communication from an Ecological Perspective*, Signaling and Communication in Plants,
DOI 10.1007/978-3-642-12162-3_8,

1 Introduction

Volatile interactions between plants and their effects on associated organisms have been studied intensively during the last 25 years, focusing mainly on signalling between damaged plants as volatile emitters and responding undamaged plants. In response to damage, plants produce specific volatile blends of organic compounds (VOCs) that may induce changes in the chemistry of undamaged plants, making them less suitable to herbivores and more attractive to herbivore natural enemies (e.g. Bruin et al. 1992; Arimura et al. 2000; Karban et al. 2000; Engelberth et al. 2004; Ruther and Kleier 2005; Baldwin et al. 2006; Dicke 2009). Recent emphasis has been placed on within-plant rather than between-plant volatile signalling, interpreted as a means for plants to react promptly after damage (Karban et al. 2006; Frost et al. 2007; Heil and Bueno 2007; Heil 2008, 2009; Heil and Ton 2008; Chapter 'Within-Plant Signalling by Volatiles Triggers Systemic Defences').

However, volatile interactions occur not only between herbivore-damaged and undamaged plants and plant parts but, undamaged plants have also been found to induce responses in their neighbours (Ninkovic et al. 2006). Since this type of chemical interaction between visibly undamaged plants may affect other organisms at higher trophic levels, i.e. herbivores and their natural enemies, it has been termed allelobiosis (Pettersson et al. 2003; Ninkovic et al. 2006; Chapters 'Volatile Interaction Between Undamaged Plants: A Short Cut to Coexistence' and 'Volatile Chemical Interaction Between Undamaged Plants: Effects at Higher Trophic Levels') in analogy with allelopathy for plant-competitive interactions (Rice 1984). That the plant volatiles induce responses in neighbouring plants, which in turn can influence the quality of plants as hosts has been shown using aphids (reviewed in this chapter) and their natural enemies (Chapter 'Volatile Chemical Interaction Between Undamaged Plants: Effects at Higher Trophic Levels') as test organisms. Other adaptive explanations for why plants may respond to their neighbours' volatiles are discussed in the chapters by Ninkovic and Heil.

In the present chapter, consequences of intra- and inter-specific plant volatile interactions for aphid host plant relations are discussed (Fig. 1). It is typical for allelobiosis that only certain combinations of volatile emitters and receivers have been found to affect aphid and natural enemy responses. The prospects for exploiting these interactions for breeding and deploying aphid resistant cultivars are discussed.

2 Aphids as Herbivores in Plant–Plant Chemical Interactions

There are many severe plant pests among the aphids, and the majority of these are specialised on crop plants within one plant family (Blackman and Eastop 1984). In the process of host plant location and selection, aphids make use of plant chemical information (Pettersson et al. 2007) and they are also very sensitive to changes in the physiological status and quality of their host plant during subsequent growth and

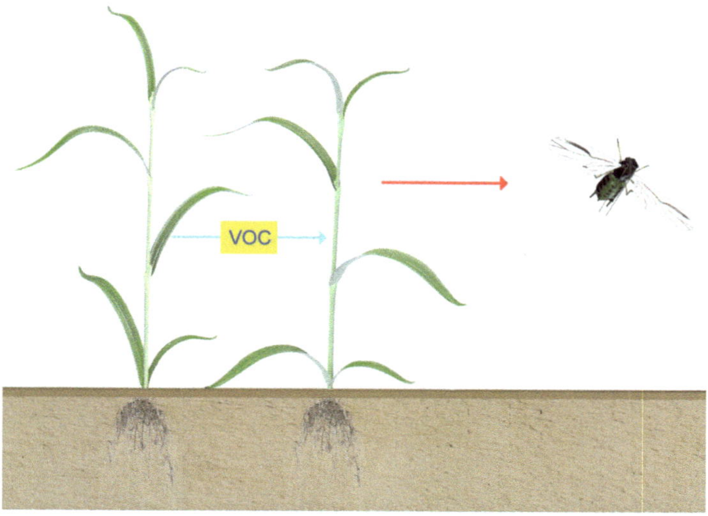

Fig. 1 Aphids can be affected by volatile interaction between undamaged plants

reproduction in successive generations (Risebrow and Dixon 1987; Douglas and van Emden 2007). Aphids feed from the phloem via their stylet. While probing, their long flexible mouthparts may puncture plant cells but compared to insects with chewing mouth parts, they inflict very little mechanical damage on their host (Walling 2000). Nevertheless, changes in plant gene regulation have been recorded as early as 2–6 h after aphid attack (Zhu-Salzman et al. 2004; Couldridge et al. 2007). Aphid-induced plants are often less acceptable and less suitable for subsequent exploitation by aphids (Prado and Tjallingii 1997; Messina et al. 2002; Messina and Bloxham 2004), and part of the 'reluctance' of new potential colonisers might be due to aphid-induced plant volatiles. Such volatiles can induce neighbouring non-infested plants to become less attractive (Pettersson et al. 1996), in a way similar to that shown for leaf-chewing herbivores (e.g. Engelberth et al. 2004). Whether the same inducing compounds or blends are inducers of both plant and aphid responses is not yet known.

3 Effects of Plant Diversity on Herbivore Abundance

In nature, plants generally grow in diverse combinations of species. Seldom is diversity as low as in farmers' fields. Mixing of species or varieties has been suggested as a means for reducing effects of pests and diseases in agriculture. An effect of plant diversity has indeed been found, both in natural (e.g. Koricheva et al. 2000; Haddad et al. 2001; Otway et al. 2005) and agricultural ecosystems (Andow 1991; Power 1991; Zhu et al. 2000; Mundt 2002; Ninkovic et al. 2002; Underwood and Rausher 2000, 2002; Cadet et al. 2007), although not in every case studied (Andow 1991;

Jokinen 1991; Mercer 2006). Various explanations for such effects have been put forward. Companion plants may favour natural enemies if they provide additional food or other resources (Root 1973), and herbivores can face problems finding their correct host plants in plant mixtures. This could in turn be explained by non-hosts functioning as physical obstacles, greater distances between suitable host plants in a mixture (Root 1973; Bach 1980), and non-host odours repelling insects or masking attractive host odours (Tahvanainen and Root 1972; Thiery and Visser 1987). Indirect effect of plant volatiles, inducing plant neighbours to become more resistant, is a further possibility. When plant genotypes grow together there is potential for interactions via plant chemicals (e.g. Uvah and Coaker 1984), both volatile and non-volatile. While differences between the plant genotypes in allelopathic ability via root exudates has been investigated intensively (e.g. Bertholdsson 2007), plant genotype-dependant aerial volatile interactions have not been widely studied.

3.1 Effects of Inter-Specific Allelobiosis on Aphids

There are relatively few studies on the effects of inter-specific interactions between plants on aphids in natural ecosystems. One example is black rush, *Juncus gerardi* Loisel, that reduces the stress effects of salinity and water logging on the marsh elder shrub, *Iva frutescens* L., increasing its photosynthetic rate and host plant quality for the aphid *Uroleucon ambrosiae* (Thomas). Elder shrub plants without rush as a neighbour were more affected by physical stress, and this correlated with lower aphid population growth rate, suggesting that interactions among plant species can have strong effects on the quality of the host plant (Hacker and Bertness 1996).

In agricultural ecosystems, plant species mixtures have been found to decrease aphid abundance in squash in maize/bean/squash mixtures compared to monoculture (Andow and Risch 1985). The explanation was lack of suitable, old, leaves in the plant mixture due to neighbour shading effects. In another example, barley acceptability to the aphid *Rhopalosiphum padi* L., (Fig. 2), was reduced when the weed, *Chenopodium album* L., was grown in mixture with barley (Ninkovic et al. 2009). This effect was at least partly attributable to volatiles released from *C. album* inducing lower plant acceptance to the aphid. Volatiles from the weeds *Cirsium* spp. (Glinwood et al. 2004) and *Solanum nigrum* L. also caused reduced aphid acceptance of barley, but volatiles from several other tested weed species had no effect (Ninkovic et al. 2009). VOCs emitted by these weeds had no direct repellence to *R. padi*, suggesting that certain weed VOCs may induce a response in barley that in turn reduces its acceptability to the aphid. Thus, the change in aphid plant acceptance is probably a result of aphid assessment of host plant status, and this may be adaptive if it coincides with changes in host quality. Indeed, aphid growth rate was lower on barley plants exposed to volatiles from *C. album* than on unexposed plants (Ninkovic et al. 2009). The evolutionary background to this type of plant communication is not clear, but aphids are unlikely to be the prime target for these induced

Fig. 2 Bird cherry-oat aphid (*Rhopalosiphum padi*)

plant changes since the presence of a neighbouring plant does not necessarily imply an imminent aphid attack. A more likely explanation may be that the plant is preparing for competition with the surrounding plants (see Chapter 'Volatile Interaction Between Undamaged Plants: A Short Cut to Coexistence'). Since neighbouring plants are likely to compete for limited resources, emitted volatiles can have an informative value for eavesdropping neighbours and act as stimuli for adaptation in order to reduce negative effects of competition with the emitter. The reduced attractiveness to aphids could then be a side effect of physiological processes in the responding plant. This theory is supported by the fact that volatile-exposed plants allocate relatively more resources to the build-up of roots (Ninkovic 2003), which might leave the shoot with a lower nutritional value. In a field experiment, the presence of *C. album* did not have any negative effect on barley yield compared to weed-free barley plots (Ninkovic et al. 2009), supporting the neighbour-adaptation interpretation. Inter-specific interactions between certain weeds and barley have also been found to affect natural enemies of aphids, resulting in greater attraction to volatile-exposed plants (see Chapter 'Volatile Chemical Interaction Between Undamaged Plants: Effects at Higher Trophic Levels').

3.2 *Effects of Intra-Specific Allelobiosis on Aphids*

Intra-specific genetic variation in plant populations can also have implications for the growth of aphid populations. This effect depends on the particular plant genotypes involved, and the variation in quality among plant genotypes can have non-additive effects on aphid population size (Underwood 2009). Another possible example of this is the significantly lower incidence of aphid-transmitted plant viruses in genetically diverse stands than in genetically homogeneous ones, likely a result of more aphid movement and less of phloem sap ingestion when plant genotypes vary from plant to plant (Power 1991).

It is now well known that volatile profiles can differ between varieties of the same crop plant (Elzen et al. 1986; Hernandez et al. 1989; Rapusas et al. 1996; Geervliet et al. 1997; Wang and Kays 2002; Scutareanu et al. 2003; Degen et al. 2004; Nissinen et al. 2005; Mewes et al. 2008; Weaver et al. 2009). The first evidence for genotype-dependent intra-specific plant interactions via specific volatile blends from apparently undamaged plants was found when barley cultivars were exposed to VOCs from other cultivars, and only certain combinations resulted in the VOC-exposed plants becoming significantly less acceptable to the aphid *R. padi* (Pettersson et al. 1999). Self-exposure, plants treated with VOCs from the same cultivar, also reduced plant acceptability to the aphid, but again only when certain cultivars were self-exposed (Pettersson et al. 1999; Ninkovic and Åhman 2009). After screening hundreds of barley genotype combinations, certain patterns have emerged. Some genotypes most often function as inducers, whereas others most often function as responders (Table 1). Another pattern observed is that older varieties have a higher tendency to respond to VOCs, whereas more recent cultivars tend to be better inducers (Martin Kellner, 2009, personal communication). A further observation is that aphid growth in a screening test for *R. padi* resistance correlated with aphid acceptance of those barley genotypes when they were induced by VOCs from Alva, a cultivar known to be able to induce other cultivars (Fig. 3; Ninkovic and Åhman 2009). In this set of plants there were 19 barley genotypes of which six responded with significantly reduced acceptance by *R. padi* after Alva-exposure, and four of these also supported significantly lower aphid growth. This set included the so far unique example of a barley genotype which became significantly more attractive to *R. padi* after barley volatile exposure. This line supported

Table 1 Varietal effects of volatile exposure in barley influencing host acceptance of the aphid *Rhopalosiphum padi* (Ninkovic et al. 2002; Ninkovic and Åhman 2009; and unpublished)

	Volatile receiver										
Volatile emitter	Alva	Lina	Kara	Barke	Scan.	Frieda	Prest.	Isabe.	Seba.	Chris.	Hulda
Alva		↓	↓	ns							ns
Lina			ns	ns	ns			ns	ns		
Kara	ns										ns
Barke		↓	ns		↓		ns	ns	ns		
Scandium		ns		ns				ns	ns		
Frieda					ns		↓	ns	ns	ns	
Prestige						ns		ns		ns	
Isabella		ns		ns	↓	ns	↓		ns	ns	
Sebastian		ns		ns	ns			ns			
Christina					ns	ns	↓	ns	ns		
Hulda	↓		ns								

Arrows indicate significantly reduced aphid acceptance and *ns* no significant effect

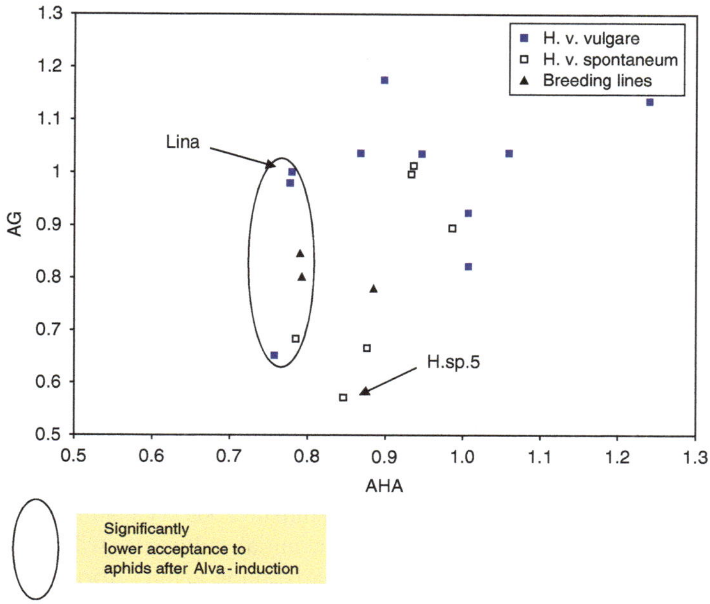

Fig. 3 Aphid growth rate relative to the control Lina (AG) and aphid host acceptance (AHA) of Alva-exposed plants relative to air-exposed plants are correlated ($r_s = 0.48$) and two out of three breeding lines (*triangles*) are similar to the BC parent Lina in AHA. Modified from Ninkovic and Åhman (2009)

the second highest aphid growth rate among the 19. Since this reaction to barley volatiles is so rare, only volatile-induced reductions in aphid acceptance are considered when the term allelobiosis is used.

In all of the above examples, young, apparently undamaged plants were combined and tested in the laboratory. When pairs of barley genotypes selected for allelobiosis effects indoors were tested in mixtures under field conditions, reduction of plant acceptance to aphids was confirmed and aphid population growth rate reduced, but not in all combinations (Ninkovic et al. 2002). Plant volatiles inducing resistance in neighbouring plants is one possible explanation why certain plant genotype mixtures may sustain less biotic damage, and often have higher yields than the mean of the individual varieties grown in monoculture (cf. e.g. Tratwal et al. 2007; Gustafsson 1953; Mundt 2002; Szumigalski and van Acker 2006).

4 Potential for Exploiting Volatile Interactions Between Undamaged Plants in Breeding Resistance to Aphids

Aphids are key pests in many crops, damaging plants both directly by feeding and indirectly by transmitting plant viruses. Efforts to breed for resistance have resulted in cultivars with resistance to Russian wheat aphid (RWA; *Diuraphis noxia*

(Mordvilko)), greenbug (*Schizaphis graminum* (Rondani)) (Berzonsky et al. 2003), lettuce aphid (*Nasonovia ribisnigri* (Mosley)) (Liu and McCreight 2006) and soybean aphid (*Aphis glycines* Matsura) (Li et al. 2007). With *R. padi* it has been more difficult to reach such a goal. *R. padi* is a pest of cereals in cold temperate regions, causing both direct plant damage and secondary damage as a vector of barley yellow dwarf virus (Blackman and Eastop 1984). It has a wide range of hosts among grasses. Since this aphid does not cause visible plant symptoms in cereals, as RWA and greenbug do, selection for resistance is more difficult and time-consuming. Some attempts to breed for resistance to *R. padi* have been made (Weibull 1994; Åhman et al. 2000), and in an ongoing effort to breed for resistance to *R. padi* in barley, the selection method is based on measurements of aphid growth, using a wild barley (*H. vulgare* spp. *spontaneum*) accession as resistance source. Nymphal growth on this wild barley is approximately half of that on cultivated barley (Delp et al. 2009). Successive generations of back-crossing (BC) and selection for reduced aphid growth have resulted in barley lines with lower resistance levels, but still with significantly lower aphid growth rate than on the susceptible BC-parent (Ninkovic and Åhman 2009). Quantitative trait loci (QTL) analysis of the F1-population of the initial cultivar x wild barley cross revealed a QTL explaining ca. 20% of the variation in aphid growth rates (Louise O'Donoughue, 1994, personal communication). A marker for this QTL is now used as a primary selection criterion; only lines carrying the marker are further tested for aphid growth rate.

In most of the BC, the recurrent parent was the cultivar Lina. Coincidently, this parent is amenable to induction by certain barley volatiles such as those from cvs. Alva and Barke (Table 1), and two out of three tested resistant BC-lines carrying Lina-genome responded with significantly reduced acceptance to *R. padi* after Alva-induction (Ninkovic and Åhman 2009), indicating that the trait is heritable (Fig. 3).

Even though the allelobiosis-induced type of resistance from Lina and the type of resistance from wild barley was incidentally combined in some of the breeding lines, it is highly desirable to be able to combine different types of resistance traits with the allelobiosis type in a more controlled manner. There are several reasons to strive for multigenic resistance to aphids. One is that strong monogenic resistance to aphids such as RWA, greenbug (Berzonsky et al. 2003) and lettuce aphid has been rapidly overcome. Another is that only a moderate level of resistance to *R. padi* has been identified in barley that is possible to use in crosses (Weibull 1987). Thus it is probably necessary to combine more than one resistance factor in order to obtain efficient, durable resistance to this aphid. This encourages further efforts to exploit allelobiosis in the ongoing breeding programme for resistance to *R. padi*.

However, since it would be even more time consuming to screen specifically for the allelobiosis type of resistance than screening with the aphid growth test, it is of utmost importance that molecular tools are developed as understanding of the mechanisms of allelobios is increases. In a QTL-test using a breeding population with variation in the aphid growth trait but lacking the previously mentioned QTL marker, a further QTL was detected on a chromosome segment originating from the parent Lina (Cheung et al. 2010). One possibility is that this chromosomal region is important for the allelobiosis responder function in that cultivar. Currently, a microarray analysis of cv. Lina exposed to cv. Alva is followed up in various ways, to narrow down the number of candidate genes (Karpinska, 2009, personal communication).

Monoculture is the norm for growing crops, whereas screens for resistance to pests, as well as to diseases, in breeding programmes are normally carried out with mixtures of many different plant genotypes. Thus there is a risk for induced effects via allelobiosis in the screenings, which are then not realised in monoculture in the field (Ninkovic and Åhman 2009). On the other hand, knowledge about allelobiosis suggests that it may be favourable to grow combinations of cultivars known to induce resistance to a key pest, and enhance attraction of its natural enemies (Chapter 'Volatile Chemical Interaction Between Undamaged Plants: Effects at Higher Trophic Levels'). Although allelobiotic responding ability is rarer among more modern than among older cultivars (Martin Kellner, 2009, personal communication), it does exist in modern ones (Table 1), which would enable immediate adoption. Part of the seed industry in Sweden is now using the opportunity to have their marketed barley varieties screened for allelobiotic abilities in our laboratory and subsequently tested as cultivar mixtures in field tests for pest abundance, yield and other agronomic characteristics. Further, if allelobiosis-eliciting volatile blends become available as formulations, it will also be possible to grow a responding cultivar in monoculture and apply these formulations, with the timing based on forecasts of pest attack.

5 Conclusions and Future Prospects

Allelobiosis is a promising mechanism to exploit for developing new means of aphid control in agriculture and horticulture. Timing of induced resistance is critical for a sufficient resistance effect in terms of reduced plant damage (Åhman 2009). A rapid and systemic induced plant response is essential since *R. padi*, like many other pests, invades crops during a relatively short time period, after which the population increases exponentially. Thus it seems as aphid-induced within- and between-plant responses are too slow to have a significant effect on population build-up in monocultures. The advantage of allelobiosis is that plants are already induced when migrant aphids arrive and attempt to settle in the crop, either via plant mixtures or via well-timed application of volatile formulations. It is still unknown if the effect of allelobiosis alone is enough to combat all levels of aphid infestations, but even if it is not, decreased frequency of pesticide use could still be achieved.

The results obtained with barley may be exploited in breeding for further improved allelobiotic abilities. The screening study performed by Martin Kellner (2009, personal communication) showed that allelobiotic responding properties are becoming increasingly rare in modern cultivars. Therefore it is important to identify the genes responsible for this trait and preserve them in the gene pool.

In the present chapter mostly barley/barley/aphid allelobiosis has been discussed, but if the theory that allelobiosis is related to plant competition is correct, it is likely to be a widespread feature in the plant kingdom. More pest/crop combinations should be investigated for allelobiotic abilities, and favourable interactions should be exploited in new cropping systems.

Acknowledgements This work was financially supported by the Swedish Foundation for Strategic Environmental Research (Mistra) through the PlantComMistra programme and by the Swedish Research Council for Environment, Agricultural Sciences and Spatial Planning (FORMAS). Lisbeth Jonsson, Robert Glinwood and Martin Kellner are thanked for reviewing previous versions of the manuscript.

References

Åhman I (2009) Breeding for inducible resistance against insects – applied plant breeding aspects. In: Schmitt A, Mauch-Mani B, Birch N, Dicke M (eds) Proceedings working group "Induced resistance in plants against insects and diseases", Heraklion, Crete (Greece), Apr 27–29 2006. ISBN 978-92-9067-218-0. IOBC/WPRS Bull 44:121–130

Åhman I, Tuvesson S, Johansson M (2000) Does indole alkaloid gramine confer resistance in barley to aphid *Rhopalosiphum padi*? J Chem Ecol 26:233–255

Andow DA (1991) Vegetational diversity and arthropod population response. Ann Rev Entomol 36:561–586

Andow DA, Risch SJ (1985) Predation in diversified agroecosystems: relations between a coccinellid predator *Coleomegilla maculata* and its food. J Appl Ecol 22:357–372

Arimura GI, Ozawa R, Shimoda T, Nishioka T, Boland W, Takabayashi J (2000) Herbivory-induced volatiles elicit defence genes in lima bean leaves. Nature 406:512–515

Bach CE (1980) Effect of plant diversity and time of colonization on an herbivore–plant interaction. Oecologia 44:319–326

Baldwin IT, Halitschke R, Paschold A, von Dahl CC, Preston CA (2006) Volatile signaling in plant–plant interactions: "talking trees" in the genomics era. Science 311:812–815

Bertholdsson NO (2007) Varietal variation in allelopathic activity in wheat and barley and possibilities for use in plant breeding. Allelopathy J 19:193–201

Berzonsky WA, Ding H, Haley SD, Harris MO, Lamb RJ, Mckenzie RIH, Ohm HW, Patterson FL, Peairs FB, Porter DR, Ratcliffe RH, Shanower TG (2003) Breeding wheat for resistance to insects. Plant Breed Rev 22:221–296

Blackman RL, Eastop VF (1984) Aphids on the world's crops: an identification guide. Wiley, Chichester

Bruin J, Dicke M, Sabelis MW (1992) Plants are better protected against spider-mites after exposure to volatiles from infested conspecifics. Experientia 48:525–529

Cadet E, Fried G, Chauvel B (2007) Floristic diversity in wildlife-friendly set-aside: what risk for invasive species to appear? In: 20eme Conference du COLUMA. Journées Internationales sur la Lutte contre les Mauvaises Herbes, Dijon, France, Dec 11–12 2007, pp 382–392

Cheung WY, Di Giorgio L, Åhman I (2010) Mapping resistance to the bird cherry-oat aphid (*Rhopalosiphum padi*) in barley. Plant Breed 129:

Couldridge C, Newbury HJ, Ford-Lloyd B, Bale J, Pritchard J (2007) Exploring plant responses to aphid feeding using a full *Arabidopsis* microarray reveals a small number of genes with significantly altered expression. Bull Entomol Res 97:523–532

Degen T, Dillmann C, Marion-Poll F, Turlings TCJ (2004) High genetic variability of herbivore-induced volatile emission within a broad range of maize inbred lines. Plant Physiol 135:1928–1938

Delp G, Gradin T, Åhman I, Jonsson LMV (2009) Microarray analysis of the interaction between the aphid *Rhopalosiphum padi* and host plants reveals both differences and similarities between susceptible and partially resistant barley lines. Mol Genet Genomics 281:233–248

Dicke M (2009) Behavioural and community ecology of plants that cry for help. Plant Cell Environ 32:654–665

Douglas AE, van Emden HF (2007) Nutrition and symbiosis. In: van Emden H, Harrington R (eds) Aphids as crop pest. CAB International, Wallingford, pp 115–134

Elzen GW, Williams HJ, Vinson SB (1986) Wind tunnel flight responses by hymenopterous parasitoid *Campoletis sonorensis* to cotton cultivars and lines. Entomol Exp Appl 42:285–289

Engelberth J, Alborn HT, Schmelz EA, Tumlinson JH (2004) Airborne signals prime plants against insect herbivore attack. Proc Natl Acad Sci USA 101:1781–1787

Frost CJ, Appel HM, Carlson JE, De Moraes CM, Mescher MC, Schultz JC (2007) Within-plant signalling via volatiles overcomes vascular constraints on systemic signalling and primes responses against herbivores. Ecol Lett 10:490–498

Geervliet JBF, Posthumus MA, Vet LEM, Dicke M (1997) Comparative analysis of headspace volatiles from different caterpillar-infested or uninfested food plants of *Pieris* species. J Chem Ecol 23:2935–2954

Glinwood R, Ninkovic V, Pettersson J, Ahmed E (2004) Barley exposed to aerial allelopathy from thistles (*Cirsium* spp.) becomes less acceptable to aphids. Ecol Entomol 29:188–195

Gustafsson Å (1953) The cooperation of genotypes in barley. Hereditas 39:1–18

Hacker SD, Bertness MD (1996) Trophic consequences of a positive plant interaction. Am Nat 148:559–575

Haddad NM, Tilman D, Haarstad J, Ritchie M, Knops JMH (2001) Contrasting effects of plant richness and composition on insect communities: a field experiment. Am Nat 158:17–35

Heil M (2008) Indirect defence via tritrophic interactions. New Phytol 178:41–61

Heil M (2009) Damaged-self recognition in plant herbivore defence. Trends Plant Sci 14:356–363

Heil M, Bueno JCS (2007) Within-plant signaling by volatiles leads to induction and priming of an indirect plant defense in nature. Proc Natl Acad Sci USA 104:5467–5472

Heil M, Ton J (2008) Long-distance signalling in plant defence. Trends Plant Sci 13:264–272

Hernandez HP, Hsieh TC-Y, Smith CM, Fischer NH (1989) Foliage volatiles of two rice cultivars. Phytochemistry 28:2959–2962

Jokinen K (1991) Competition and yield advantage in barley–barley and barley–oats mixtures. J Agric Sci Finl 63:255–285

Karban R, Baldwin IT, Baxter KJ, Laue G, Felton GW (2000) Communication between plants: induced resistance in wild tobacco plants following clipping of neighboring sagebrush. Oecologia 125:66–71

Karban R, Shiojiri K, Huntzinger M, McCall AC (2006) Damage-induced resistance in sagebrush: volatiles are key to intra- and interplant communication. Ecology 87:922–930

Koricheva J, Mulder CPH, Schmid B, Joshi J, Huss-Danell K (2000) Numerical responses of different trophic groups of invertebrates to manipulations of plant diversity in grasslands. Oecologia 125:271–282

Li Y, Hill CB, Carlson SR, Diers BW, Hartman GL (2007) Soybean aphid resistance genes in the soybean cultivars Dowling and Jackson map to linkage group M. Mol Breed 19:25–34

Liu YB, McCreight JD (2006) Responses of *Nasonovia ribisnigri* (Homoptera: Aphididae) to susceptible and resistant lettuce. J Econ Entomol 99:972–978

Mercer PC (2006) Growing organic cereals in Northern Ireland – disease and weed problems. Asp Appl Biol 79:229–232

Messina FJ, Bloxham AJ (2004) Plant resistance to the Russian wheat aphid: effects on a nontarget aphid and the role of induction. Can Entomol 136:129–137

Messina FJ, Taylor R, Karren ME (2002) Divergent responses of two cereal aphids to previous infestation of their host plant. Entomol Exp Appl 16:43–50

Mewes S, Kruger H, Pank F (2008) Physiological, morphological, chemical and genomic diversities of different origins of thyme (*Thymus vulgaris* L.). Genet Res Crop Evol 55:1303–1311

Mundt CC (2002) Use of multiline cultivars and cultivar mixtures for disease management. Ann Rev Phytopathol 40:381–410
Ninkovic V (2003) Volatile communication between barley plants affects biomass allocation. J Exp Bot 54:1931–1939
Ninkovic V, Åhman I (2009) Aphid acceptance of *Hordeum* genotypes is affected by plant volatile exposure and is correlated with aphid growth. Euphytica 169:177–185
Ninkovic V, Glinwood R, Dahlin I (2009) Weed–barley interactions affect plant acceptance by aphids in laboratory and field experiments. Entomol Exp Appl 133:38–45
Ninkovic V, Glinwood R, Pettersson J (2006) Communication between undamaged plants by volatiles: the role of allelobiosis. In: Baluška F, Mancuso S, Volkmann D (eds) Communication in plants: neuronal aspects of plant life, vol 28. Springer, Heidelberg, pp 421–434
Ninkovic V, Olsson U, Pettersson J (2002) Mixing barley cultivars affects aphid host plant acceptance in field experiments. Entomol Exp Appl 102:177–182
Nissinen A, Ibrahim M, Kainulainen P, Tiilikkala K, Holopainen JK (2005) Influence of carrot psyllid (*Trioza apicalis*) feeding or exogenous limonene or methyl jasmonate treatment on composition of carrot (*Daucus carota*) leaf essential oil and headspace volatiles. J Agric Food Chem 53:8631–8638
Otway SJ, Hector A, Lawton JH (2005) Resource dilution effects on specialist insect herbivores in a grassland biodiversity experiment. J Anim Ecol 74:234–240
Pettersson J, Ninkovic V, Ahmed E (1999) Volatiles from different barley cultivars affect aphid acceptance of neighbouring plants. Acta Agric Scand B Soil Plant Sci 49:152–157
Pettersson J, Ninkovic V, Glinwood R (2003) Plant activation of barley by intercropped conspecifics and weeds: allelobiosis. BCPC Crop Sci Technol 2:1135–1144
Pettersson J, Quiroz A, Fahad AE (1996) Aphid antixenosis mediated by volatiles in cereals. Acta Agric Scand B Soil Plant Sci 46:135–140
Pettersson J, Tjallingii WF, Hardie J (2007) Host-plant selection and feeding. In: van Emden H, Harrington R (eds) Aphids as crop pest, vol 4. CAB International, Wallingford, pp 87–113
Power AG (1991) Virus spread and vector dynamics in genetically diverse plant populations. Ecology 72:232–241
Prado E, Tjallingii WF (1997) Effects of previous plant infestation on sieve element acceptance by two aphids. Entomol Exp Appl 82:189–200
Rapusas HR, Bottrell DG, Coll M (1996) Intraspecific variation in chemical attraction of rice to insect predators. Biol Control 6:394–400
Rice EL (1984) Allelopathy, 2nd edn. Academic Press, New York, 1989
Risebrow A, Dixon AFG (1987) Nutritional ecology of phloem-feeding insects. In: Slansky F Jr, Rodriguez JG (eds) Nutritional ecology of insects, mites, spiders, and related invertebrates. Wiley Interscience, USA, pp 421–448
Root RB (1973) Organization of plant-arthropod association in simple and diverse habitats: the fauna of collards (*Brassica oleracea*). Ecol Monogr 43:95–124
Ruther J, Kleier S (2005) Plant–plant signaling: ethylene synergizes volatile emission in *Zea mays* induced by exposure to (Z)-3-hexen-1-ol. J Chem Ecol 31:2217–2222
Scutareanu P, Bruin J, Posthumus MA, Drukker B (2003) Constitutive and herbivore-induced volatiles in pear, alder and hawthorn trees. Chemoecology 13:63–74
Szumigalski AR, van Acker RC (2006) The agronomic value of annual plant diversity in crop-weed systems. Can J Plant Sci 86:865–874
Tahvanainen JO, Root RB (1972) The influence of vegetational diversity on the population ecology of a specialized herbivore, *Phyllotreta cruciferae* (Coleoptera: Chrysomelidae). Oecologia 10:321–346
Thiéry D, Visser JH (1987) Misleading the Colorado potato beetle with an odor blend. J Chem Ecol 13:1139–1146
Tratwal A, Law J, Philpott H, Horwell A, Garner J (2007) The possibilities of reduction of winter barley chemical protection by growing variety mixtures. Part II. Effect on yield. J Plant Prot Res 47:79–86

Underwood N (2009) Effect of genetic variance in plant quality on the population dynamics of a herbivorous insect. J Anim Ecol 78:839–847

Underwood N, Rausher MD (2000) The effects of host-plant genotype on herbivore population dynamics in a model system. Ecology 81:1565–1576

Underwood N, Rausher MD (2002) Comparing the consequences of induced and constitutive plant resistance for herbivore population dynamics. Am Nat 160:20–30

Uvah III, Coaker TH (1984) Effect of mixed cropping on some insect pests of carrots and onions. Entomol Exp Appl 36:159–167

Walling LL (2000) The myriad plant responses to herbivores. J Plant Growth Regul 19:195–216

Wang Y, Kays SJ (2002) Sweetpotato volatile chemistry in relation to sweetpotato weevil (*Cylas formicarius*) behaviour. J Am Soc Hortic Sci 127:656–662

Weaver DK, Buteler M, Hofland ML, Runyon JB, Nansen C, Talbert LE, Lamb P, Carlson GR (2009) Cultivar preferences of ovipositing wheat stem sawflies as influenced by the amount of volatile attractant. J Econ Entomol 102:1009–1017

Weibull J (1987) Screening for resistance against *Rhopalosiphum padi* (L.). 2. *Hordeum* species and interspecific hybrids. Euphytica 36:571–576

Weibull J (1994) Resistance to *Rhopalosiphum padi* L. (Homoptera: Aphididae) in *Hordeum vulgare* subsp. *spontaneum* and in hybrids with *H. vulgare* subsp. *vulgare*. Euphytica 78:97–101

Zhu YY, Chen HR, Fan JH, Wang YY, Li Y, Chen JB, Fan JX, Yang SS, Hu LP, Leung H, Mew TW, Teng PS, Wang ZH, Mundt CC (2000) Genetic diversity and disease control in rice. Nature 406:718–722

Zhu-Salzman K, Salzman RA, Ahn JE, Koiwa H (2004) Transcriptional regulation of sorghum defense determinants against a phloem-feeding aphid. Plant Physiol 134:420–431

Communication in Ant–Plant Symbioses

Rumsaïs Blatrix and Veronika Mayer

Abstract Plant communication abilities are the subject of intensive research. They have been particularly investigated in the context of signalling herbivore activity and responding to these signals. In this chapter, we review the current knowledge on communication between plants and ants in ant–plant symbioses. Chemistry is the preponderant channel in ant–plant communication. Communication is identified in five contexts: the selection of seeds by ants to sow ant-gardens, the detection of the host plant by founding queens, the discrimination of the host plant by the inhabiting ants to prune exogenous vegetation, the selective continuous patrolling on young shoots by workers and the damage-induced ant-mediated plant protection. Implications of communication for the evolutionary ecology of ant–plant symbioses are discussed and directions for future research are given.

1 Introduction

Plants have long been considered as passive organisms devoid of sensory activity and it is only during the last few decades that plant-to-plant communication and the communication of plants with other organisms have been widely acknowledged. These types of communication involve plant emission and perception of signals that are regulated by the same biosynthetic pathways as systemic cell signalling within plants. It is thus not surprising that they are so widespread. Our long scepticism

R. Blatrix (✉)
Centre d'Ecologie Fonctionnelle et Evolutive, UMR 5175, CNRS 1919 route de Mende, 34293 Montpellier cedex 5, France
e-mail: blatrix@cefe.cnrs.fr

V. Mayer
Department of Structural and Functional Botany, Faculty Centre of Biodiversity, Faculty of Life Sciences, University of Vienna, Rennweg 14, 1030 Vienna, Austria
e-mail: veronika.mayer@univie.ac.at

F. Baluška and V. Ninkovic (eds.), *Plant Communication from an Ecological Perspective*, Signaling and Communication in Plants,
DOI 10.1007/978-3-642-12162-3_9,

about plant communication is probably due to the nature of the most important channels of human communication, the visual and auditory channels, which biased the way we first considered biotic interactions. Plants lack conspicuous visual and auditory organs, even though they do perceive light. The plant sensory world is rather driven by chemicals. Our poor capacities in olfaction have blurred our perception of plant sensitivity until recent technological advances. A major breakthrough in the acknowledgement of plant communication occurred in the early 1990s when it was discovered that the emission of plant volatiles could be triggered by herbivore feeding. The compounds emitted were shown to act as a "cry for help", attracting predatory and parasitic species that attack herbivores (Dicke et al. 1990a, b; Turlings et al. 1990).

Communication can be defined as the exchange of information between individuals, wherein both the signaller and receiver may expect to benefit from the exchange (Greenfield 2002). A stimulus can be considered to be a signal if (1) individuals derive expected net benefits from emitting and receiving the stimulus and (2) the stimulus has undergone evolutionary modifications that enhance the benefits derived from providing information to receivers and influencing their behaviour (Greenfield 2002). Otherwise, the stimulus is considered to be a cue. The semantic distinction is important for understanding the evolution of communication. Plant cues used by parasitoids to locate their hosts are beneficial to the plant. Such cues are expected to evolve towards greater efficiency, and become signals, if the benefit of attracting parasitoids compensates for the cost of the modification. These costs may be the direct energetic cost of producing the compounds, and also indirect costs occasioned by the use of the signal by eavesdroppers, such as herbivores searching for mates or a host plant (Kalberer et al. 2001). Moreover, differentiating *signals* from *cues* in biological systems may not be straightforward because a continuum exists between the two and distinguishing traits that result from selection from traits that are simply by-products may prove difficult. It is thus important to take into consideration plant characteristics that may at first appear to be mere cues because they could in fact be true communication signals or represent early stages in signal evolution. In this chapter, we focus on plant stimuli that are used by their symbiotic ants, regardless of whether they eventually prove to be *cues* or *signals*.

Plants are able to use volatile cues to increase their own fitness either through direct deterrence of herbivores (by effects on survivorship of their eggs, or oviposition rate, De Moraes et al. 2001; Kessler and Baldwin 2001) or through communication between different parts of the same plant and between conspecific or heterospecific individuals. Undamaged leaves on the same stem of *Phaseolus lunatus* (lima bean) plants showed an increase in extrafloral nectar secretion when they were placed near damaged leaves, but not when volatile organic compounds were removed from the system (Heil and Silva Bueno 2007). Tobacco plants primed their defence response against herbivory in reaction to volatile organic compounds emitted by nearby artificially damaged sagebrush (*Artemisia tridentata*) and experienced less damage than control plants established next to undamaged *Artemisia* plants (Karban et al. 2000, 2003). Lima bean leaves infested by mites produce volatile compounds that induce the expression of defence genes in conspecific non-infested plants (Arimura et al. 2000, 2001). Moreover, those compounds are different from those that are released as a consequence of mere mechanical damage, and only mite-induced compounds trigger the expression of defence genes (Arimura et al. 2000). The use of volatile signals in within-plant communication may have evolved because volatile transmission from leaf to leaf is expected to be more rapid than systemic signal transfer within the vascular system, mainly because leaves close to each other may be separated by a large branch distance. Conspecific individuals, or even individuals of other species, may then use these intra-individual signals as cues to assess the risk of herbivory.

Invertebrates have evolved specific responses to plant cues that increase fitness of both the animal and the plant. The volatile organic compounds emitted by plant leaves when attacked by herbivores is a complex bouquet, including, for example, terpenoids, sesquiterpenoids, green leaf volatiles and aromatic compounds such as indole and methyl salicylate, and have been shown to be used by parasitoids or predators as cues for localising their host or their prey (Turlings et al. 1990; Choh et al. 2004; De Boer et al. 2004; Halitschke et al. 2008). Roots of maize plants attacked by the Western corn rootworm (larvae of a Chrysomelidae beetle) emit β-caryophyllene to a much greater extent than when artificially damaged (Rasmann et al. 2005). β-Caryophyllene was shown to attract an entomopathogenic nematode, reducing infestation by the rootworm. The fact that the cue used by the nematode is produced by the plant in greater quantity in response to damage by the rootworm suggests that this cue is evolving towards a signal.

Plant communication is particularly expected in plant–insect mutualistic symbioses. Symbiosis is usually defined as the interaction between individuals of different species living together and showing some kind of dependency. It ranges from parasitism to mutualism. Communication, by definition, should not evolve in parasitic interactions. Host–parasite interactions involve cues, but it is difficult to envisage a situation in which a cue emitted by a host and received by its parasite, or vice versa, would confer benefit on both players. In mutualistic interactions, however, any cue enhancing reciprocal benefits is expected to evolve into a communication signal. The shared living of symbiotic organisms makes communication very efficient because of the potential for immediate responses to signals. Spatial proximity also allows short distance signals that should be less prone to eavesdropping, and thus to potential use by enemies. Moreover, the interdependency of such organisms should rapidly lead to the evolution of an efficient communication system.

Ant–plant symbioses are good biological models to study plant–insect communication. The exocrine system is far more developed in social than solitary insects (Billen and Morgan 1998), and the ant worker has been described as a "walking battery of exocrine glands" (Hölldobler and Wilson 1990). It is, thus, not surprising that the communication system of ants is based on chemical compounds. Ants use a diverse range of volatile compounds as chemical signals, for example acids, alcohols, aldehydes, esters, hydrocarbons, ketones, pyrazines and terpenoids, to organise alertness, recruitment, defence, and foraging behaviour (Hölldobler and

Wilson 1990; Hölldobler 1995; Passera and Aron 2005; Morgan 2009). Furthermore, cuticular hydrocarbons of the body surface serve for discrimination of nestmates from non-nestmates (Lahav et al. 1999; Greene and Gordon 2007; Guerrieri et al. 2009), for identification of worker eggs and assessment of queen fertility (Endler et al. 2004; Howard and Blomquist 2005; Endler et al. 2006) and for recognition of their dead (Choe et al. 2009). Although ants and plants belong to different kingdoms, they use the same alphabet as a base for communication and cooperation.

Ant–plant symbioses can be categorised into two types: ant-gardens and ant–myrmecophyte interactions. Both types display a series of interesting examples of ant–plant communication. They are only known in the tropics, and have evolved many times independently. Ant–myrmecophyte interactions involve more than 100 genera of angiosperms and 40 genera of ants, and they are important components of tropical communities (Davidson and McKey 1993).

Ant gardens are associations of epiphytes and arboricolous ants, in which the ants build carton nests rich in organic material. They incorporate seeds of epiphytes that then grow on the nest. The growing epiphytes constitute what is called an "ant-garden", because the ants literally sow seeds and thus control the composition of the plant community. Myrmecophytes, or ant–plants, are plants that offer shelter and food -directly or indirectly- to host ants. They usually have specialised preformed hollow structures, called "domatia", of various morphological origins, including stem, petiole, leaf blade, stipule, root or rhizome (Jolivet 1996). Many ant–plants additionally provide food in extrafloral nectaries (numerous species), glandular trichomes (e.g. *Clidemia*, *Tococa*) or specialised food bodies derived from emergences (*Cecropia*, *Macaranga*, *Piper*) or leaflet tips (genus *Vachellia* = former American *Acacia* subg. *Acacia*, McNeill et al. 2005).

Nesting sites are one important benefit that ants receive in their interaction with plants. In ant gardens the roots of the epiphytes penetrate into the carton nest and constitute a solid structural framework for the nest. In ants associated with myrmecophytes, at least part of the colony inhabits the hollow structures of the plant. In specific associations a single ant colony occupies one to several individuals. In many cases, the workers restrict their foraging activity to their host plant and do not even go to the ground.

The ants also benefit from the nutrition offered by the plants. Ant-garden epiphytes can produce floral and extrafloral nectar as well as elaiosomes and fruit pulp that the ants consume (Kleinfeldt 1978). Many myrmecophytes produce extrafloral nectar, consisting primarily of aqueous solutions of mono- and disaccharides (fructose, glucose and sucrose) and free amino acids. Extrafloral nectar is an important resource for their resident ants (Gonzalez-Teuber and Heil 2009). Some also provide their ant symbionts with food bodies which contain various nutritionally valuable compounds (Heil et al. 1998; Fischer et al. 2002; Heil et al. 2004a). Even if no direct food rewards are produced, the ants get nutrients indirectly by rearing hemipteran trophobionts. The *Tetrathylacium macrophyllum*/*Azteca brevis*, *Cordia*/*Azteca* and *Tachigali*/*Pseudomyrmex* associations are some of the numerous systems in which plant–ants tend hemipterans (coccids or pseudococcids) (see Gullan 1997, Table 1.3.5.3), even if some of them receive direct plant food rewards. Hemipteran honeydew provides an important source of carbohydrates, and some plant–ants seem to harvest the insects themselves for proteins and lipids (Carroll and Janzen 1973).

Plant benefits are also of two types: (1) Nutrition. The roots of ant-garden epiphytes penetrating into the carton nest find nutrients there. Those nutrients come from the carton itself and from refuse accumulated by the ants. Ant refuse contains ant faeces, discarded pieces of arthropod prey, dead colony members, and other wastes. In a number of studies of myrmecophytes, nutrient transfer from the ant refuse to the host plant has been demonstrated (reviewed in Rico-Gray and Oliveira 2007). First thought to be restricted to epiphytic myrmecophytes, nutritional mutualism now appears to be a widespread phenomenon in free-standing myrmecophytes (Fischer et al. 2003; Rico-Gray and Oliveira 2007). (2) Plants benefit from protection. As ants are fierce predators they provide to their host plants anti-herbivore protection (Gaume et al. 1997; Gaume and McKey 1998; Heil et al. 2001; Rosumek et al. 2009). The protective behaviour of ants has been demonstrated in numerous cases by comparing herbivory in ant exclusion experiments, showing that ant–plants with artificially removed ants suffer a considerably higher degree of herbivory than plants with their mutualistic ants (Moraes and Vasconcelos 2009; see also reviews in Davidson and McKey 1993; Heil and McKey 2003; Rosumek et al. 2009). Ants are efficient at discovering and removing eggs of herbivores (Letourneau 1983; Fiala et al. 1989). Some of them are even efficient against mammalian herbivores (McKey 1974a; Madden and Young 1992). They also protect their host plants from fungal pathogens (Letourneau 1998) and competing vegetation through the pruning of encroaching vines and surrounding plants (Janzen 1969; Davidson et al. 1988; Renner and Ricklefs 1998). The protection provided to hosts, however, differs depending on the occupant ant species (Djieto-Lordon et al. 2004; Dejean et al. 2006).

In the following text, we will review and discuss the two major contexts in which communication has evolved in ant–plant symbioses: (1) in the discrimination of host versus non-host plants, including the choice of suitable epiphyte seeds in assembling ant-gardens, the location of myrmecophyte hosts by founding queens and the pruning of exogenous vegetation; and (2) in the optimisation of protection against herbivores by synchronising ant activity with herbivore presence and by directing ant attention to more valuable and/or more vulnerable organs and to attacked plant parts.

2 Host-Plant Discrimination

2.1 Sowing the Right Partner

Ant-gardens involve obligate and specific associations between certain species of ants and plants. To initiate and grow the garden, ant workers have to choose among a large set of seeds because they use only a few species. The plant would benefit from facilitating the task of the ants. Thus, the ant is under selective pressure for choosing the right seeds to bring back to the nest, and the plant is also under selective pressure for displaying a signal easily recognised by the ants. This makes ant-gardens prone to the evolution of a specialised signal for host-plant recognition. This expectation was pointed out by Davidson (1988), who showed that *Camponotus femoratus*, the inhabiting ant species, actively collected seeds of the epiphyte species with which they are associated. Ants were even observed collecting seeds directly from the infructescences. Subsequent work identified candidate compounds for signalling but behavioural tests with synthetic compounds gave ambiguous results (Davidson et al. 1990; Seidel et al. 1990). Youngsteadt et al. (2008) combined chemical analysis, electroantennography and behavioural tests to demonstrate that *Camponotus femoratus* workers were attracted to seeds from a distance by a blend of five volatile organic compounds. However the synthetic blend alone did not elicit seed transport, indicating that other cues are involved in the complete behavioural sequence. Seeds from unrelated ant-garden plants displayed similar compounds (Seidel et al. 1990; Youngsteadt et al. 2008), one of which, methyl 6-methylsalicylate (6-MMS), is not known in other plants. Moreover, 6-MMS is known to act as a semiochemical of several ant species (Morgan et al. 1990; Kohl et al. 2000; Greenberg et al. 2007), and was found in heads of *Camponotus femoratus* males as well as in heads of other *Camponotus* species. (Seidel et al. 1990). Several other facts indicate that a specific communication signal has evolved between *Camponotus femoratus* and its epiphyte associates: (1) Seeds of *Peperomia macrostachya*, one of the most common epiphytes in Neotropical ant-gardens, are very rarely dispersed by organisms other than *Camponotus femoratus* (Youngsteadt et al. 2009), (2) *Camponotus femoratus* does not seem to be attracted to seeds of plants other than its hosts (Davidson 1988; Youngsteadt et al. 2008), and (3) the food reward of the elaiosome is not responsible (in terms of proximate mechanism) for the attractiveness of the diaspore. Indeed, behavioural tests showed that ant-garden ants were still attracted to seeds from associated epiphytes, even when the elaiosomes were removed (Davidson 1988; Orivel and Dejean 1999; Youngsteadt et al. 2008). Besides a specific communication signal, we may expect compounds repelling non-symbiotic ants to have evolved. This remains to be investigated, but the facts that a compound present in *Peperomia macrostachya* seeds is toxic to various ants and that diaspores are not collected by other ants despite the presence of a fleshy aril attached to the seed (Lemaire et al. 1990; Youngsteadt et al. 2009) give some clues.

2.2 *Finding the Right Host Plant: Better to Look for a Needle in a Haystack?*

Horizontally transmitted symbioses, such as ant–plant symbioses, involve organisms that reproduce and disperse independently. Horizontal transmission has an important consequence for the ecology and evolution of ant–myrmecophyte associations: the partners have to find each other anew at each generation. Over evolutionary time, this is expected to have led to the use of plant cues by the founding queens to find a suitable host plant. This is particularly relevant in the cases where the ant and the plant are obligate and specific associates. Mature colonies of many ant species are known to produce hundreds of males and winged females that mate during massive synchronised swarming flights (Hölldobler and Wilson 1990). This strategy has probably been selected in response to a high predation risk. In fact, a very low proportion of queens succeed in founding a new colony. In *Allomerus octoarticulatus*, a plant–ant colonising *Cordia nodosa*, only 3% of the young queens survive from nuptial flight to 1 year after colony founding (Frederickson 2006). Nuptial flights of plant–ants are rarely observed, and are thus not well known, but from *Pseudomyrmex* sp. living on *Vachellia* (=*Acacia*) and from *Crematogaster* sp. living on *Macaranga* it is known that some plant–ants produce and disperse sexuals rather continuously (Janzen 1967; Fiala and Maschwitz 1990), whereas *Allomerus octoarticulatus* on *Cordia nodosa* and *Pheidole minutula* living on *Maieta guianensis* showed a seasonal pattern in their reproduction (Vasconcelos 1993; Frederickson 2006). Both queen mortality during mating and searching for a suitable host plant, and the mortality due to competition during colony founding (Janzen 1973; Fiala and Maschwitz 1990; Federle et al. 1998; Frederickson 2006) are factors influencing access to reproduction. Since woody plants have lifespan and generation times usually longer than those of ants, only few will establish and reach the stage for ant colonisation even if seeds can be produced in massive numbers. On the other hand, ant queens are ready to found soon after leaving their natal colony. As a consequence, nest sites are limited and host-plant populations are saturated with foundresses of the symbiotic ant (Fonseca 1999; Frederickson 2006). Traits enabling host-plant recognition are thus expected to be under strong selective pressure because queen success depends on the rapidity of the foundress in locating and colonising an available host plant. In the host plants, the strength of selection pressure on signals to attract symbiotic queens should vary with the ratio between densities of unoccupied plants and of foundresses. If founding queens outnumber available plants, specific signals are not expected to evolve because the plants will be colonised in any event. Ant queens should thus be able to recognise their host plant on the basis of cues that have not necessarily been selected for this function. In this respect, some plant–ants could be faced with problems similar to those of specialised herbivores: they must rely on unspecialised cues to find the suitable host within the tangle of greenery. On the other hand, when there is a reasonable chance that founding queens will not saturate the population of available plants, it is expected that plants will evolve a specific signal that facilitates host-plant discrimination. Even if foundresses are super-abundant, selection could still favour emission by the plant of a strong and specific signal. Plants could conceivably benefit from attracting numerous foundresses, thereby engendering a contest for occupancy, in which the most vigorous and protective colony wins.

Studying the behaviour of newly mated winged females during founding is the best way to determine whether queens use cues to locate their host plant. However, such observations remain scarce. In a study on the South-East Asian ant–plant *Macaranga puncticulata* associated with *Camponotus* sp., Federle et al. (1998) were able to observe queens in the field right after the nuptial flight. They saw that winged females fly from plant to plant, alighting on leaves for short durations, until they arrive on their host-plant species. Then they run to the stem and start digging an entrance hole into a domatium. This behaviour indicates that the ants recognise their host plant by contact, and that cues detectable at a distance, such as shape or volatile compounds, are much less important. However, the authors also observed that a plant protected by a plastic cover was colonised. The queen had to access the plant from the ground, suggesting that ant–plant queens do not find their host plant by chance but that volatile compounds might also be involved as cues. Plant–ant queens and butterflies specialised on host plants that are scattered and inconspicuous face the same problem of locating their host plants. Not surprisingly, they display similar behaviours. To find a host plant suitable for laying eggs, female butterflies fly from plant to plant, landing on each until they encounter the host species (Wiklund 1984). The female butterfly recognises the host plant by contact with the gustatory organs located in the forelegs.

In ants, the use of choice experiments in an olfactometer demonstrated that volatile compounds are used by mated queens of *Peidole minutula*, *Allomerus* cf. *octoarticulatus* and four species of *Azteca* to orient towards their host plant (Edwards et al. 2006; Dattilo et al. 2009). But contact cues have also been shown to be important in the process of host-plant choice (Inui et al. 2001; Jürgens et al. 2006; Dattilo et al. 2009). Jürgens et al. (2006) observed that foundress queens of *Crematogaster* associated with *Macaranga* plants "began walking up and down the stem several times and sometimes also scrutinised leaves" after landing on a young plant. They repeated this behaviour with several saplings before choosing their host plant. Preference experiments using ethyl acetate extracts from the stem surfaces of *Macaranga* seedlings showed that such choice could also be triggered by chemical compounds of low volatility located on the stem surfaces (Inui et al. 2001).

The paleotropical genus *Macaranga* is one of the few plant genera in which a sizable radiation of myrmecophyte species has occurred. The genus includes many myrmecophytic and some non-myrmecophytic species. The interaction between plants and ants ranges from a loose, facultative non-specific relationship to obligate myrmecophytic associations (Fiala et al. 1994). This allows comparative studies of ant–plant signalling in associations involving related species. In a study site where two *Macaranga* species were each associated with a unique *Crematogaster* ant species, Inui et al. (2001) showed that the queens of each *Crematogaster* were able to distinguish their own host among four *Macaranga* species, on the basis of stem surface extracts offered to them. Jürgens et al. (2006), who investigated the volatile compounds from intact leaves of 11 *Macaranga* species, found clear differences in the volatile odour profile of each species, and ant queens were able to discriminate their usual host plant in choice experiments, even when confronted with closely related myrmecophytic taxa. All these experiments with *Macaranga* support the hypothesis that both volatile cues and contact cues of low volatility are crucial for host-plant recognition.

But how do ant symbionts associated with several plant species deal with host recognition? Results in this direction could give great insight into the factors constraining the evolution of host-plant recognition but they are still very scarce. In a choice experiment with a generalist ant species of *Crematogaster*, known from several myrmecophytic *Macaranga*, some queens choose an unsuitable *Macaranga* species instead of their habitual host, a mistake that was not observed with a more specialised ant species (Jürgens et al. 2006). This suggests that the decision rule in host discrimination is relaxed in ants with multiple hosts. The benefit of having a wider spectrum of plant use may however outweigh the cost of recognition errors.

2.3 *Discriminating the Host Plant: Self Versus Non-Self?*

Workers of several plant–ant species from fewer than a dozen genera protect their host plant against competing vegetation (see review in Davidson and McKey 1993). They do it by pruning parts of non-host plants in contact with their host or by pruning young individuals of non-host plants growing nearby. Pruning is performed either mechanically with the mandibles through cutting of parts of the intruding plant parts such as tendrils or shoots, or through destroying the terminal meristem of the shoot tips of saplings by chewing it to death. Pruning can also be performed chemically by spraying poison from the venom gland (Morawetz et al. 1992; Renner and Ricklefs 1998; Frederickson et al. 2005). The most dramatic examples of plant pruning by plant–ants are the so-called "devil's gardens" in the Amazonian rainforest. A devil's garden consists of a pure stand of ant–plants in the forest understory, for example *Duroia hirsuta*, *Cordia nodosa*, *Tococa guianensis* or *Clidemia heterophylla*, occupied by a single polygynous, polydomous colony of *Myrmelachista* ants. Frederickson et al. (2005) report that the largest devil's garden they observed consisted of 351 *Duroia hirsuta* plants that were tended by a single *Myrmelachista schumanni* colony comprising as many as three million workers and 15,000 queens. These myrmecophyte monocultures contrast so strongly with the highly diverse surrounding tropical vegetation that the native people believe they are gardened by evil spirits of the forest (see Edwards et al. 2009). The selective destruction of non-host plants is performed by the ants, which bite small holes in meristematic tissues and leaf veins and release formic acid into these tissues. This treatment induces leaf necrosis only a few hours after the attack (Morawetz et al. 1992; Renner and Ricklefs 1998; Frederickson et al. 2005).

As plants compete for light and/or nutrients, the behaviour of pruning competing vegetation is highly beneficial to the host plant. Its more rapid growth in turn has a positive effect on the growth of the ant colony, which benefits from the production of new domatia and food (Frederickson and Gordon 2009). In devil's gardens, pruning facilitates the establishment of new host plants (seedlings and vegetative shoots), allowing expansion of the colony. Pruning also reduces the access of other ants to the host plant, thereby limiting ant-ant competition, which is particularly fierce in tropical environments. In fact, this last factor could be one of the most important for explaining the selection of pruning behaviour (Davidson et al. 1988; Davidson and McKey 1993; Yumoto and Maruhashi 1999; Federle et al. 2002). Federle et al. (2002) found that *Crematogaster* species inhabiting *Macaranga* host plants with a slippery, waxy stem surface (which functions as a mechanical barrier against generalist ant competitors) pruned significantly less than those ants inhabiting plants without a protective wax barrier. The hypothesis that the principal advantage of pruning is reducing accessibility to intruders is further supported by the fact that not only intruding vegetation but also the canopy of the ants' own host plant may be pruned to minimise the risk of contact with competitors (Stanton et al. 1999). The latter authors observed repeated severe conflicts between plant–ant species competing for the same host tree species, the African *Acacia drepanolobium*. Competitively dominant plant–ants from nearby trees attempt to dislodge workers and brood of a less competitive species from their nests inside the swollen stipular thorns. This less competitive plant–ant species was also observed to prune the canopy of trees it occupies.

Pruning of competing vegetation means that workers are able to discriminate their own host plant from non-host-plant species. In devil's gardens, Frederickson et al. (2005) observed that host plants introduced from another stand, and from which domatia had been removed, were not pruned, suggesting that discrimination was at the species level, and not based on the suitability of the plant for nesting. Plant signals facilitating this discrimination are expected to evolve because of the high reciprocal benefits conferred by pruning. Despite this expectation and the fact that pruning behaviour has been known for several decades, investigation of the signals that mediate this behaviour is only beginning. Cosio (2009) and co-workers found that the chemical components on the surface of leaves and stems are important for the recognition of *Triplaris americana* trees by their host ants *Pseudomyrmex triplarinus*. Filter papers impregnated with cuticular extracts of *Triplaris* species pinned to the trunk, received significantly less attention and damage from the workers than untreated filter papers. This indicates that the plant–ants are able to recognise the chemicals found on the surfaces of their host plant.

Ants discriminate non-nestmates from the chemical profile they bear on the cuticle (Lenoir et al. 1999; Howard and Blomquist 2005). A typical ant cuticular profile is composed of hydrocarbons, linear and methyl-branched molecules (alkanes) and sometimes unsaturated molecules (alkenes), with a chain length ranging generally from 20 to 40 carbon atoms (Guerrieri et al. 2009). All colony members have a similar chemical profile called the "colony visa". Variation in the identity and relative proportions of those hydrocarbons allows for each colony to have its own visa. Nestmate discrimination is possible via antennation or at a short distance (Cuvillier-Hot et al. 2005; Brandstaetter et al. 2008). Moreover, the colony visa is learned, and is also partly acquired by individual ants at emergence (Lenoir et al. 1999). The primary function of both insect and plant cuticular lipids is to limit water loss (Hadley 1994; Raven and Edwards 2004). In social insects the need for communication has led to the use of cuticular hydrocarbon blends as a communication signal. As ant imagos emerge in contact with the plant surface, it could be that they learn the plant chemical profile along with the colony visa. However, this still remains to be demonstrated.

3 Signalling for Anti-Herbivore Protection

3.1 *Induced Defence: Signalling Herbivore Activity to Defending Ants*

Plant defences can be classified as direct or indirect. Direct defences are traits produced by the plant itself that act against herbivores or pathogens. Indirect defences, the subject of this paragraph, involve defence of the plant by a third party attracted by plant traits. Both direct and indirect defences can be either constitutive or induced. Constitutive defences are those that are displayed continuously, while induced defences are produced in response to particular stimuli such as those resulting from herbivory or volatile organic compounds emitted by neighbouring plants. Induced indirect defences involve the production of food rewards, the emission of volatile organic compounds, or both. The attraction of predators or parasitoids by herbivore-induced plant volatiles has been widely documented and results have been summarised in several reviews (Turlings and Wäckers 2004; Arimura et al. 2005; Heil 2008). In a number of cases it has been shown that the production of extrafloral nectar can be induced by herbivory or by signals from neighbouring plants (review in Agrawal and Rutter 1998; Heil 2008). An increase in nectar production leads to higher levels of attendance by ants (Passera et al. 1994) which deter or prey upon herbivores when patrolling the plant for food rewards. Wäckers et al. (2001) showed that the inducing effect of herbivory on extrafloral nectar production is often restricted to the damaged leaf and helps in actively guiding ants to the site of attack. Nevertheless, extrafloral nectar by itself is a rather imbalanced diet. Low amino acid levels, or the absence of certain essential amino acids, forces nectar-consuming ants to seek out supplementary protein sources. In a series of baiting experiments it could be shown that access to relatively carbohydrate-rich baits increased the effectiveness of ants as predators compared to colonies with access to protein (Ness et al. 2009). The plants' increase of extrafloral nectar production has, therefore, a double effect: not only is ant presence augmented, but also their aggressiveness as predators.

Damage to plants is known to induce immediate recruitment by ants resident in or patrolling on the plant (reviewed in Rico-Gray and Oliveira 2007; Heil 2008). This was shown in host-generalist arboreal ants that build carton nests in various tree species (Dejean et al. 2008b, c), but it has been best studied in ant–plant symbioses. Herbivore-induced defence by ants seems to be more efficient in myrmecophytes than in ant-gardens (Vantaux et al. 2007). We will thus focus on myrmecophytes.

The two most important cues that seem to be used by plant–ants to locate herbivores are mechanical vibrations and chemical compounds. Although the first one is the more obvious, it has never been studied in detail. There is ample evidence that shaking an ant–plant, even gently, results in ant workers swarming out of the domatia (see for example Madden and Young 1992; Agrawal and Rutter 1998; Federle et al. 1998; Lapola et al. 2003; Dejean et al. 2008a, 2009). However, the precise nature of the mechanical signal has not been investigated. Plant movements provoked by wind do not affect ant activity, and ant response to physical disturbance varies according to its intensity and among ant species. There is thus great potential for the discovery of herbivore-produced mechanical cues that could be used by ants to defend their host plant.

Herbivore-induced plant volatiles are other important cues used by ants in the defence of their host. Many studies have shown that artificial damage to myrmecophytes attracts their resident ants (Fig. 1) (Fiala and Maschwitz 1990; Jolivet 1996; Agrawal 1998; Federle et al. 1998; Agrawal and Dubin-Thaler 1999; Lapola et al. 2003; Bruna et al. 2004, 2008; Christianini and Machado 2004; Romero and Izzo 2004; Inui and Itioka 2007; Dejean et al. 2008b; Gianoli et al. 2008; Grangier et al. 2008; Mayer et al. 2008; Schatz et al. 2009). These studies have often demonstrated that ant attraction is similarly triggered by plant sap or by solvent extracts of plant parts, showing the role of volatile compounds (Fiala and Maschwitz 1990; Agrawal 1998; Agrawal and Dubin-Thaler 1999; Lapola et al. 2003; Bruna et al. 2004; Christianini and Machado 2004; Romero and Izzo 2004; Bruna et al. 2008). The emission of volatile compounds for defence is not restricted to myrmecophytic plants, but there is a big difference in predator-mediated induced defence between plants – such as myrmecophytes – that attract generalist predators and plants attracting specialist predators and parasitoids. In the latter, elicitors from herbivore regurgitants are often required to induce a specific chemical blend that can vary according to the herbivore species or developmental stage. Natural enemies of these herbivores readily detect these differences and respond accordingly (Takabayashi et al. 1995; Du et al. 1996; De Moraes et al. 1998; Powell et al. 1998). Maize plants emit different blends of volatiles according to the developmental stage of *Pseudaletia separata* caterpillars, and the parasitoid *Cotesia kariyai* is only attracted by blends emitted under attack by suitable stages (Takabayashi et al. 1995). The specificity of the volatile compounds emitted by plants attacked by different herbivores may play

Fig. 1 Anti-herbivore activity of *Pheidole bicornis* on stems of *Piper* ant-plants: (**a**) intact, unwounded internode and (**b**) recently wounded internode (the wound was made with a scalpel and is visible as a dark spot in the middle of the internode)

a key role in prey location by some parasitoid species (Turlings et al. 1990; De Moraes et al. 1998; Turlings and Wäckers 2004; Halitschke et al. 2008). Plant–arthropod communication appears in some cases to be highly specialised, and probably results from strong selective pressures. Indeed, a communication signal attracting predators or parasitoids without offering a reward (i.e. high probability of encountering their specific prey or hosts) would lead to rapid breakdown of the reciprocal benefits upon which mutualistic specialisation depends.

In myrmecophytes, odour emissions resulting from artificial mechanical damage or extracts of leaves are sufficient to induce ant response and recruitment; elicitors from herbivore regurgitants seem not to be necessary (*Macaranga*/*Crematogaster* Fiala and Maschwitz 1990, *Cecropia obtusifolia*/*Azteca* sp. Agrawal 1998, *Maieta guianense*/*Pheidole minutula* Lapola et al. 2003, *Hirtella myrmecophila*/*Allomerus octoarticulatus* Romero and Izzo 2004, *Piper* spp./*Pheidole bicornis* Mayer et al. 2008, *Leonardoxa africana*/*Petalomyrmex phylax* Schatz et al. 2009). This peculiarity of myrmecophytes is probably linked to the fact that ants are generalist predators, and that the ant mutualists of myrmecophytes are selected to be aggressive in defending their plants against all comers. There is thus no selective pressure on the plant to fine-tune signalling the identity of herbivores. This suggests that the ant–plant communication signals should not be threat-specific. It could also explain why the actual presence of the herbivore seems not to be needed to trigger plant–ant defence behaviour. However, it has never been tested in ant–myrmecophyte associations whether artificial damage is as efficient in inducing ant patrolling as is damage by actual herbivores. It would be interesting to determine to what extent blends induced by artificial damage and by real herbivores differ in myrmecophytes, and to test whether they differ less than in other plants.

The ants' response is influenced by the age and vulnerability of the damaged part of the host plant, as shown in the *Hirtella myrmecophila*/*Allomerus octoarticulatus* association. The response to extracts of young leaves was much stronger than to extracts of mature and old ones (Romero and Izzo 2004). The inhabitants of myrmecophytic *Piper* species, *Pheidole bicornis*, reacted much more strongly to wounding of the stem than to damage to leaves (Fig. 1) (Mayer et al. 2008). The reaction to the plant signal may thus depend more on tissue identity and relative value to the plant (and, by extension, to its resident ant colony as well) than on the herbivore species. In associations with multiple ant species the response to cues emitted by the host plant varies with identity of the inhabiting ants. The two plant–ant occupants of *Maieta guianensis*, *Pheidole minutula* and *Crematogaster laevis*, reacted in different ways to host-plant damage and leaf extract (Lapola et al. 2003).

Plant–ants usually show a stronger response to damaged leaves or leaf extracts of their host species than to those of related myrmecophytes (Fiala and Maschwitz 1990; Inui and Itioka 2007; Bruna et al. 2008), indicating some degree of specificity in ant response. Only a few studies have characterised the nature of volatile blends induced by damage in myrmecophytes and they only considered artificial damage (Inui and Itioka 2007; Mayer et al. 2008; Schatz et al. 2009). All volatiles detected fall into classes of compounds that are well known from induced blends in other plants: green leaf volatiles, aromatic compounds and terpenoids. A comparison of blends emitted from related myrmecophytic and non-myrmecophytic species indicates that the myrmecophytes emit specific blends (Brouat et al. 2000; Mayer et al. 2008). In the bouquet of volatile compounds from wounded stem bark of Central American *Piper* myrmecophytes, emission of the sesquiterpenes β-caryophyllene and germacrene D, and the green leaf volatiles hexanal, *cis*-3-hexene-1-ol, 2-hexenal, 2-undecanone and 1-hexanol is characteristic of all myrmecophytic species investigated so far (Mayer et al. 2008). These compounds were not detected in any of the investigated non-myrmecophytic *Piper*. In field experiments *Pheidole* colonies of a specific host plant reacted to stem pieces of related sympatric myrmecophytes when these were experimentally attached to the host plant but not to stem pieces of non-myrmecophytes growing in the same area (Mayer et al. 2008). This indicates that the signal is not plant species-specific within the group of myrmecophytic *Piper*. Large *P. bicornis* colonies are polydomous and it was observed that they can occupy more than one species of *Piper*. The defence of related myrmecophytes from attacks by herbivores is therefore important in the *Piper–Pheidole* association.

While selection is not expected to lead to herbivore-specific signals in myrmecophytes, it should favour emission of a strong generalised signal communicating attack, and rapid and marked response of the ants to the signal. Have plant signal and ant response coevolved? The next step in research on ant–myrmecophyte communication is the identification of the precise nature of the communication signal. All the above studies have characterised chemical blends whose emission is induced by damage in myrmecophytes, but only one has tested response of the ants to synthetic compounds and mixtures of them (Schatz et al. 2009). This case is enlightening. *Leonardoxa africana* subsp. *africana* is a myrmecophyte endemic to coastal rainforests of Cameroon (McKey 2000). It is obligatorily associated with the ant *Petalomyrmex phylax*, which protects the tree against herbivores (Gaume et al. 1997). The plant's young leaves are patrolled continuously (McKey 1984). The very closely related subspecies, *Leonardoxa africana* subsp *gracilicaulis*, does not have domatia. Analysis of brief hexane washes of intact leaves intended to collect molecules only from leaf surfaces, highlighted a specificity of young leaves of *Leonardoxa africana*: they bear a high proportion of methyl salicylate, which was found only as traces on young *Leonardoxa africana gracilicaulis* leaves and not at all on mature leaves of the myrmecophyte (Brouat et al. 2000). In addition, Schatz et al. (2009) showed that in the myrmecophyte ant recruitment was induced by damage to mature leaves, which produce methyl salicylate among other damage-induced volatiles. Moreover, synthetic methyl salicylate alone produced the same effect. Methyl salicylate is thus the communication signal for ant-mediated defence in the myrmecophyte *Leonardoxa africana*. This compound is widely found in the plant kingdom, where it often functions as a herbivore- and pathogen-induced volatile attracting parasitoids and predators (Dicke et al. 1990b; Shulaev et al. 1997; Zhu and Park 2005; Sasso et al. 2009). This illustrates our expectation that signals between myrmecophytes and ants to guide patrolling behaviour of the latter do not need to be specific.

A characteristic of ant–myrmecophyte symbioses is the short lag time – often spectacularly short in comparison to other tritrophic plant–herbivore–predator interactions – between plant damage and ant behavioural response. This short lag time is due not solely to the constant presence of ants on these plants (and thus do not have to be attracted from a distance), but also to the more rapid emission of the signal by the plant. In non-myrmecophytes the induced production and release of volatiles lasts typically for several hours (Metlen et al. 2009). Many ant species display efficient recruitment behaviour in the context of nest defence or food source exploitation. Once a worker has detected a herbivore on its host plant, there is no need for the myrmecophyte to continue to "cry for help". In ant–plant symbioses we predict a negative correlation between the duration of the emission of herbivore-induced plant volatiles and the efficiency of recruitment of the inhabiting ant. Moreover, ants have good learning capabilities. We may expect them to respond more quickly to the plant signal if they associate the signal with the presence of herbivores. In such a case, there is no need for a long-lasting signal. The natural enemies of herbivores attracted by non-myrmecophytes are at an unpredictable distance from the plant. In contrast, myrmecophytes have their bodyguards already on site. This is a great advantage for the rapidity of the protective response. All studies reported ant attraction within less than five minutes (Fiala and Maschwitz 1990; Agrawal 1998; Federle et al. 1998; Agrawal and Dubin-Thaler 1999; Lapola et al. 2003; Bruna et al. 2004; Christianini and Machado 2004; Romero and Izzo 2004; Inui and Itioka 2007; Dejean et al. 2008b; Grangier et al. 2008; Mayer et al. 2008; Schatz et al. 2009), and the signal appears to be produced immediately upon damage. Lag times for emission of herbivore-induced volatiles by non-myrmecophytic plants are much longer; lag times of a few hours are considered rapid (Metlen et al. 2009). Lag time greatly depends on compound class, green leaf volatiles being the first released, and terpenoids coming hours later (Turlings and Tumlinson 1992; Turlings et al. 1995). It also depends on whether the compounds are stored in plant tissues or whether they are synthesised de novo. Non-myrmecophytic plants may need to produce blends that are herbivore-specific. Storage of several specific compounds is likely to be more expensive than storage of a single generalised signal, as in myrmecophytes. There are thus two explanations, not mutually exclusive, for the particularly rapid induction in myrmecophytes: the signal is made of green leaf volatiles that can be emitted rapidly after damage, and/or of other compounds that are stored structurally.

3.2 *Ants as a Constitutive Defence*

In addition to functioning as an induced defence against herbivores, ants also provide constitutive defence in ant–myrmecophyte interactions. Evidence for ants as a constitutive defence is of two types: (1) ants live permanently on the plant and many species do not leave their plant to forage for food, (2) certain plant parts are continuously patrolled by ants in the absence of a food reward.

Myrmecophytes are characterised by the development of domatia, even in the absence of ants. Many species also produce food bodies and extra-floral nectar regardless of the presence of ants. Thus, domatia, food bodies and extrafloral nectar are constitutive traits, allowing ants to take up permanent residence in the host plant. Moreover, many plant–ants are obligate inhabitants of their host plant. Ants are fierce predators, and their presence alone is dissuasive for many herbivores. This is shown by the fact that non-myrmecophytic plants that bear extrafloral nectaries, and thus attract ants opportunistically, are better protected against herbivores (review in Heil and McKey 2003): the purpose of the ant visit is to collect nectar but the consequence is herbivore deterrence. From an evolutionary perspective, domatia, food bodies and extrafloral nectar in myrmecophytes can be considered as communication features whose function is to provide the plant with constitutive defence (i.e. ants). Extrafloral nectar in non-myrmecophytic plants that attract ants opportunistically can be induced by herbivory. Interestingly, in the highly diversified genus *Vachellia* (=*Acacia*), extrafloral nectar production is inducible in species attracting opportunistic ants, and constitutive in myrmecophytes (Heil et al. 2004c). The phylogeny of the genus shows that nectar inducibility is the ancestral state, providing evidence that the switch from highly inducible to constitutive nectar flow has evolved along with myrmecophytism.

The other type of evidence that ants provide constitutive defence is the constant patrolling of young shoots (Fig. 2). In many ant–myrmecophyte symbioses, ants patrol mostly young shoots, around-the-clock in some cases, even though these plant parts often do not produce food rewards (Downhover 1975; McKey 1984; Fiala and Maschwitz 1990; Moog et al. 1998; Gaume and McKey 1999; Christianini and Machado 2004; Edwards et al. 2007; Grangier et al. 2008). Moreover, the myrmecophytic *Crypteronia* and *Tachigali* do not produce direct food rewards at all, but their respective *Cladomyrma* and *Pseudomyrmex* symbionts still patrol young shoots (Moog et al. 1998, Blatrix personal observation, Fig. 2). However, new shoots are often associated with the production of a new domatium. As nesting space is a limiting factor for colony development (Fonseca 1993; Edwards et al. 2009), it is advantageous for the ants to protect new shoots. So it is very likely that a signal characterising new shoots has evolved in myrmecophytic species. However, it has rarely been investigated. Edwards et al. (2007) detected compounds that could be similar between ant larvae of *Allomerus octoarticulatus* and new shoots of their host plant *Cordia nodosa*. However, the brood-tending caste was attracted to brood extracts but not to young leaf extracts, and the patroller caste was attracted to young leaf extracts but not to brood extracts. This showed that any compounds which might be shared between brood and new leaves were not responsible for ant attraction. Thus, attraction to new shoots is an adaptation of this ant to its host. The only case where the signal that elicits patrolling specifically of young shoots has been identified is the ant–plant *Leonardoxa africana* and its obligate ant mutualist *Petalomyrmex phylax* (Fig. 2). Young leaves produce a large amount of methyl salicylate, which is barely detectable in the brief hexane washes of intact mature leaves (Brouat et al. 2000). Moreover, the ants were shown to react strongly to synthetic methyl salicylate (Schatz et al. 2009). It is noteworthy that the

Fig. 2 Constitutive patrolling of young shoots of the ant-plants *Tachigali* (**a**) and *Leonardoxa africana* (**b**) by *Pseudomyrmex* and *Petalomyrmex phylax* ants respectively

production of this compound is induced by herbivory in mature leaves and that the ants react much less to the other compounds emitted by damaged leaves. Thus, the same compound is used as an induced signal in mature leaves, and as a constitutive signal in young shoots. To understand the evolution of such signals, a comparative analysis of cues displayed by mature leaves and new leaves should be conducted in a diversified genus containing both myrmecophytes and non-myrmecophytes and match the divergence of cues used in mature and young leaves, respectively, with plant phylogeny.

3.3 How Specificity Affects Efficiency

Because of the spatial proximity between the ants and the plant in ant–myrmecophyte symbioses, the ants do not have to discriminate the right signal from a noisy background expected in a complex environment. Even if cues from surrounding plants may reach the ants, the strongest ones are likely to be those of the host plant. Thus, the signals for constitutive and induced defences in myrmecophytes are not expected to be specific to the plant or the ant species. In this context, the same signals could function with several ant species, and different plants could share the same signals. Once the ant colony occupies the plant, compounds that are commonly found in plants could work as signals as soon as the ants are able to perceive them. Defence signals are thus not expected to be specific to myrmecophytes because the specificity of the ant–plant association, if any, is established at the colony founding stage.

The degree of specificity shows great variability among ant–myrmecophyte associations. Jürgens et al. (2006) found even in *Macaranga* species sharing the same obligate *Crematogaster* morphospecies no obvious similarities in the scent profiles of the investigated plants. On the other hand, myrmecophytes can also be associated with various ant species, not necessarily related to each other. For instance, in a myrmecophytic *Tachigali*, each individual plant can potentially be inhabited successively by several ant species belonging to different subfamilies according to an ontogenetic succession process (Fonseca and Benson 2003). It would be interesting to investigate the nature of the signal at work with the different ant partners. We may expect similar compounds to be used by several ant species. Unfortunately, very little is known in this field. *Maieta guianensis* can be inhabited by two ants, *Pheidole minutula* and *Crematogaster laevis*. Both are similarly attracted by leaf damage, but only the first is attracted by leaf extracts, showing that the cues used may differ (Lapola et al. 2003). The type of cue used by ants may depend on their life history. Two ant species associated with *Tococa bullifera* showed different reactions to cues induced by leaf damage (Bruna et al. 2004). The polydomous *Azteca* colonies, rearing brood and tending coccids also on neighbouring plants (usually plants other than *Tococa bullifera*), responded strongly to volatiles of a non-myrmecophytic species (Bruna et al. 2008). To defend the colony they must be able to recognise damage to non-myrmecophytes. In contrast, *Pheidole minutula* is an obligate dweller of *Tococa bullifera* and responded to chemical cues from potential host plants, but not to non-myrmecophytic species. Contrary to the predictions, this very preliminary information suggests that the myrmecophytes do not display one signal that can be used by any ant symbiont, but rather that each ant species uses different cues. It has been proposed that ant communication involves multi-component signals resulting from a mix of "private" signals addressing specific members of a group or organisational level, and "anonymous" signals that are uniform throughout a group or organisational level (Hölldobler and Carlin 1987; Hölldobler 1995). Adapted to ant–plant interactions, this means that there are "private" host species-specific signals which can be only "understood" by the ant symbiont, whereas anonymous ones can be "understood" by all other potential receivers. The modulation of signals as anonymous or private may cause a variety of possible answers and interactions. In ant–plant associations with low specificity, "anonymous" signals should be predominant because the signals should be understood by various ant species. However, "private" signals may have evolved in highly specific ant–plant associations because the plant would benefit from selecting the most efficient partner.

An unusual case is found in the ant–plant genus *Piper*, where one ant species, *Pheidole bicornis*, is obligatorily associated with five myrmecophytic species in the genus. Mayer et al (2008) showed that the ant responded strongly to chemical cues from myrmecophytes tested, but not (or only very weakly) to those from non-myrmecophytes. The analysis of the volatile organic compounds emitted after damage showed that some compounds were specific to the myrmecophytes. Those were green leaf volatiles and terpenoids, which are common in inter-plant signalling (Dicke 2009). The *Piper* case is a good model to test whether ants react to a unique signal displayed by all myrmecophytes and whether potential specific signals of host plant could influence defence efficiency.

The nature of plant cues used by ants for plant protection seems to result from complex interactions between life histories of plants and ants. This results in variability of protection efficiency. Low specificity in ant–myrmecophyte interactions could impair the coevolution between signal emission by plants and response by ants. Specificity would be favoured if it results in better protection. More comparative analyses of communication signals and defence efficiency should be performed to gain a better understanding of their role in the evolution of "filters" in some myrmecophytes that secure the specificity of the ant partner at the founding stage. Communication may thus have an important role in community ecology and evolution.

3.4 Optimal Defence Theory

The optimal defence theory (McKey 1974b; Rhoades 1979) postulates that defences against herbivores are costly and that natural selection must have favoured plants that maximise the benefits of defence. One prediction is that plants should avoid overlap of functionally redundant defences. Induced defence is expected to be less costly than constitutive defence, and to be favoured when the probability of herbivory is low but variable. Thus, the theory predicts that constitutive defences should occur in organs on which the risk and predictability of herbivory are high. How do theoretical predictions fit with the natural history of myrmecophytes? Supporting an ant colony is costly to the plant because it often involves producing specialised hollow structures and supplying the ants with food, either directly with extrafloral nectar and/or food bodies, or indirectly through the farming of hemipterans by the ants. If overlapping of functionally redundant defences should be avoided, mechanisms should have been selected to focus ant patrolling on young shoots rather than on mature leaves because the latter are usually protected by toughness and the accumulation of secondary metabolites. Young leaves suffer more from herbivory because they are easier to consume and digest. As herbivores are choosy, the probability of herbivory is higher on young shoots. Thus we expect the plant to use ants as a constitutive defence on young leaves and as an induced defence on old ones. In summary, the optimal defence theory predicts the evolution of plant features that concentrate ant activity continuously on young shoots and that induce ant activity on mature leaves only when they are actually attacked. Not surprisingly, this is exactly the pattern observed in myrmecophytes whose ant-mediated defence mechanisms have been investigated (see above). Additional examples further confirm the predictions of the optimal defence theory. The ant *Allomerus octoarticulatus* is attracted by aqueous leaf extracts from its host plant *Hirtella myrmecophila*, but more when leaves are young than when they are mature (Romero and Izzo 2004). The intensity of the chemical signal inducing ant recruitment seems to be negatively correlated with leaf age. Gianoli et al. (2008) found support for the optimal defence theory by testing the trade-off between constitutive and induced defences according to the intensity of herbivory pressure across trees in the ant–plant *Cecropia obtusifolia* occupied by *Azteca* ants. They used the percentage of area lost by old leaves to assess the risk of herbivory and found that it was positively correlated with the level of constitutive patrolling whereas it was negatively correlated with the level of induced recruitment to leaf damage.

The constitutive production of food bodies and/or extrafloral nectar in myrmecophytes matches with the optimal defence theory. Extrafloral nectar, for example, is usually produced on young leaves which need particular protection (Fiala and Maschwitz 1991; Raine et al. 2002). However, food rewards are spatially separated from plant parts that need most protection (e.g. young leaves) in several myrmecophytes, such as *Cecropia* (Downhover 1975), *Barteria* (Janzen 1972), *Leonardoxa* (McKey 1984) and *Macaranga* (Heil et al. 2004b). This may have occurred in myrmecophytes because ants are provided with a reliable food source from the plant (extrafloral nectar, food bodies and hemipteran trophobionts) setting the conditions for ant constitutive defence behaviour not triggered by immediate direct food rewards (Heil 2008).

Plant defence strategies can be very complex, involving various types of defence categories, and it is often difficult to get the full picture. Incomplete information on the entire suite of defence traits can lead to questionable interpretations (Webber and McKey 2009). The role of ants in the evolution of plant defence strategies should thus be considered with caution.

4 The Evolution of Plant–Ant Communication

4.1 How Can Plant–Ant Communication Signals Have Evolved so Many Times Independently?

Myrmecophytism occurs in many plant genera (over 100) distributed over a large diversity of Angiosperm families (c. 36). The same kind of pattern holds true for plant–ants (c. 40 genera, Davidson and McKey 1993). Moreover, very few ant–plant and plant–ant genera have produced considerable species radiations (exceptions are *Vacellia* = *Acacia*, and *Macaranga*). The scattered taxonomic distribution of myrmecophytism indicates that it has evolved many times independently (Davidson and McKey 1993; McKey and Davidson 1993). Therefore, the key feature in the functioning of ant–plant symbioses, ant–plant communication, has also evolved concomitantly with myrmecophytism many times independently, suggesting that this feature is not too constrained. Within both types of organisms, ants and plants, communication is achieved through the channel of chemistry. They both make a widespread use of blends of volatile compounds for their own communication system independently from each other (in ants: alarm and defence releaser, trail marker, nestmate recognition, etc.; in plants: attraction of parasitoids, plant-to-plant signalling). Moreover, similar compounds are found in the arsenal of both groups. Terpenoids such as citronellal, citral or farnesene homologues, and aromatic compounds such as indole or methyl 6-methyl salicylate, are plant volatiles also found in ant glands (Regnier and Wilson 1968; Attygalle and Morgan 1985, see review in Morgan 2009). Those compounds are usually trail pheromones (Morgan 2009). The so-called green leaf volatiles hexanal, 1-hexanol and 3-undecanone, substances which are usually emitted from plants when they are under biotic stress, are produced by the weaver ant *Oecophylla longinoda* as components of the alarm pheromone (Bradshaw et al. 1975). Ants and plants thus have predispositions facilitating information exchange, which could account for the apparent ease of evolution of ant–plant communication.

4.2 The Sensory Trap Hypothesis

Sensory traps are defined by Edwards and Yu (2007) as "signal mimics that elicit out-of-context behaviours by exploiting the adaptive neuronal responses of signal receivers". They are particularly well illustrated by Mediterranean *Ophrys* orchids, the flowers of which mimic the odour, shape and texture of female bees in order to attract males for pollination, without producing food rewards (Borg-Karlson 1990). Mimicry has been suggested to be an important mechanism in the evolution of animal communication (Wickler 1965). Courtship behaviour has been the focus of most attention because the signals involved can be strikingly similar to cues used to find food (Christy 1995). Recently it was argued that "sensory traps" could be involved in mutualisms (Edwards and Yu 2007). However, the term "trap" suggests something detrimental to the receiver of the signal, which we did not find in myrmecophytes so far. The original term "mimicry" is preferred as it is more neutral.

The "mimicry hypothesis" applied to ant–plant symbioses predicts that cues displayed by plants have evolved to mimic the communication signals used by ants in their social activities. In ant-gardens built by the ant *Camponotus femoratus*, a compound from seeds of several unrelated symbiotic epiphytes is responsible for ant attraction (Youngsteadt et al. 2008). It is noteworthy that this compound is not found in other plants. However, it is known from various ant species and was found in the mandibular glands of *Camponotus femoratus* males (Seidel et al. 1990).

The constitutive defences provided by ants to myrmecophytes are also contexts where mimicry could occur. According to the mimicry hypothesis, continuous patrolling of young shoots by ants could be achieved through emission of brood odours. Edwards et al. (2007) specifically tested the hypothesis that mimicry could have a role in ant–plant communication. They found similarities between chemical profiles of young shoots of the ant–plant *Cordia nodosa* and brood of its ant *Allomerus octoarticulatus*. But the hypothesis was invalidated because ant patrollers were attracted to extracts of young leaves but not of brood, and nurses were more attracted by extracts of brood. Moreover, only patrollers were more attracted to extracts of young leaves than mature leaves. Most plant–ants must be able to recognise their host plant because they prune encroaching vines and intruding vegetation. The mimicry hypothesis would predict that the plant's chemical profile could imitate the odour of the colony. However, this is highly unlikely because colony odour is unpredictable and specific for each colony. Imitating it would thus require detection and synthesis of the blend by each individual plant. To explain host-plant discrimination by ants, it is much more parsimonious to invoke either learning of the plant odour or incorporation of plant chemicals into the colonial odour, two mechanisms that could be by-products of common ant features, or innate recognition. Indeed, learning of plant odour by arboreal ants was shown to explain attraction of some ants to their nesting site plants (Djieto-Lordon and Dejean 1999). The second possible explanation remains to be tested. As plant–ants are fed by food derived from the host plant, the sequestration of plant chemicals resulting in a specific odour is possible (see also Liang and Silverman 2000; Buczkowski et al. 2005).

The similarity between plant volatiles and ant trail pheromones may appear more relevant to the mimicry hypothesis. Some terpenoids and aromatic compounds known from plants have been described as ant trail pheromones (Morgan 2009). It could thus be tempting to propose that volatiles of myrmecophytes may have been selected to match trail pheromones of some pre-symbiotic ant ancestor. However, an alternative explanation is convergent evolution. Indeed, both trail pheromones and volatiles emitted by myrmecophytes are under similar selective pressure: they must attract ant congeners at relatively short distances, the signal they convey should be produced quickly and vanish rapidly if not reinforced, and it should be cheap to produce.

Except for ant-garden seeds, the compounds identified as potential communication signals in ant–plant symbioses were more typical for plants than for ants, ranging over the three categories displayed by plants in the context of defence, i.e. green leaf volatiles, terpenoids and aromatic compounds. They have thus obviously not evolved to match ant odours. An alternative explanation to account for similarities between ants and myrmecophyte chemicals would be that ants incorporate compounds from their nesting environment, i.e. their host plant. Indeed, ant chemical profiles are known to vary according to diet and nest material (Heinze et al. 1996; Liang and Silverman 2000; Sorvari et al. 2008). The hypothesis that mimicry plays a role in the origin and/or maintenance of ant–plant symbioses is difficult, if not impossible, to validate. Finding similarities between plant and ant odours does not mean that plants mimic ants. Although so far only very few studies have addressed the role of mimicry in the evolution of ant–plant communication, the available data do not support this hypothesis.

4.3 Processes in the Evolution of Communication Signals

Improving our understanding of the evolution of communication in ant–plant symbioses will require determining what may constrain the signal. Plant biosynthetic pathways for volatile production are considered to be highly constrained by phylogeny. They have even been proposed as characters for phylogenetic reconstruction (Barkman 2001). However, hierarchical clustering of blends of volatiles proved to be mostly incongruent with DNA-based phylogeny (Levin et al. 2003). Production of secondary metabolites by plants is indeed known to depend not only on taxonomy but also on evolutionary interactions with herbivores or pollinators (Jürgens 2004; Agrawal et al. 2009). The nature of plant–invertebrate communication signals thus seems to result from both phylogenetic (*sensus* McKitrick 1993) and ecological constraints.

What are the respective roles of these constraints in the evolution of ant–plant communication? No study has addressed this question yet, but ant and plant traits suggest some hypotheses for future investigations.

The fact that very few biosynthetic pathways are involved in the numerous cases of emission of herbivore-induced plant volatiles attractive to natural enemies (Turlings and Wäckers 2004) suggests that the biosynthetic pathways may constitute the most important phylogenetic constraint on the nature of the communication signal between plants and ants. Insect olfactory receptor neurons exhibit high selectivity and respond to specific compounds, but the identity of these compounds can be very different across related insect species (Todd and Baker 1999). Thus, antennal receptors appear to impose little phylogenetic constraint on the evolution of signals. Plants are known to produce more than 1,000 volatile compounds (Pichersky et al. 2006), and more than 1,500 compounds with a pheromonal action have been identified in social insects (Passera and Aron 2005). Moreover, we saw that there is structural overlap between ant pheromones and plant volatiles. Thus, the few phylogenetic constraints, i.e. mostly those affecting plant biosynthetic pathways, allow for a wide range of compounds to become potential signals.

The ecological selective pressures exerted on the signal depend on the type of interaction between the two partners. We have identified five settings where communication is expected to have evolved in ant–plant symbioses: the selection of ant-garden seeds to sow the garden, the discrimination of the host plant by pruning plant–ants, the detection of the host myrmecophyte by founding queens, the selective continuous patrolling on young shoots of myrmecophytes, and the damage-induced ant-mediated protection of myrmecophytes. The specificity of the signal is expected to be high in the first two settings because competition with non-symbiotic plants is expected. Therefore, cues reducing ant choice errors should be favoured by selection. The third setting may select for a high specificity for the same reason, except when the density of founding queens is constantly higher than that of the available host plants. In this last case, competition between plants for occupants is relaxed because myrmecophytes have a high probability of being colonised under high ant density. In these three settings, i.e. dispersal of ant-garden seeds, pruning of non-host-plants and host-plant recognition by founding queens, ecological constraints may be expected to shape a more or less specific signal. However, in the last two settings, i.e. patrolling of young shoots and damage-induced protection, the ants are permanently in close proximity to the plant and the signal does not need to be so specific. Constraints are thus very low, and we may expect any compound with the appropriate physical properties to be used as signal. Nevertheless, efficiency of the communication process should be favoured. The selection of one particular signal and further coevolution of the production by the plant and perception/integration by the ants would result in increased efficiency. Reciprocal adaptation would lead, over evolutionary time, to a higher production of the signal by the plant and a stronger ant behavioural response to it. Since the nature of the signal is not expected to be particularly constrained, the compounds which evolve as signals may be the outcome of a random or unpredictable chain of events, each step contingent on the history of previous steps. Historical contingency has been demonstrated in experimental evolution with bacteria (Blount et al. 2008). Such demonstrations are of course impossible in biological systems such as ant–plant symbioses, but a detailed analysis of communication signals in symbioses in which both partner species have undergone evolutionary radiation could give some clues about the processes underlying the evolution of communication signals. Under contingent evolution, many different signals are expected to have been selected across species, with low repeatability. Under phylogenetic constraints, the pattern of signal variation should be congruent with phylogeny either of the plants or the ants, respectively, depending on whether emission or reception is constrained. Under ecological constraints, a predictable and reduced set of signals should occur, irrespective of phylogeny. Signal evolution probably results from interactions between those constraints. Ant–plant associations that show radiation in both lineages may thus constitute good models for understanding the respective roles of the two kinds of constraints in the evolution of communication signals in these and other plant–insect mutualisms.

5 Conclusions

The emission of communication signals by plants addressed to arthropods is largely documented. However, there is a huge gap in the investigation of communication signals from arthropods to plants. Despite the many examples of signals emitted by plants for reception by ants, there is hardly any work on signals emitted by ants to which plants respond. An indication that such signals exist is found in myrmecophytic *Piper*, where food body production is induced by the presence of the ant symbiont (Risch and Rickson 1981; Letourneau 1983). The plant probably uses ant cues to match the production of myrmecophytic traits with ant presence, and thus avoids wasting resources when ants are absent. Ant–plant symbioses involve partners that live together during most of their respective life-spans. This means that there is a great potential for mutual exchange of information. Communication signals in both directions are more likely to have evolved in ant–plant symbioses than in any other plant–insect relationship. The search for communication from ants to plants may constitute a promising field for future research.

Communication in ant–plant symbioses remains understudied and poorly understood. This chapter provides research directions, which we hope will stimulate further investigations. The discovery a few decades ago that plants attract natural enemies of herbivores led to a better understanding of plant–arthropod communication and opened the door to a new research area. This area highlighted the complexity of inter-specific interactions and led to the view that communication between organisms as different as plants and arthropods has tremendous implications for the structure and functioning of ecological communities. Ant–plant symbioses set various contexts for inter-specific communication to evolve. Because of their particular life history, myrmecophytes provide interesting comparisons with other plants that "cry for help". Comparing the various settings of plant-arthropod communication constitute an interesting framework for a better understanding of evolutionary community ecology.

Acknowledgments We thank Doyle McKey for discussions and proof-reading the manuscript, and Finn Kjellberg for stimulating discussions. Studies of ant–plant interactions were funded by two grants from the French Agence Nationale de la Recherche to R.B. ("Young scientists" programme, research agreement no. ANR-06-JCJC-0127, and "Biodiversity" programme, IFORA project).

References

Agrawal AA (1998) Leaf damage and associated cues induce aggressive ant recruitment in a neotropical ant–plant. Ecology 79:2100–2112

Agrawal AA, Dubin-Thaler BJ (1999) Induced responses to herbivory in the Neotropical ant-plant association between *Azteca* ants and *Cecropia* trees: response of ants to potential inducing cues. Behav Ecol Sociobiol 45:47–54

Agrawal AA, Rutter MT (1998) Dynamic anti-herbivore defense in ant-plants: the role of induced responses. Oikos 83:227–236

Agrawal AA, Salminen JP, Fishbein M (2009) Phylogenetic trends in phenolic metabolism of milkweeds (*Asclepias*): evidence for escalation. Evolution 63:663–673

Arimura G, Ozawa R, Shimoda T, Nishioka T, Boland W, Takabayashi J (2000) Herbivory-induced volatiles elicit defence genes in lima bean leaves. Nature 406:512–515

Arimura G, Ozawa R, Horiuchi J, Nishioka T, Takabayashi J (2001) Plant–plant interactions mediated by volatiles emitted from plants infested by spider mites. Biochem Syst Ecol 29:1049–1061

Arimura G, Kost C, Boland W (2005) Herbivore-induced, indirect plant defences. Biochim Biophys Acta 1734:91–111

Attygalle AB, Morgan ED (1985) Ant trail pheromones. Adv In Insect Phys 18:1–30

Barkman TJ (2001) Character coding of secondary chemical variation for use in phylogenetic analyses. Biochem Syst Ecol 29:1–20

Billen J, Morgan ED (1998) Pheromone communication in social insects: sources and secretions. In: Vander Meer RK, Breed MD, Espelie KE, Winston ML (eds) Pheromone communication in social insects: ants, wasps, bees, and termites. Westview, Boulder, pp 3–33

Blount ZD, Borland CZ, Lenski RE (2008) Historical contingency and the evolution of a key innovation in an experimental population of *Escherichia coli*. Proc Natl Acad Sci USA 105:7899–7906

Borg-Karlson AK (1990) Chemical and ethological studies of pollination in the genus *Ophrys* (Orchidaceae). Phytochemistry 29:1359–1387

Bradshaw JW, Baker R, Howse PE (1975) Multicomponent alarm pheromones of the weaver ant. Nature 258:230–231

Brandstaetter AS, Endler A, Kleineidam CJ (2008) Nestmate recognition in ants is possible without tactile interaction. Naturwissenschaften 95:601–608

Brouat C, McKey D, Bessière JM, Pascal L, Hossaert-McKey M (2000) Leaf volatile compounds and the distribution of ant patrolling in an ant–plant protection mutualism: preliminary results on *Leonardoxa* (Fabaceae: Caesalpinioideae) and *Petalomyrmex* (Formicidae: Formicinae). Acta Oecon 21:349–357

Bruna EM, Lapola DM, Vasconcelos HL (2004) Interspecific variation in the defensive responses of obligate plant–ants: experimental tests and consequences for herbivory. Oecologia 138:558–565

Bruna EM, Darrigo MR, Pacheco AMF, Vasconcelos HL (2008) Interspecific variation in the defensive responses of ant mutualists to plant volatiles. Biol J Linn Soc 94:241–249

Buczkowski G, Kumar R, Suib SL, Silverman J (2005) Diet-related modification of cuticular hydrocarbon profiles of the Argentine ant, *Linepithema humile*, diminishes intercolony aggression. J Chem Ecol 31:829–843

Carroll CR, Janzen DH (1973) Ecology of foraging by ants. Annu Rev Ecol Syst 4:231–257

Choe DH, Millar JG, Rust MK (2009) Chemical signals associated with life inhibit necrophoresis in Argentine ants. Proc Natl Acad Sci USA 106:8251–8255

Choh Y, Shimoda T, Ozawa R, Dicke M, Takabayashi J (2004) Exposure of lima bean leaves to volatiles from herbivore-induced conspecific plants results in emission of carnivore attractants: active or passive process? J Chem Ecol 30:1305–1317

Christianini AV, Machado G (2004) Induced biotic responses to herbivory and associated cues in the Amazonian ant-plant *Maieta poeppigii*. Entomol Exp Appl 112:81–88

Christy JH (1995) Mimicry, mate choise, and the sensory trap hypothesis. Am Nat 146:171–181

Cosio E (2009) Chemical recognition in an ant–plant mutualistic interaction. In: Joint meeting of the society for tropical ecology and the association for tropical biology and conservation. Marburg, Germany, p 580

Cuvillier-Hot V, Renault V, Peeters C (2005) Rapid modification in the olfactory signal of ants following a change in reproductive status. Naturwissenschaften 92:73–77

Dattilo WFC, Izzo TJ, Inouye BD, Vasconcelos HL, Bruna EM (2009) Recognition of host plant volatiles by *Pheidole minutula* Mayr (Myrmicinae), an Amazonian ant–plant specialist. Biotropica 41:642–646

Davidson DW (1988) Ecological studies of neotropical ant gardens. Ecology 69:1138–1152

Davidson DW, McKey D (1993) The evolutionary ecology of symbiotic ant–plant relationships. J Hymenopt Res 2:13–83

Davidson DW, Longino JT, Snelling RR (1988) Pruning of host plant neighbors by ants – an experimental approach. Ecology 69:801–808

Davidson DW, Seidel JL, Epstein WW (1990) Neotropical ant garden II. Bioassays of seed compounds. J Chem Ecol 16:2993–3013

De Boer JG, Posthumus MA, Dicke M (2004) Identification of volatiles that are used in discrimination between plants infested with prey or nonprey herbivores by a predatory mite. J Chem Ecol 30:2215–2230

De Moraes CM, Lewis WJ, Pare PW, Alborn HT, Tumlinson JH (1998) Herbivore-infested plants selectively attract parasitoids. Nature 393:570–573

De Moraes CM, Mescher MC, Tumlinson JH (2001) Caterpillar-induced nocturnal plant volatiles repel conspecific females. Nature 410:577–580

Dejean A, Delabie JHC, Cerdan P, Gibernau M, Corbara B (2006) Are myrmecophytes always better protected against herbivores than other plants? Biol J Linn Soc 89:91–98
Dejean A, Djieto-Lordon C, Orivel J (2008a) The plant ant *Tetraponera aethiops* (Pseudomyrmecinae) protects its host myrmecophyte *Barteria fistulosa* (Passifloraceae) through aggressiveness and predation. Biol J Linn Soc 93:63–69
Dejean A, Grangier J, Leroy C, Orivel J (2008b) Host plant protection by arboreal ants: looking for a pattern in locally induced responses. Evol Ecol Res 10:1217–1223
Dejean A, Grangier J, Leroy C, Orivel J, Gilbernau M (2008c) Nest site selection and induced response in a dominant arboreal ant species. Naturwissenschaften 95:885–889
Dejean A, Grangier J, Leroy C, Orivel J (2009) Predation and aggressiveness in host plant protection: a generalization using ants from the genus *Azteca*. Naturwissenschaften 96:57–63
Dicke M (2009) Behavioural and community ecology of plants that cry for help. Plant Cell Environ 32:654–665
Dicke M, Sabelis MW, Takabayashi J (1990a) Do plants cry for help? Evidence related to a tritrophic system of predatory mites, spider mites and their host plants. Symp Biol Hung 39:127–134
Dicke M, Vanbeek TA, Posthumus MA, Bendom N, Vanbokhoven H, Degroot AE (1990b) Isolation and identification of volatile kairomone that affects acarine predator–prey interactions: involvement of host plant in its production. J Chem Ecol 16:381–396
Djieto-Lordon C, Dejean A (1999) Tropical arboreal ant mosaics: innate attraction and imprinting determine nest site selection in dominant ants. Behav Ecol Sociobiol 45:219–225
Djieto-Lordon C, Dejean A, Gibernau M, Hossaert-McKey M, McKey D (2004) Symbiotic mutualism with a community of opportunistic ants: protection, competition, and ant occupancy of the myrmecophyte *Barteria nigritana* (Passifloraceae). Acta Oecon 26:109–116
Downhover JF (1975) The distribution of ants on *Cecropia* leaves. Biotropica 7:59–62
Du YJ, Poppy GM, Powell W (1996) Relative importance of semiochemicals from first and second trophic levels in host foraging behavior of *Aphidius ervi*. J Chem Ecol 22:1591–1605
Edwards DP, Yu DW (2007) The roles of sensory traps in the origin, maintenance, and breakdown of mutualism. Behav Ecol Sociobiol 61:1321–1327
Edwards DP, Hassall M, Sutherland WJ, Yu DW (2006) Assembling a mutualism: ant symbionts locate their host plants by detecting volatile chemicals. Insectes Soc 53:172–176
Edwards DP, Arauco R, Hassall M, Sutherland WJ, Chamberlain K, Wadhams LJ, Yu DW (2007) Protection in an ant – plant mutualism: an adaptation or a sensory trap? Anim Behav 74:377–385
Edwards DP, Frederickson ME, Shepard GH, Yu DW (2009) A plant needs ants like a dog needs fleas: *Myrmelachista schumanni* ants gall many tree species to create housing. Am Nat 174:734–740
Endler A, Liebig J, Schmitt T, Parker JE, Jones GR, Schreier P, Hölldobler B (2004) Surface hydrocarbons of queen eggs regulate worker reproduction in a social insect. Proc Natl Acad Sci USA 101:2945–2950
Endler A, Liebig J, Hölldobler B (2006) Queen fertility, egg marking and colony size in the ant *Camponotus floridanus*. Behav Ecol Sociobiol 59:490–499
Federle W, Maschwitz U, Fiala B (1998) The two-partner ant-plant system of *Camponotus* (*Colobopsis*) sp 1 and *Macaranga puncticulata* (Euphorbiaceae): natural history of the exceptional ant partner. Insectes Soc 45:1–16
Federle W, Maschwitz U, Hölldobler B (2002) Pruning of host plant neighbours as defence against enemy ant invasions: *Crematogaster* ant partners of *Macaranga* protected by "wax barriers" prune less than their congeners. Oecologia 132:264–270
Fiala B, Maschwitz U (1990) Studies on the south east Asian ant–plant association *Crematogaster borneensis*/*Macaranga* – adaptations of the ant partner. Insectes Soc 37:212–231
Fiala B, Maschwitz U (1991) Extrafloral nectaries in the genus *Macaranga* (Euphorbiaceae) in Malaysia: comparative studies of their possible significance as predispositions for myrmecophytism. Biol J Linn Soc 44:287–305

Fiala B, Maschwitz U, Pong TY, Helbig AJ (1989) Studies of a South East Asian ant–plant association: protection of *Macaranga* trees by *Crematogaster borneensis*. Oecologia 79:463–470
Fiala B, Linsenmair KE, Maschwitz U (1994) Diversität von Interaktionen zwischen Ameisen und Pflanzen im südostasiatischen Regenwald. Andrias 13:169–178
Fischer RC, Richter A, Wanek W, Mayer V (2002) Plants feed ants: food bodies of myrmecophytic *Piper* and their significance for the interaction with *Pheidole bicornis* ants. Oecologia 133:186–192
Fischer RC, Wanek W, Richter A, Mayer V (2003) Do ants feed plants? A 15 N labelling study of nitrogen fluxes from ants to plants in the mutualism of *Pheidole* and *Piper*. J Ecol 91:126–134
Fonseca CR (1993) Nesting space limits colony size of the plant–ant *Pseudomyrmex concolor*. Oikos 67:473–482
Fonseca CR (1999) Amazonian ant–plant interactions and the nesting space limitation hypothesis. J Trop Ecol 15:807–825
Fonseca CR, Benson WW (2003) Ontogenetic succession in Amazonian ant trees. Oikos 102:407–412
Frederickson ME (2006) The reproductive phenology of an Amazonian ant species reflects the seasonal availability of its nest sites. Oecologia 149:418–427
Frederickson ME, Gordon DM (2009) The intertwined population biology of two Amazonian myrmecophytes and their symbiotic ants. Ecology 90:1595–1607
Frederickson ME, Greene MJ, Gordon DM (2005) "Devil's gardens" bedevilled by ants. Nature 437:495–496
Gaume L, McKey D (1998) Protection against herbivores of the myrmecophyte *Leonardoxa africana* (Baill.) Aubrèv. T3 by its principal ant inhabitant *Aphomomyrmex afer* Emery. C R Acad Sci Ser III Sci Vie 321:593–601
Gaume L, McKey D (1999) An ant–plant mutualism and its host-specific parasite: activity rhythms, young leaf patrolling, and effects on herbivores of two specialist plant–ants inhabiting the same myrmecophyte. Oikos 84:130–144
Gaume L, McKey D, Anstett MC (1997) Benefits conferred by "timid" ants: active anti-herbivore protection of the rainforest tree *Leonardoxa africana* by the minute ant *Petalomyrmex phylax*. Oecologia 112:209–216
Gianoli E, Sendoya S, Vargas F, Mejia P, Jaffe R, Rodriguez M, Gutierrez A (2008) Patterns of *Azteca* ants' defence of *Cecropia* trees in a tropical rainforest: support for optimal defence theory. Ecol Res 23:905–908
Gonzalez-Teuber M, Heil M (2009) The role of extrafloral nectar amino acids for the preferences of facultative and obligate ant mutualists. J Chem Ecol 35:459–468
Grangier J, Dejean A, Male PJ, Orivel J (2008) Indirect defense in a highly specific ant–plant mutualism. Naturwissenschaften 95:909–916
Greenberg L, Troger AG, Francke W, McElfresh JS, Topoff H, Aliabadi A, Millar JG (2007) Queen sex pheromone of the slave-making ant, *Polyergus breviceps*. J Chem Ecol 33:935–945
Greene MJ, Gordon DM (2007) Structural complexity of chemical recognition cues affects the perception of group membership in the ants *Linephithema humile* and *Aphaenogaster cockerelli*. J Exp Biol 210:897–905
Greenfield MD (2002) Signalers and receivers: mechanisms and evolution of arthropod communication. Oxford University Press, Oxford
Guerrieri FJ, Nehring V, Jorgensen CG, Nielsen J, Galizia CG, d'Ettorre P (2009) Ants recognize foes and not friends. Proc R Soc B 276:2461–2468
Gullan PJ (1997) Relationships with ants. In: Ben-Dov Y, Hodgson CJ (eds) Soft scale insects: their biology, natural enemies and control. Elsevier Science, Amsterdam, pp 351–373
Hadley NF (1994) Water relations of terrestrial arthropods. Academic, San Diego
Halitschke R, Stenberg JA, Kessler D, Kessler A, Baldwin IT (2008) Shared signals – 'alarm calls' from plants increase apparency to herbivores and their enemies in nature. Ecol Lett 11:24–34
Heil M (2008) Indirect defence via tritrophic interactions. New Phytol 178:41–61

Heil M, McKey D (2003) Protective ant–plant interactions as model systems in ecological and evolutionary research. Annu Rev Ecol Evol Syst 34:425–453

Heil M, Silva Bueno JC (2007) Within-plant signaling by volatiles leads to induction and priming of an indirect plant defense in nature. Proc Natl Acad Sci USA 104:5467–5472

Heil M, Fiala B, Kaiser W, Linsenmair KE (1998) Chemical contents of *Macaranga* food bodies: adaptations to their role in ant attraction and nutrition. Funct Ecol 12:117–122

Heil M, Fiala B, Maschwitz U, Linsenmair KE (2001) On benefits of indirect defence: short- and long-term studies of antiherbivore protection via mutualistic ants. Oecologia 126:395–403

Heil M, Baumann B, Kruger R, Linsenmair KE (2004a) Main nutrient compounds in food bodies of Mexican *Acacia* ant–plants. Chemoecology 14:45–52

Heil M, Feil D, Hilpert A, Linsenmair KE (2004b) Spatiotemporal patterns in indirect defence of a South-East Asian ant-plant support the optimal defence hypothesis. J Trop Ecol 20:573–580

Heil M, Greiner S, Meimberg H, Kruger R, Noyer JL, Heubl G, Linsenmair KE, Boland W (2004c) Evolutionary change from induced to constitutive expression of an indirect plant resistance. Nature 430:205–208

Heinze J, Foitzik S, Hippert A, Hölldobler B (1996) Apparent dear-enemy phenomenon and environment-based recognition cues in the ant *Leptothorax nylanderi*. Ethology 102:510–522

Hölldobler B (1995) The chemistry of social regulation: multicomponent signals in ant societies. Proc Natl Acad Sci USA 92:19–22

Hölldobler B, Carlin NF (1987) Anonymity and specificity in the chemical communication signals of social insects. J Comp Physiol A Sens Neural Behav Physiol 161:567–581

Hölldobler B, Wilson EO (1990) The ants. Belknap, Cambridge

Howard RW, Blomquist GJ (2005) Ecological, behavioral, and biochemical aspects of insect hydrocarbons. Annu Rev Entomol 50:371–393

Inui Y, Itioka T (2007) Species-specific leaf volatile compounds of obligate *Macaranga* myrmecophytes and host-specific aggressiveness of symbiotic *Crematogaster* ants. J Chem Ecol 33:2054–2063

Inui Y, Itioka T, Murase K, Yamaoka R, Itino T (2001) Chemical recognition of partner plant species by foundress ant queens in *Macaranga–Crematogaster* myrmecophytism. J Chem Ecol 27:2029–2040

Janzen DH (1967) Interaction of the bull's-horn acacia (*Acacia cornigera* L.) with an ant inhabitant (*Pseudomyrmex ferruginea* F. Smith) in Eastern Mexico. Kans Univ Sci Bull 47:315–558

Janzen DH (1969) Allelopathy by myrmecophytes: the ant *Azteca* as an allelopathic agent of *Cecropia*. Ecology 50:147–153

Janzen DH (1972) Protection of *Barteria* (Passifloraceae) by *Pachysima* ants (Pseudomyrmecinae) in a Nigerian rain-forest. Ecology 53:885–892

Janzen DH (1973) Evolution of polygynous obligate *Acacia*-ants in Western Mexico. J Anim Ecol 42:727–750

Jolivet P (1996) Ants and plants, an example of coevolution. Backhuys, Leiden

Jürgens A (2004) Flower scent composition in diurnal *Silene* species (Caryophyllaceae): phylogenetic constraints or adaption to flower visitors? Biochem Syst Ecol 32:841–859

Jürgens A, Feldhaar H, Feldmeyer B, Fiala B (2006) Chemical composition of leaf volatiles in *Macaranga* species (Euphorbiaceae) and their potential role as olfactory cues in host-localization of foundress queens of specific ant partners. Biochem Syst Ecol 34:97–113

Kalberer NM, Turlings TCJ, Rahier M (2001) Attraction of a leaf beetle (*Oreina cacaliae*) to damaged host plants. J Chem Ecol 27:647–661

Karban R, Baldwin IT, Baxter KJ, Laue G, Felton GW (2000) Communication between plants: induced resistance in wild tobacco plants following clipping of neighboring sagebrush. Oecologia 125:66–71

Karban R, Maron J, Felton GW, Ervin G, Eichenseer H (2003) Herbivore damage to sagebrush induces resistance in wild tobacco: evidence for eavesdropping between plants. Oikos 100:325–332

Kessler A, Baldwin IT (2001) Defensive function of herbivore-induced plant volatile emissions in nature. Science 291:2141–2144
Kleinfeldt SE (1978) The Interaction of *Codonanthe crassifolia* (Gesneriaceae) and *Crematogaster longispina* (Formicidae). Ecology 59:449–456
Kohl E, Hölldobler B, Bestmann HJ (2000) A trail pheromone component of the ant *Mayriella overbecki* Viehmeyer (Formicidae: Myrmicinae). Naturwissenschaften 87:320–322
Lahav S, Soroker V, Hefetz A (1999) Direct behavioral evidence for hydrocarbons as ant recognition discriminators. Naturwissenschaften 86:246–249
Lapola DM, Bruna EM, Vasconcelos HL (2003) Contrasting responses to induction cues by ants inhabiting *Maieta guianensis* (Melastomataceae). Biotropica 35:295–300
Lemaire M, Nagnan P, Clement JL, Lange C, Peru L, Basselier JJ (1990) Geranyllinalool (diterpene alcohol) an insecticidal component of pine wood and termites (Isoptera: Rhinotermitidae) in four European ecosystems. J Chem Ecol 16:2067–2079
Lenoir A, Fresneau D, Errard C, Hefetz A (1999) Individual and colonial identity in ants: the emergence of the social representation concept. In: Detrain C, Deneubourg JL, Pasteels JM (eds) Information processing in social insects. Birkhauser, Bâle, pp 219–237
Letourneau DK (1983) Passive aggression: an alternative hypothesis for the *Piper–Pheidole* association. Oecologia 60:122–126
Letourneau DK (1998) Ants, stem-borers, and fungal pathogens: experimental tests of a fitness advantage in *Piper* ant–plants. Ecology 79:593–603
Levin RA, McDade LA, Raguso RA (2003) The systematic utility of floral and vegetative fragrance in two genera of Nyctaginaceae. Syst Biol 52:334–351
Liang D, Silverman J (2000) "You are what you eat": diet modifies cuticular hydrocarbons and nestmate recognition in the Argentine ant, *Linepithema humile*. Naturwissenschaften 87:412–416
Madden D, Young TP (1992) Symbiotic ants as an alternative defense against giraffe herbivory in spinescent *Acacia drepanolobium*. Oecologia 91:235–238
Mayer V, Schaber D, Hadacek F (2008) Volatiles of myrmecophytic *Piper* plants signal stem tissue damage to inhabiting *Pheidole* ant-partners. J Ecol 96:962–970
McKey D (1974a) Ant–plants: selective eating of an unoccupied barteria by a *Colobus* monkey. Biotropica 6:269–270
McKey D (1974b) Adaptive patterns in alkaloid physiology. Am Nat 108:305–320
McKey D (1984) Interaction of the ant–plant *Leonardoxa africana* (Caesalpiniaceae) with its obligate inhabitants in rainforest in Cameroon. Biotropica 16:81–99
McKey D (2000) *Leonardoxa africana* (Leguminosae: Caesalpinioideae): a complex of mostly allopatric subspecies. Adansonia 22:71–109
McKey D, Davidson DW (1993) Ant-plant symbioses in Africa and the neotropics: history, biogeography and diversity. In: Goldblatt P (ed) Biological relationships between Africa and South America. Yale University Press, New Haven, pp 568–606
McKitrick MC (1993) Phylogenetic constraint in evolutionary theory: has it any explanatary power? Annu Rev Ecol Syst 24:307–330
McNeill J, Stuessy TF, Turland NJ, Horandl E (2005) XVII International Botanical Congress: preliminary mail vote and report of Congress action on nomenclature proposals. Taxon 54:1057–1064
Metlen KL, Aschehoug ET, Callaway RM (2009) Plant behavioural ecology: dynamic plasticity in secondary metabolites. Plant Cell Environ 32:641–653
Moog J, Drude T, Maschwitz U (1998) Protective function of the plant–ant *Cladomyrma maschwitzi* to its host, *Crypteronia griffithii*, and the dissolution of the mutualism (Hymenoptera: Formicidae). Sociobiology 31:105–129
Moraes SC, Vasconcelos HL (2009) Long-term persistence of a neotropical ant–plant population in the absence of obligate plant–ants. Ecology 90:2375–2383
Morawetz W, Henzl M, Wallnofer B (1992) Tree killing by herbicide producing ants for the establishment of pure *Tococa occidentalis* populations in the Peruvian Amazon. Biodivers Conserv 1:19–33

Morgan ED (2009) Trail pheromones of ants. Physiol Entomol 34:1–17
Morgan ED, Jackson BD, Ollett DG, Sales GW (1990) Trail pheromone of the ant *Tetramorium impurum* and model compounds: structure–activity comparisons. J Chem Ecol 16:3493–3510
Ness JH, Morris WF, Bronstein JL (2009) For ant-protected plants, the best defense is a hungry offense. Ecology 90:2823–2831
Orivel J, Dejean A (1999) Selection of epiphyte seeds by ant-garden ants. Ecoscience 6:51–55
Passera L, Aron S (2005) Les fourmis: comportement, organisation sociale et évolution. Les Presses scientifiques du CNRC, Ottawa
Passera L, Lachaud JP, Gomel L (1994) Individual food source fidelity in the neotropical ponerine ant *Ectatomma ruidum* Roger (*Hymenoptera Formicidae*). Ethol Ecol Evol 6:13–21
Pichersky E, Noel JP, Dudareva N (2006) Biosynthesis of plant volatiles: nature's diversity and ingenuity. Science 311:808–811
Powell W, Pennacchio F, Poppy GM, Tremblay E (1998) Strategies involved in the location of hosts by the parasitoid *Aphidius ervi* Haliday (Hymenoptera: Braconidae: Aphidiinae). Biol Control 11:104–112
Raine NE, Willmer P, Stone GN (2002) Spatial structuring and floral avoidance behavior prevent ant-pollinator conflict in a Mexican ant-*Acacia*. Ecology 83:3086–3096
Rasmann S, Kollner TG, Degenhardt J, Hiltpold I, Toepfer S, Kuhlmann U, Gershenzon J, Turlings TCJ (2005) Recruitment of entomopathogenic nematodes by insect-damaged maize roots. Nature 434:732–737
Raven JA, Edwards D (2004) Physiological evolution of lower embryophytes: adaptations to the terrestrial environment. In: Hemsley AR, Poole I (eds) The evolution of plant physiology. Elsevier, Amsterdam, pp 17–41
Regnier FE, Wilson EO (1968) The alarm-defence system of the ant *Acanthomyops claviger*. J Insect Physiol 14:955–970
Renner SS, Ricklefs RE (1998) Herbicidal activity of domatia-inhabiting ants in patches of *Tococa guianensis* and *Clidemia heterophylla*. Biotropica 30:324–327
Rhoades DF (1979) Evolution of plant chemical defense against herbivores. In: Rosenthal GA, Janzen DH (eds) Herbivores. their interaction with secondary plant metabolites. Academic, New York, pp 1–55
Rico-Gray V, Oliveira PS (2007) The ecology and evolution of ant–plant interactions. The University of Chicago Press, Chicago
Risch SJ, Rickson FR (1981) Mutualismin which ants must be present before plants produce fodd bodies. Nature 291:149–150
Romero GQ, Izzo TJ (2004) Leaf damage induces ant recruitment in the Amazonian ant–plant *Hirtella myrmecophila*. J Trop Ecol 20:675–682
Rosumek FB, Silveira FAO, Neves FD, Barbosa NPD, Diniz L, Oki Y, Pezzini F, Fernandes GW, Cornelissen T (2009) Ants on plants: a meta-analysis of the role of ants as plant biotic defenses. Oecologia 160:537–549
Sasso R, Iodice L, Woodcock CM, Pickett JA, Guerrieri E (2009) Electrophysiological and behavioural responses of *Aphidius ervi* (Hymenoptera: Braconidae) to tomato plant volatiles. Chemoecology 19:195–201
Schatz B, Djieto-Lordon C, Dormont L, Bessiere JM, McKey D, Blatrix R (2009) A simple nonspecific chemical signal mediates defence behaviour in a specialised ant–plant mutualism. Curr Biol 19:R361–R362
Seidel JL, Epstein WW, Davidson DW (1990) Neotropical ant gardens. I. Chemical constituents. J Chem Ecol 16:1791–1816
Shulaev V, Silverman P, Raskin I (1997) Airborne signalling by methyl salicylate in plant pathogen resistance. Nature 385:718–721
Sorvari J, Theodora P, Turillazzi S, Hakkarainen H, Sundström L (2008) Food resources, chemical signaling, and nest mate recognition in the ant *Formica aquilonia*. Behav Ecol 19:441–447
Stanton ML, Palmer TM, Young TP, Evans A, Turner ML (1999) Sterilization and canopy modification of a swollen thorn acacia tree by a plant–ant. Nature 401:578–581

Takabayashi J, Takahashi S, Dicke M, Posthumus MA (1995) Developmental stage of herbivore *Pseudaletia separata* affects production of herbivore induced synomone by corn plants. J Chem Ecol 21:273–287
Todd JL, Baker TC (1999) Function of peripheral olfactory organs. In: Hansson BS (ed) Insect olfaction. Springer, Berlin, pp 67–96
Turlings TCJ, Tumlinson JH (1992) Systemic release of chemical signals by herbivore-injured corn. Proc Natl Acad Sci USA 89:8399–8402
Turlings TCJ, Wäckers F (2004) Recruitment of predators and parasitoids by herbivore-injured plants. In: Cardé RT, Millar JG (eds) Advances in insect chemical ecology. Cambridge University Press, Cambridge, pp 21–75
Turlings TCJ, Tumlinson JH, Lewis WJ (1990) Exploitation of herbivore-induced plant odors by host-seeking parasitic wasps. Science 250:1251–1253
Turlings TCJ, Loughrin JH, McCall PJ, Rose USR, Lewis WJ, Tumlinson JH (1995) How caterpillar-damaged plants protect themselves by attracting parasitic wasps. Proc Natl Acad Sci USA 92:4169–4174
Vantaux A, Dejean A, Dor A, Orivel J (2007) Parasitism versus mutualism in the ant-garden parabiosis between *Camponotus femoratus* and *Crematogaster levior*. Insectes Soc 54:95–99
Vasconcelos HL (1993) Ant colonization of *Maieta guianensis* seedlings, an Amazon ant–plant. Oecologia 95:439–443
Wäckers FL, Zuber D, Wunderlin R, Keller F (2001) The effect of herbivory on temporal and spatial dynamics of foliar nectar production in cotton and castor. Ann Bot 87:365–370
Webber BL, McKey D (2009) Cyanogenic myrmecophytes, redundant defence mechanisms and complementary defence syndromes: revisiting the neotropical ant-acacias. New Phytol 182:792–794
Wickler W (1965) Mimicry and evolution of animal communication. Nature 208:519–521
Wiklund C (1984) Egg-laying patterns in butterflies in relation to their phenology and the visual apparency and abundance of their host plants. Oecologia 63:23–29
Youngsteadt E, Nojima S, Haberlein C, Schulz S, Schal C (2008) Seed odor mediates an obligate ant-plant mutualism in Amazonian rainforests. Proc Natl Acad Sci USA 105:4571–4575
Youngsteadt E, Alvarez Baca J, Osborne J, Schal C (2009) Species-specific seed dispersal in an obligate ant–plant mutualism. Plos One 4:e4335
Yumoto T, Maruhashi T (1999) Pruning behavior and intercolony competition of *Tetraponera* (*Pachysima*) *aethiops* (Pseudomyrmecinae, Hymenoptera) in *Barteria fistulosa* in a tropical forest, Democratic Republic of Congo. Ecol Res 14:393–404
Zhu JW, Park KC (2005) Methyl salicylate, a soybean aphid-induced plant volatile attractive to the predator *Coccinella septempunctata*. J Chem Ecol 31:1733–1746

Photosensory Cues in Plant–Plant Interactions: Regulation and Functional Significance of Shade Avoidance Responses

Diederik H. Keuskamp and Ronald Pierik

Abstract Plants growing in dense vegetations compete with their neighbors for resources such as water, nutrients, and light. Particularly, competition for light has been thoroughly studied, both for fitness consequences as well as the adaptive behaviors that plants display to win the battle for light interception. Aboveground, plants detect their competitors through photosensory cues, notably the red:far-red light ratio (R:FR). The R:FR is a very reliable indicator of future competition as it decreases in a plant-specific manner through red light absorption for photosynthesis and is sensed with the phytochrome photoreceptors. In addition, also blue light depletion is perceived for neighbor detection. As a response to these light signals, plants display a suite of phenotypic traits defined as the shade avoidance syndrome (SAS). The SAS helps to position the photosynthesizing leaves in the higher zones of a canopy where light conditions are more favorable. In this chapter, we discuss the physiological control mechanisms through which the photosensory signals are transduced into the adaptive phenotypic responses that make up the SAS. Using this mechanistic knowledge as a starting point, we discuss how the SAS functions in the context of the complex multifacetted environments that plants usually grow in. Special attention is paid to trade-offs between SAS and defense against attackers, since recent breakthrough papers have elucidated some of the mechanisms behind this interaction.

1 Introduction

Plants usually grow in dynamic environments with oftentimes severe competition over limited resources with surrounding neighbors. To deal with the limitations in

D.H. Keuskamp and R. Pierik (✉)
Plant Ecophysiology, Institute of Environmental Biology, Utrecht University, Padualaan 8, Utrecht 3584 CH, The Netherlands
e-mail: r.pierik@uu.nl

F. Baluška and V. Ninkovic (eds.), *Plant Communication from an Ecological Perspective*, Signaling and Communication in Plants,
DOI 10.1007/978-3-642-12162-3_10,

resources such as water, nutrients, and light, plants display phenotypic responses to neighboring plants to maximize resource capture in dense canopies. To do so, plants have to sense the vicinity of neighbors. Perception of neighbors and/or the abiotic stresses, such as altered light levels that come with them, may lead to different types of behavior: shade tolerance, shade avoidance, or confrontation (e.g., allelopathy in roots) (Novoplansky 2009). The ability of plants to develop different phenotypes in response to environmental cues of (future) selective conditions is an important determinant of plant performance and ultimately plant fitness. To respond appropriately to neighboring, plants use reliable external signals for the presence of neighbors and internal receptor systems to perceive and process these signals. Perception of such neighbor detection signals can induce rapid changes in gene expression and physiological processes, which regulate the phenotypic plasticity required to competitively acquire resources.

Phenotypic plasticity is thought to be an essential feature of plants in response to their environment, which is usually dynamic for various aspects simultaneously. In summary, plants use a wide variety of external cues and internal perception mechanisms, which are subsequently integrated through cross-talk at the signal transduction level, leading to an integral phenotypic outcome. Examples of environmental signals that can be used as input signals for the presence of neighboring competitors are light quality (Vandenbussche et al. 2005; Franklin 2008), plant-produced volatile organic compounds (Pierik et al. 2004a; Kegge and Pierik 2010), nutrients (Schenk 2006; de Kroon 2007), and root exudates (Bais et al. 2006). The mechanistic regulation and functional importance of plant neighbor detection through light quality signals is probably the best studied example of plant phenotypic plasticity. In this chapter, we review the current knowledge on what is perhaps the most important and wide-spread behavior of plants under competition: the Shade Avoidance Syndrome (SAS). We review the recent progress on molecular and physiological regulation of the SAS and the current understanding of its adaptive significance for plant competition.

2 The Shade Avoidance Syndrome

Light is the main source of energy for plants. When plants grow side by side, leaves will eventually overlap and shade each other, leading to competition for light. Competition for light is also an important factor determining the biodiversity of dense plant communities (Hautier et al. 2009).

Next to being a source of energy, light is also a source of information that plants can respond to. Most plants are able to react to the direction, intensity, composition, and duration of light. These light components regulate such features as seed germination, photomorphogenesis, flowering time and the SAS (Borthwick et al. 1952; Terzaghi and Cashmore 1995; Guo et al. 1998). The SAS encompasses various phenotypic traits (Fig. 1), including elongation of internodes, petioles and hypocotyls, apical dominance, early flowering and upward leaf movement (hyponasty)

Fig. 1 Shade avoidance responses in *Petunia X hybrida*, *Nicotiana tabacum* CV Samsun NN, *Rumex palustris* and *Arabidopsis thaliana* accession Col-0. For each species, the plant on the left represents a control plant and the plant on the right represents a plant under low R:FR conditions. Note that low R:FR-exposed plants display more vertical leaf angles, elongated stems and/or petioles and early flowering

(Franklin 2008). These growth and developmental responses help plants to outgrow the shade imposed by neighbors, thus allowing them to position the photosynthesizing young leaves in the better lit, upper parts of the vegetation. Phenotypic traits that are also observed in shaded plants, but that are not part of the SAS, include an increase of Specific Leaf Area (SLA) and a decrease of the chlorophyll a/b ratio (Evans and Poorter 2001). These latter two responses are thought to maximize light harvesting under shaded conditions and constitute shade tolerance, rather than shade avoidance. For plants to display the SAS under the appropriate conditions, reliable cues for (future) crowding are essential.

Importantly, in a canopy there is not only a reduction in the light availability, but also a dramatic change in light composition. Unlike any other object, plants typically do not absorb far-red (FR) light ($\lambda = 700$–800 nm) whilst strongly absorbing red (R: $\lambda = 600$–700 nm) and blue (B: $\lambda = 400$–500 nm) light for photosynthesis (Franklin 2008). The photon fluence rate of B light is a reliable indicator of light intensity, whereas the red to far-red ratio (R:FR) declines as more R light is absorbed by leaves. Plants have the capacity to respond to both these light signals. A long-standing paradigm in plant biology predicts that the SAS is induced by detection of a reduced R:FR, indicative of proximate vegetation (reviewed in Franklin (2008)). However, studies on a variety of species have shown that also reduction of B light photon fluence rates in the incident light can induce pronounced shade avoidance responses (Ballaré et al. 1991; Casal and Sanchez 1994; Pierik et al. 2004a; Djakovic-Petrovic et al. 2007).

Leaves even reflect a relatively large amount of FR light. The lowering of the R:FR is therefore an accurate and early indicator of neighbor proximity even in stages of vegetation development where leaf overlap and shading is not yet occurring (Ballaré et al. 1991). A low R:FR is, therefore, considered an early warning signal for upcoming competition for light. It is likely that a simultaneous occurrence of low R:FR and low B is used to evaluate actual shade, and thus competition, by neighbors (e.g., Sasidharan et al. 2008).

3 Photoreceptors

All higher plant species studied so far have photoreceptors tailored to detecting B, R and FR light cues. The model plant *Arabidopsis thaliana* for example has three families of photoreceptors. The cryptochrome and phototropin families of photoreceptors are sensitive to B light fluence rates, whereas phytochromes are mostly sensitive to R and FR light (Ahmad and Cashmore 1993; Quail et al. 1995; Briggs et al. 2001).

3.1 Cryptochromes

Cryptochromes are the major blue light receptors involved in stimulating hypocotyl elongation of light-grown seedlings exposed to B light depletion (Pierik et al. 2009), but they also regulate de-etiolation of seedling and entrainment of the circadian clock (Somers et al. 1998; Lin and Shalitin 2003). Cryptochromes are not only sensitive to B light but they also perceive UV/A light (Lin and Shalitin 2003). It has been shown for *A. thaliana* cryptochromes 1 (CRY1) and 2 (CRY2), that these proteins are phosphorylated upon blue light exposure and this autophosporylation affects both their activity and stability (Lin and Shalitin 2003). CRY2 is localized to the nucleus, but the exposure of *A. thaliana* seedlings to B light leads to a rapid degradation of CRY2, whereas CRY1 is much more light-stable and acts at higher fluence rates (Lin and Shalitin 2003). In a microarray study on de-etiolation in *A. thaliana* by Ma et al. (2001) it was observed that most of the genes that were regulated in wild-type plants upon B light exposure were not differentially regulated anymore in the *cry1 cry2* double mutant under the same light conditions. In either of the single mutants there were still pronounced B-induced changes in gene expression (Ohgishi et al. 2004) indicating that the two cryptochromes are partially redundant but are both responsible for B light-mediated de-etiolation. Pierik et al. (2009) showed that light-grown *Arabidopsis* seedlings display increased hypocotyl elongation upon B light depletion of the incident light, and that, similar to de-etiolation, this response is abolished only in the *cry1 cry2* double mutant. This confirms that for the B light-mediated shade avoidance the two cryptochromes are the main photoreceptors that modify the shade avoidance response.

Upon activation by B light CRYs have been shown to be able to bind to the downstream factor constitutively photomorphogenic 1 (COP1) (Wang et al. 2001). COP1 acts as an E3 Ub ligase and is a key repressor of photomorphogenesis. COP1 is located in the nucleus where it can interact with the transcription factor long hypocotyls 5 (HY5), and this interaction will lead to the ubiquitination and subsequent degradation of HY5 (Osterlund et al. 2000). In this way cry can regulate light responses through HY5 abundance. In addition, cryptochromes can also directly interact with the phytochromes (Mas et al. 2000)

3.2 Phototropins

Phototropins are photoreceptors which are sensitive to blue (390–500 nm) and ultraviolet-A (320–390 nm) light. They are involved in a set of responses to optimize light harvesting and growth promotion, like phototropism, chloroplast movements, B-induced inhibition of hypocotyl elongation, cotyledon expansion, leaf expansion and light-stimulated leaf movement (reviewed in Christie (2007)). Of these responses, particularly phototropism is an important phenomenon for optimization of light capture in dense stands, as it will guide plant organs to grow differentially towards better lit conditions, for example at the border of a dense field.

Phototropins belong to the AGC family of kinases, containing two very similar photosensory domains named LOV1 and LOV2 and a kinase domain. The photo-excitation of these LOV domains upon blue light leads to autophosphorylation of phototropin. In darkness phototropin is not phosphorylated and inactive, but the LOV domains cycle between active (LOV_{390}), which has an absorption peak at 390 nm, and the inactive (LOV_{447}) state depending on the light condition. Phototropins are hydrophilic and in darkness are situated in the plasma membrane. Upon B light radiation Phototropin1 (PHOT1) and PHOT2 are rapidly internalized or transported to the golgi apparatus, respectively. PHOT1 is active at low fluence rates ($<1\ \mu mol\ m^{-2}\ s^{-1}$ blue light), whereas PHOT2 functions at higher B photon fluencies (reviewed in Christie (2007)).

Little is known about the targets of the kinase activity of phototropin but with the use of mutations specifically in the kinase domain it was found that the kinase activity is essential (Huala et al. 1997). An example of a phototropin interacting protein is NPH3 (nonphototropic hypocotyls 3), which is involved in auxin distribution during phototropism, and NPH3 function depends on B-induced PHOT phosphorylation (Pedmale and Liscum 2007). More recently, studies demonstrated that the phytochrome signaling component PKS (phytochrome kinase substrate) protein is required for hypocotyl phototropism in Arabidopsis and PKS1 can form a complex with PHOT1 and NPH3 (Lariguet et al. 2006). Since phytochromes (and cryptochromes) influence phototropic curvature in *A. thaliana* as well (Janoudi et al. 1997), PKS proteins may constitute a link between the photoreceptor families.

3.3 Phytochromes

As mentioned above, phytochromes are sensitive to R and FR light. Phytochromes exist in two different conformation states, the inactive Pr form and the active Pfr form, with somewhat different light absorption characteristics (Quail et al. 1995). The biologically inactive form of phytochrome (Pr) has an absorption peak at 666 nm and is therefore sensitive for R light. When R light is absorbed this will lead to a photoconversion of Pr into Pfr, which is the active form of phytochrome. The Pfr has an absorption peak at 730 nm (FR light) and upon absorbing FR light it will photoconvert back into the inactive Pr form. In this way the R:FR of light reaching the plant will determine the equilibrium between Pr and Pfr forms of phytochrome in plants. Relative phytochrome activity (Pfr:Pr) thus is a direct function of the R:FR. Upon activation, Pfr translocates to the nucleus where it can interact with its molecular regulatory partners. In Arabidopsis five different phytochromes have been characterized, which are named PHYA-E. Of these, PHYB is the main regulator for low R:FR-mediated shade avoidance, although PHYD and PHYE are also involved (Robson et al. 1993; Franklin 2008). In various species, such as *A. thaliana*, *Brassica rapa* (turnip mustard) and *Lycopersicon esculentum* (tomato), mutants which are deficient in phyB display a constitutive shade avoidance phenotype. (Somers et al. 1991; Devlin et al. 1992; Robson et al. 1993; Reed et al. 1993). An important phytochrome-mediated regulatory pathway is through interaction with PIF (Phytochrome Interacting Factors) proteins. PIF and PIF-like (PIL) proteins are a subfamily of the bHLH (basic helix–loop-helix) family of transcription factors that bind DNA to regulate gene transcription as part of the phytochrome signal transduction (Duek and Fankhauser 2005). At least PIF3, 4, 5 and 7 are important regulators of phytochrome-mediated light responses (Huq and Quail 2002; Fujimori et al. 2004; Nozue et al. 2007; Leivar et al. 2008). A subset of these are even specific to shade avoidance responses, such as shown for PIL1 (PIF3-like 1) and HFR1 (long hypocotyl in far-red 1) (Fairchild et al. 2000; Fankhauser and Chory 2000; Salter et al. 2003). In addition to the interaction with PIFs, PHYB can also bind directly to CRY2 (Mas et al. 2000), mediating cross-talk with the B light signaling pathway.

4 Hormonal Regulation

PIFs seem to be an important player in plant responses to the environment and the hormonal regulation of these responses. They play a key role in modulating developmental responses to both light and temperature (Lorrain et al. 2008; Koini et al. 2009). As indicated earlier, phytochromes migrate to the nucleus when activated by R light, where PHYB can interact with PIFs (Martinez-Garcia et al. 2000). PIF4, for example, is in control of genes mediating cell elongation and is targeted for ubiquitination and subsequent degradation by the binding of PHYB (Quail 2002). It has recently been shown that PIFs can also interact with DELLA proteins (Feng et al. 2008; de Lucas et al. 2008). The DELLA proteins are growth-repressing proteins and a subfamily of the GRAS domain family of transcriptional regulators (Alvey and Harberd 2005). The regulation of these DELLA proteins appears to be key to photomorphogenic responses, including shade avoidance responses (Achard et al. 2007; Djakovic-Petrovic et al. 2007). DELLA abundance is down regulated during shade avoidance in low R:FR and in dense stands, which is essential to prevent DELLAs from inhibiting the SAS (Djakovic-Petrovic et al. 2007). Interestingly, DELLA stability is primarily controlled by the plant hormone gibberellin (GA), thus connecting phytochrome signaling to hormone action. However, DELLA stability is also affected by other hormones, such as auxin and ethylene (Achard et al. 2003; Fu and Harberd 2003; Pierik et al. 2009) and these hormones all are essential regulators of the SAS as well (Morelli and Ruberti 2000; Pierik et al. 2004a, 2009) (Fig. 2). This suggests that DELLA proteins are not only playing a key role in integrating the regulatory effect of PIFs and GA, but are also an integrator of several hormonal signal transduction pathways.

4.1 Gibberellin

Gibberellin (GA) is a key regulator of cell elongation, and for that alone it would be a good candidate for regulation of the SAS. Transgenic tomatoes expressing high

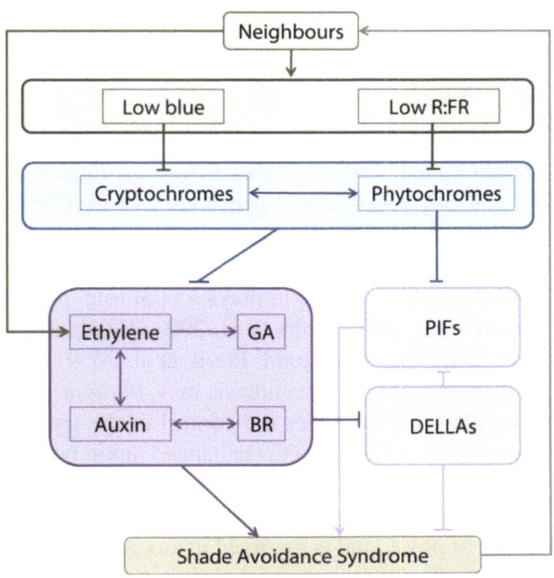

Fig. 2 Schematic representation of some of the signal transduction steps in the process of SAS induction upon detection of neighbors in dense stands

levels of oat phytochrome A, which results in an inhibition of the SAS phenotype (Boylan and Quail 1989), are remarkably similar in phenotype to tomato mutants defective in GA biosynthesis (Koornneef et al. 1990). Both types of plants display a shortened stature, curled leaves, and increased leaf and fruit pigmentation. In contrast to all phytochrome family members, PHYA undergoes rapid light-induced proteolysis and therefore accumulates to high levels only in etiolated seedlings (Clough and Vierstra 1997). A study with tobacco that over expresses oat PHYA confirmed the strong inhibition of the SAS (Robson et al. 1996), including inhibition of internode and petiole elongation, and showed that the SAS could be restored by external GA application (Jordan et al. 1995). In *A. thaliana*, low R:FR promotes the expression of GA-related genes (Devlin et al. 2003) and the constitutively elongated phenotype of *phyb* mutants is suppressed by GA deficiency and GA insensitivity (Peng and Harberd 1997). A direct link between GA and the SAS has been shown for *A. thaliana*, since low R:FR treatment appears to enhance both GA biosynthesis (Hisamatsu et al. 2005) and responsiveness (Reed et al. 1996). In addition, GA-related mutants are less responsive to low B and low R:FR treatment to induce the SAS (Djakovic-Petrovic et al. 2007).

When GA is present, it will bind to its receptor GID1 and will facilitate direct interaction with DELLA proteins, which are subsequently ubiquitinated and targeted for proteasome-mediated degradation (Ueguchi-Tanaka et al. 2005). As mentioned above, DELLA protein abundance is reduced in plants grown in dense stands (Djakovic-Petrovic et al. 2007), confirming earlier data on R:FR controlled GA-levels and DELLA stability. In the absence of GA, DELLA proteins will accumulate to higher levels and interact with PIF3 and PIF4 and prevent these PIFs from regulating gene expression (Feng et al. 2008; de Lucas et al. 2008) associated with cell elongation.

4.2 Auxin

As mentioned above, not only GA-related genes are under the control of R:FR and B light, but auxin related genes are as well (Devlin et al. 2003; Folta et al. 2003; Pierik et al. 2009). Auxin is associated with several processes like embryogenesis, stem cell niche, cell division and cell elongation (Teale et al. 2006) and several studies propose that auxin plays a vital role in the SAS (Morelli and Ruberti 2000; Devlin et al., 2003; Folta et al. 2003; Vandenbussche et al. 2003; Roig-Villanova et al. 2007; Tao et al. 2008; Pierik et al. 2009). Tao et al. (2008) showed a rapid up regulation of auxin biosynthesis in *A. thaliana* seedlings through a dedicated auxin biosynthesis route under the control of the tryptophan aminotransferase of Arabidopsis1 (TAA1) gene. This enhanced auxin production is needed for the hypocotyl to respond to a FR-enrichment treatment to lower R:FR. Another way for light to regulate auxin biosynthesis is through red elongated 1 (RED1), which acts downstream of phyB, and is involved in auxin homeostasis (Hoecker et al. 2004). Morelli and Ruberti (2000) proposed almost 10 years ago that polar auxin transport (PAT) could play an important role in the redistribution of auxin and thereby induce elongation responses, as part of the SAS. This was partly confirmed by Devlin et al. (2003) who showed that not only auxin-related genes like several AUX/IAAs are regulated but also pin-formed3 (PIN3) and PIN7 are up regulated upon far-red light enrichment. The PINs are facilitators of auxin efflux and can thereby determine the direction of PAT (Teale et al 2006). The role of the cellular (re)localization of these PINs under the control of light was shown by Friml et al. (2002) and Blakeslee et al. (2004). It was shown that phototropism in response to B light induces intracellular lateral relocalization of PIN1 and PIN3 proteins, thus producing a differential auxin gradient which induces differential cell elongation. As a result, there will be hypocotyl bending towards the light. The importance of this redistribution of auxin has also been shown for unidirectional shoot elongation as part of the SAS in hypocotyls and petioles of *A. thaliana*. With the use of an auxin activity reporter line *pIAA19::GUS*, it was shown there is an induction of the expression pattern induced upon different light treatments throughout the elongated hypocotyls (Pierik et al. 2009) or petioles. When these plants where treated with the auxin efflux inhibitor NPA (1-naphthylphthalamic), not only the *pIAA19*::GUS pattern was abolished, but also the shade-induced elongation response (Pierik et al. 2009). These data indicate that not only auxin biosynthesis plays an important role in the SAS but also an intact PAT is needed for *A. thaliana* to respond properly to shade.

4.3 Brassinosteroids

Auxin and BR are linked to many of the same growth processes, including vascular differentiation, flower and fruit development, root growth and elongation. Furthermore auxin and BR show a large overlap in genes regulated, suggesting crosstalk between BR and IAA, (Nakamura et al; 2003; Goda et al. 2004; Nemhauser et al. 2004). Luccioni et al. (2002) suggest that BR may fine tune phytochrome-mediated responses. Although this has not been studied yet in a SAS context, light is involved in BR biosynthesis (Kang et al. 2001) and/or BR inactivation in a phytochrome- and cryptochrome-dependent manner (Neff et al. 1999). In addition, BR-related mutants are dark green, slow-growing dwarfs with epinastic leaves and short stems and petioles (Neff et al. 1999), which is the opposite of a SAS phenotype. There are also indications that BR is involved in auxin and ethylene responses in *A. thaliana* (de Grauwe et al. 2005). It is suggested that BR might affect auxin transport in response to light (de Grauwe et al. 2005), which is consistent with the fact that BR can induce the expression of some of the PINs (Nakamura et al. 2004). The role of auxin in the SAS is more clear than the role of BR, but there is a great overlap in the regulatory pathway between these two hormones (Nakamura et al. 2003; Goda et al. 2004; Halliday 2004; Nemhauser et al. 2004) and PIF3 expression is at least partly under the control of BR (Goda et al. 2002) indicating that a role for BR in the SAS is at least possible.

4.4 Ethylene

In several species, ethylene production is stimulated by low R:FR (Finlayson et al. 1999; Pierik et al. 2004b; reviewed in Kegge and Pierik 2010). It was shown that ethylene can accumulate within dense stands of cultivated tobacco in a greenhouse to threefold the ambient concentration and can therefore be a signal for neighbor detection (Pierik et al. 2004a). The elevated ethylene levels reached up to 20 ppb (parts per billion) which was sufficiently high to induce stem elongation and hyponastic leaf growth, two SAS components, in wild-type plants. Ethylene-insensitive transgenic tobacco plants show a reduced and delayed response to neighboring plants and are therefore out competed by wild-type neighbors (Pierik et al. 2003). Although the ethylene-insensitive tobacco plants did show a reasonable response to low R:FR, the response to a reduction in B light fluence rates was entirely absent in these plants (Pierik et al. 2004a). For *A. thaliana*, it seems to be the opposite, where the response to low R:FR light is ethylene dependent, the response of seedlings to B light depletion appears to be independent of ethylene (Pierik et al. 2009).

Ethylene can stimulate auxin production and transport in *A. thaliana* (e.g., Ruzicka et al. 2007). Consistently, for ethylene to stimulate hypocotyl elongation in *A. thaliana*, intact auxin signaling is required (Pierik et al. 2009). Interestingly, auxin can also enhance ethylene production by stimulating the activity of ACC synthase, a precursor of ethylene biosynthesis (Yi et al. 1999).

4.5 SAS Regulation Downstream of the Hormones

Shoot elongation as part of the SAS, will be primarily driven by cellular expansion and therefore by cell wall modifications (Sasidharan et al. 2008). Some of these modifications are brought about by the cell wall protein families of expansins and XTHs (Xyloglucan endotransglucosylases/hydrolases) (Rose et al. 2002; Cosgrove 2005). The activity and gene expression of these protein families are indeed regulated during responses to biotic and abiotic stresses, including shade (Ma et al., 2001; Sessa et al. 2005; Roig-Villanova et al. 2007; Sasidharan et al. 2008; Hornitschek et al. 2009). Furthermore, it has been shown in various species that several members of these protein families are under the control of auxin, GA and BR (Potter and Fry 1993; Zurek et al. 1994; Catala et al. 1997; Goda et al. 2004), ethylene (Rose et al. 1997) and PIFs (Hornitschek et al. 2009). Auxin plays a role in the acidification of the apoplast, where auxin can lower the pH in the cell wall within minutes (Kutschera 1994). This rapid acidification of the apoplast probably sets the optimal pH for expansions (Cosgrove 2005) and XTHs (Fry 1998). We suggest that these protein families are activated during SAS as the downstream targets of the interacting network of hormones mentioned above.

5 Adaptive Value of the SAS in Ecological Context

The actual adaptive value of the SAS can be derived from research where it was shown that elongated plants (thus showing the SAS) have increased fitness compared to non-elongated plants growing at a high density, but a reduced fitness at low densities (Schmitt et al. 1995; Dudley and Schmitt 1995). At the same time, from an agronomic viewpoint, inhibition of elongation responses to neighbors in crop monocultures may actually enhance the harvest index since more carbon will be allocated harvestable organs, rather than to be invested in nonharvestable stems (Robson and Smith 1997).

Not all species show a similarly strong SAS. A relatively steep vertical light gradient in a canopy makes it more likely that a plant can benefit from enhanced light interception by elongation. It will therefore be more effective to show this adaptive behavior in, for example dense grasslands than underneath an over story canopy, where the reduction in light occurs at a greater height and the SAS cannot enable understory plants to escape from these shaded conditions. This is consistent with the observation that in general forest understory plants do not show strong shade avoidance responses to low R:FR (Morgan and Smith 1979; Dudley and Smith 1995; Weinig 2000). The adaptiveness of the SAS is, therefore, contingent upon local conditions of competition. Since plants will usually also experience more than one stress at a time, it is likely that these potential other stresses can also affect the adaptiveness of SAS induction.

Thus, for the SAS to be beneficial to the plant, it should be induced only by light signals that reliably signal crowding, should result in a more favorable place within a canopy leading to higher rates of photosynthesis, and SAS-inducing signals should also be integrated with other stress signals. In this signal integration plant hormones can play an important role as many of the hormones mentioned above are involved in responses to stress, like drought (Leymarie et al. 1996), temperature (Gray et al. 1998), submergence (Bailey-Serres and Voesenek 2008), herbivory and pathogen infection (Robert-Seilaniantz et al. 2007).

5.1 *Local Adaptation in Ecotypes*

Comparative studies have revealed reduced responsiveness to R:FR in species or populations from forested areas, compared to more open habitats (Morgan and Smith 1979; Dudley and Schmitt 1995; Weinig 2000). For example *Impatiens capensis* genotypes from a clearing population display greater elongation responses to low R:FR (Dudley and Schmitt, 1995) and crowding (Donohue et al. 2000) than woodland genotypes under common garden conditions, suggesting local adaptive differentiation in shade avoidance traits. The woodland population of this annual species cannot outgrow the trees and displaying the SAS would thus be maladaptive. Studies on Canadian prairie and alpine ecotypes of *Stellaria longipes*, constitute another interesting example of ecotypic variation for the SAS. The prairie ecotype is naturally subjected to crowding as it grows in dense, competitive vegetations. In contrast, the dwarfed alpine ecotype grows in alpine regions of the Rocky Mountains with very little vegetation in which above-ground competition is almost absent. It was shown by Alokam et al. (2002) and in more detail by Sasidharan et al. (2008) that there is a clear variation in response to different light signals, between the two ecotypes (Fig. 3). Alpine plants show no response to low R:FR, consistent with the lack of competing neighbors in its native habitat, whereas the prairie ecotype displays the classic SAS in low R:FR. When these plants were exposed to real shade, both ecotypes showed a clear internode elongation response, in the alpine ecotype probably representing a response to grow away from the deep shade created by rocky surfaces from which the shoots need to grow into the light (R. Pierik, personal observation). These differences between ecotypes indicate that there is variation in SAS properties between ecotypes within species, probably representing specialization to particular environments.

Fig. 3 Ecotypic variation in *Stellaria longipes* (alpine versus prairie). *Panels A & B*: Native habitat for the alpine (rocky, virtually no competitors) and prairie (dense grassland) ecotype in the Canadian Rocky Mountains. *Panel C*: Shoot elongation responses in the alpine and prairie ecotype to low R:FR and true shade, Copyright American Society of Plant Biologists, www.plantphysiol.org, reproduced with permission from Sasidharan et al. (2008), picture kindly provided by R. Sasidharan

5.2 Environmental Complexity: Integrating SAS and Defense

Plants in their natural environments encounter, in addition to neighboring competitors, a vast array of other biotic stresses imposed by pathogens and herbivores. Interestingly there seems to be a trade-off between resistance against these attackers and the SAS. Herbivory treatment affected plant height, but since this occurred during both control and a SAS-inducing green shade treatment it is possible that herbivory mostly constrained general growth rather than SAS (Kurashige and Agrawal 2005). The SAS, however, does strongly affect plant defense against attackers. For example, growth of pathogens is enhanced in constitutively shade avoiding mutants (Genoud et al. 2002; McGuire and Agrawal 2005; Griebel and Zeier 2008). Furthermore, plants are less resistant against herbivory when grown in crowded stands or when exposed to low R:FR conditions mimicking dense stands (Kurashige and Agrawal 2005; Izaguirre et al. 2006; Moreno et al. 2009). Defense responses to these biotic stresses are primarily not only controlled by the hormones salicylic acid (SA) and jasmonic acid (JA) but also by ethylene. Interestingly, these hormones also affect development and growth (Martínez et al. 2004; Wasternack 2007), and their signaling pathways are partially under the control of the earlier mentioned DELLA proteins (Navarro et al. 2008). In addition, the hormones mentioned earlier to be involved in SAS regulation also interact with plant defense against microbial pathogenesis and herbivory (Robert-Seilaniantz et al. 2007). This reinforces the idea that the interplay between different stresses is due to cross-talk between different hormones (Potters et al. 2007).

It is shown by Navarro et al. (2008)) that DELLA proteins are involved in plant defense regulation. DELLAs appeared to promote the susceptibility to biotrophic bacteria, which is SA-dependent and enhances resistance against necrotrophic fungi, which is JA-dependent. Furthermore it has been shown that during infection DELLA quadruple knockout mutants have higher SA levels and that DELLAs can stimulate the JA signaling pathway (Navarro et al. 2008). Since DELLAs are degraded during low R:FR-induced SAS (Djakovic-Petrovic et al. 2007) it is possible that this would contribute to the earlier mentioned weakened defense against herbivory upon low R:FR (Moreno et al. 2009).

Auxin has also been shown to modulate plant defense responses and can repress the expression of pathogenesis-related (*PR*) genes (e.g., Jouanneau et al. 1991). Recent evidence indicates that SA inhibits pathogen growth partly through the suppression of auxin signaling (reviewed in Kazan and Manners 2009). Interestingly, SA and auxin signaling seem to be mutually antagonistic because auxin suppresses SA-dependent defenses, such as *PR1* expression (e.g., Wang et al. 2007; Park 2007) and SA-deficient plants show increased IAA levels (Abreu and Munne-Bosch 2009). An ultimate test to study if the trade-off between SAS and herbivore defense, involves auxin was performed by Moreno et al. (2009). It was shown, however, that the pronounced trade-off observed was not through auxin, nor through a direct competition for resource allocation between defense and SAS. It appeared that phytochrome signaling affects JA sensitivity in an auxin-independent manner.

The volatile hormone ethylene not only controls shade avoidance, but is also involved in defense responses against pathogens and herbivores (van Loon et al. 2006; von Dahl and Baldwin 2007). It has been shown that attackers lead to enhanced production of ethylene through control of ACC synthase, the rate limiting enzyme in ethylene biosynthesis (de Vos et al. 2005). Exogenous application of ethylene induced the expression of specific genes which are also regulated during attack by necrotrophic pathogens (Penninckx et al. 1998). JA and ethylene together activate specific defense signaling pathways and elevated levels of both JA and ethylene were observed upon, e.g., wounding, herbivore attack or colonization with rhizobacteria. Under these conditions these elevated hormone levels correlate with a reduction in disease symptoms (Rojo et al. 1999). Based on these observations it could be suggested that low R:FR-induced ethylene production would facilitate rather than inhibit defense against herbivores.

6 Conclusions

Plants perceive the threat of competing neighbors through various signals. Aboveground, the most important one seems to be the reduced R:FR, which signals upcoming competition well before actual shading occurs. Plants carry sophisticated photoreceptor systems to signal this and subsequently activate an interacting network of various hormones and transcriptional regulators. The complete signal detection and signal transduction network together defines the SAS to be expressed. This complicated network of interacting molecular and physiological regulators allows for fine-tuned modifications of the response by additional inputs, such as signals coming from defense pathways. The relative advantage of expressing the SAS during competition will thus depend on the presence of additional threats to plant performance, such as herbivory. The current understanding of the physiological and molecular regulation of the SAS is instrumental to understanding how plant behavior in dense, competitive vegetations will be adjusted to additional biotic or abiotic stress factors.

Acknowledgements We thank Rens Voesenek, Mieke de Wit and Rashmi Sasidharan for helpful comments on a draft of this manuscript. The authors are financed by the Netherlands Organization for Scientific Research (Veni grant 863.06.01 to RP) and Utrecht University.

References

Abreu ME, Munne-Bosch S (2009) Salicylic acid deficiency in NahG transgenic lines and sid2 mutants increases seed yield in the annual plant Arabidopsis thaliana. J Exp Bot 60:1261–1271

Achard P, Vriezen WH, Van der Straeten D, Harberd NP (2003) Ethylene regulates *Arabidopsis* development via the modulation of DELLA protein growth repressor function. Plant Cell 15:2816–2825

Achard P, Liao LL, Jiang CF, Desnos T, Bartlett J, Fu XD, Harberd NP (2007) DELLAs contribute to plant photomorphogenesis. Plant Physiol 143:1163–1172
Ahmad M, Cashmore AR (1993) Hy4 gene of *A. thaliana* encodes a protein with characteristics of a blue-light photoreceptor. Nature 366:162–166
Alokam S, Chinnappa CC, Reid DM (2002) Red/far-red light mediated stem elongation and anthocyanin accumulation in *Stellaria longipes*: differential response of alpine and prairie ecotypes. Can J Bot 80:72–81
Alvey L, Harberd NP (2005) DELLA proteins: integrators of multiple plant growth regulatory inputs? Physiol Plant 123:153–160
Bailey-Serres J, Voesenek LACJ (2008) Flooding stress: acclimations and genetic diversity. Annu Rev Plant Biol 59:313–339
Bais HP, Weir TL, Perry LG, Gilroy S, Vivanco JM (2006) The role of root exudates in rhizosphere interactions with plants and other organisms. Annu Rev Plant Biol 57:233–266
Ballaré CL, Casal JJ, Kendrick RE (1991) Responses of light-grown wild-type and long-hypocotyl mutant cucumber seedlings to natural and simulated shade. Photochem Photobiol 54:819–826
Blakeslee JJ, Bandyopadhyay A, Peer WA, Makam SN, Murphy AS (2004) Relocalization of the PIN1 auxin efflux facilitator plays a role in phototropic responses. Plant Physiol 134(1):28–31
Borthwick HA, Hendricks SB, Parker MW, Toole EH, Toole VK (1952) A reversible photoreaction controlling seed germination. Proc Natl Acad Sci USA 38:662–666
Boylan MT, Quail PH (1989) Oat phytochrome is biologically active in transgenic tomatoes. Plant Cell 1:765–773
Briggs WR, Beck CF, Cashmore AR, Christie JM, Hughes J, Jarillo JA, Kagawa T, Kanegae H, Liscum E, Nagatani A, Okada K, Salomon M, Rudiger W, Sakai T, Takano M, Wada M, Watson JC (2001) The phototropin family of photoreceptors. Plant Cell 13:993–997
Casal JJ, Sanchez RA (1994) Impaired stem-growth responses to blue-light irradiance in light-growth transgenic tobacco seedlings overexpressing *Avena* phytochrome A. Physiol Plant 91:268–272
Catala C, Rose JKC, Bennett AB (1997) Auxin regulation and spatial localization of an endo-1, 4-beta-D-glucanase and a xyloglucan endotransglycosylase in expanding tomato hypocotyls. Plant J 12:417–426
Christie J (2007) Phototropin blue light receptors. Annu Rev Plant Biol 58:21–45
Clough RC, Vierstra RD (1997) Phytochrome degradation. Plant Cell Environ 20:713–721
Cosgrove DJ (2005) Growth of the plant cell wall. Nat Rev Mol Cell Biol 6:850–861
de Grauwe L, Vandenbussche F, Tietz O, Palme K, Van der Straeten D (2005) Auxin, ethylene and brassinosteroids: tripartite control of growth in the *Arabidopsis* hypocotyl. Plant Cell Physiol 46:827–836
de Kroon H (2007) Ecology – how do roots interact? Science 318:1562–1563
de Lucas M, Daviere JM, Rodriguez-Falcon M, Pontin M, Iglesias-Pedraz JM, Lorrain S, Fankhauser C, Blazquez MA, Titarenko E, Prat S (2008) A molecular framework for light and gibberellin control of cell elongation. Nature 451:480–491
de Vos M, Van Oosten VR, Van Poecke RMP, Van Pelt JA, Pozo MJ, Mueller MJ, Buchala AJ, Metraux JP, van Loon LC, Dicke M, Pieterse CMJ (2005) Signal signature and transcriptome changes of *Arabidopsis* during pathogen and insect attack. Mol Plant Microbe Interact 18:923–937
Devlin PF, Rood SB, Somers DE, Quail PH, Whitelam GC (1992) Photophysiology of the elongated internode (ein) mutant of *Brassica rapa* – Ein mutant lacks a detectable phytochrome B-like polypeptide. Plant Physiol 100:1442–1447
Devlin PF, Yanovsky MJ, Kay SA (2003) A genomic analysis of the shade avoidance response in Arabidopsis. Plant Physiol 133:1617–1629
Djakovic-Petrovic T, de Wit M, Voesenek LACJ, Pierik R (2007) DELLA protein function in growth responses to canopy signals. Plant J 51:117–126
Donohue K, Messiqua D, Pyle EH, Heschel MS, Schmitt J (2000) Evidence of adaptive divergence in plasticity: density- and site-dependent selection on shade-avoidance responses in *Impatiens capensis*. Evolution 54:1956–1968

Dudley SA, Schmitt J (1995) Genetic differentiation in morphological responses to simulated foliage shade between populations of Impatiens capensis from open and woodland sites. Funct Ecol 9:655–666

Duek PD, Fankhauser C (2005) bHLH class transcription factors take centre stage in phytochrome signalling. Trends Plant Sci 10:51–54

Evans JR, Poorter H (2001) Photosynthetic acclimation of plants to growth irradiance: the relative importance of specific leaf area and nitrogen partitioning in maximizing carbon gain. Plant Cell Environ 24:755–767

Fairchild CD, Schumaker MA, Quail PH (2000) HFR1 encodes an atypical bHLH protein that acts in phytochrome A signal transduction. Genes Dev 14:2377–2391

Fankhauser C, Chory J (2000) RSF1, an *Arabidopsis* locus implicated in phytochrome A signaling. Plant Physiol 124:39–45

Feng SH, Martinez C, Gusmaroli G, Wang Y, Zhou JL, Wang F, Chen LY, Yu L, Iglesias-Pedraz JM, Kircher S, Schafer E, Fu XD, Fan LM, Deng XW (2008) Coordinated regulation of *Arabidopsis thaliana* development by light and gibberellins. Nature 451:475–4U9

Finlayson SA, Lee IJ, Mullet JE, Morgan PW (1999) The mechanism of rhythmic ethylene production in sorghum. The role of phytochrome B and simulated shading. Plant Physiol 119:1083–1089

Folta KM, Pontin MA, Karlin-Neumann G, Bottini R, Spalding EP (2003) Genomic and physiological studies of early cryptochrome 1 action demonstrate roles for auxin and gibberellin in the control of hypocotyl growth by blue light. Plant J 36:203–214

Franklin KA (2008) Shade avoidance. New Phytol 179:930–944

Friml J, Wisniewska J, Benkova E, Mendgen K, Palme K (2002) Lateral relocation of auxin efflux regulator PIN3 mediates tropism in *Arabidopsis*. Nature 415:806–809

Fry SC (1998) Oxidative scission of plant cell wall polysaccharides by ascorbate-induced hydroxyl radicals. Biochem J 332:507–515

Fu XD, Harberd NP (2003) Auxin promotes *Arabidopsis* root growth by modulating gibberellin response. Nature 421:740–743

Fujimori T, Yamashino T, Kato T, Mizuno T (2004) Circadian-controlled basic/helix–loop-helix factor, PIL6, implicated in light-signal transduction in *Arabidopsis thaliana*. Plant Cell Physiol 45:1078–1086

Genoud T, Buchala AJ, Chua NH, Metraux JP (2002) Phytochrome signalling modulates the SA-perceptive pathway in *Arabidopsis*. Plant J 31:87–95

Goda H, Shimada Y, Asami T, Fujioka S, Yoshida S (2002) Microarray analysis of brassinosteroid-regulated genes in *Arabidopsis*. Plant Physiol 130:1319–1334

Goda H, Sawa S, Asami T, Fujioka S, Shimada Y, Yoshida S (2004) Comprehensive comparison of auxin-regulated and brassinosteroid-regulated genes in *Arabidopsis*. Plant Physiol 134:1555–1573

Gray WM, Ostin A, Sandberg G, Romano CP, Estelle M (1998) High temperature promotes auxin-mediated hypocotyl elongation in *Arabidopsis*. Proc Natl Acad Sci USA 95:7197–7202

Griebel T, Zeier J (2008) Light regulation and daytime dependency of inducible plant defenses in arabidopsis: phytochrome signaling controls systemic acquired resistance rather than local defense. Plant Physiol 147:790–801

Guo HW, Yang H, Mockler TC, Lin CT (1998) Regulation of flowering time by *Arabidopsis* photoreceptors. Science 279:1360–1363

Halliday KJ (2004) Plant hormones: the interplay of brassinosteroids and auxin. Curr Biol 14:1008–1010

Hautier Y, Niklaus AN, Hector A (2009) Competition for light causes plant biodiversity loss after eutrophication. Science 324:636–638

Hisamatsu T, King RW, Helliwell CA, Koshioka M (2005) The involvement of gibberellin 20-oxidase genes in phytochrome-regulated petiole elongation of *Arabidopsis*. Plant Physiol 138:1106–1116

Hoecker U, Toledo-Ortiz G, Bender J, Quail PH (2004) The photomorphogenesis-related mutant red1 is defective in CYP83B1, a red light-induced gene encoding a cytochrome P450 required for normal auxin homeostasis. Planta 219:195–200

Hornitschek P, Lorrain S, Zoete V, Michielin O, Fankhauser C (2009) Inhibition of the shade avoidance response by formation of non-DNA binding bHLH heterodimers. EMBO J 28:3893–3902

Huala E, Oeller PW, Liscum E, Han IS, Larsen E, Briggs WR (1997) *Arabidopsis* NPH1: a protein kinase with a putative redox-sensing domain. Science 278:2120–2123

Huq E, Quail PH (2002) PIF4, a phytochrome-interacting bHLH factor, functions as a negative regulator of phytochrome B signaling in *Arabidopsis*. EMBO J 21:2441–2450

Izaguirre MM, Mazza CA, Biondini M, Baldwin IT, Ballare CL (2006) Remote sensing of future competitors: impacts on plant defenses. Proc Natl Acad Sci USA 103:7170–7174

Janoudi AK, Gordon WR, Wagner D, Quail P, Poff KL (1997) Multiple phytochromes are involved in red-light-induced enhancement of first-positive phototropism in *Arabidopsis thaliana*. Plant Physiol 113:975–979

Jordan ET, Hatfield PM, Hondred D, Talon M, Zeevaart JAD, Vierstra RD (1995) Phytochrome-A overexpression in transgenic tobacco – correlation of dwarf phenotype with high-concentrations of phytochrome in vascular tissue and attenuated gibberellin levels. Plant Physiol 107:797–805

Jouanneau JP, Lapous D, Guern J (1991) In plant-protoplasts, the spontaneous expression of defense reactions and the responsiveness to exogenous elicitors are under auxin control. Plant Physiol 96:459–466

Kang JG, Yun J, Kim DH, Chung KS, Fujioka S, Kim JI, Dae HW, Yoshida S, Takatsuto S, Song PS, Park CM (2001) Light and brassinosteroid signals are integrated via a dark-induced small G protein in etiolated seedling growth. Cell 105:625–636

Kazan K, Manners JM (2009) Linking development to defense: auxin in plant–pathogen interactions. Trends Plant Sci 14:373–382

Kegge W, Pierik R (2010) BVOCs and plant competition. Trends Plant Sci (in press)

Koini MA, Alvey L, Allen T, Tilley CA, Harberd NP, Whitelam GC, Franklin KA (2009) High temperature-mediated adaptations in plant architecture require the bHLH transcription factor PIF4. Curr Biol 19:408–413

Koornneef M, Bosma TDG, Hanhart CJ, van der Veen J, Zeevaart JAD (1990) The isolation and characterization of gibberellin-deficient mutants in tomato. Theor Appl Genet 80:852–857

Kurashige NS, Agrawal AA (2005) Phenotypic plasticity to light competition and herbivory in *Chenopodium album* (Chenopodiaceae). Am J Bot 92:21–26

Kutschera U (1994) The current status of the acid-growth hypothesis. New Phytol 126:549–569

Lariguet P, Schepens I, Hodgson D, Pedmale UV, Trevisan M, Kami C, de Carbonnel M, Alonso JM, Ecker JR, Liscum E, Fankhauser C (2006) Phytochrome kinase substrate 1 is a phototropin 1 binding protein required for phototropism. Proc Natl Acad Sci USA 103:10134–10139

Leivar P, Monte E, Al-Sady B, Carle C, Storer A, Alonso JM, Ecker JR, Quail PH (2008) The *Arabidopsis* phytochrome-interacting factor PIF7, together with PIF3 and PIF4, regulates responses to prolonged red light by modulating phyB levels. Plant Cell 20:337–352

Leymarie J, Damerval C, Marcotte L, Combes V, Vartanian N (1996) Two-dimensional protein patterns of *Arabidopsis* wild-type and auxin insensitive mutants, *axr1*, *axr2*, reveal interactions between drought and hormonal responses. Plant Cell Physiol 37:966–975

Lin CT, Shalitin D (2003) Cryptochrome structure and signal transduction. Annu Rev Plant Biol 54:469–496

Lorrain S, Allen T, Duek PD, Whitelam GC, Fankhauser C (2008) Phytochrome-mediated inhibition of shade avoidance involves degradation of growth-promoting bHLH transcription factors. Plant J 53:312–323

Luccioni LG, Oliverio KA, Yanovsky MJ, Boccalandro HE, Casal JJ (2002) Brassinosteroid mutants uncover fine tuning of phytochrome signaling. Plant Physiol 128:173–181

Ma LG, Li JM, Qu LJ, Hager J, Chen ZL, Zhao HY, Deng XW (2001) Light control of *Arabidopsis* development entails coordinated regulation of genome expression and cellular pathways. Plant Cell 13:2589–2607
Martinez-Garcia JF, Huq E, Quail PH (2000) Direct targeting of light signals to a promoter element-bound transcription factor. Science 288:859–863
Martínez C, Pons E, Prats G, Leon J (2004) Salicylic acid regulates flowering time and links defence responses and reproductive development. Plant J 37:209–217
Mas P, Devlin PF, Panda S, Kay SA (2000) Functional interaction of phytochrome B and cryptochrome 2. Nature 408:207–211
McGuire R, Agrawal AA (2005) Trade-offs between the shade-avoidance response and plant resistance to herbivores? Tests with mutant *Cucumis sativus*. Funct Ecol 19:1025–1031
Morelli G, Ruberti I (2000) Shade avoidance responses. Driving auxin along lateral routes. Plant Physiol 122:621–626
Moreno JE, Tao Y, Chory J, Ballare CL (2009) Ecological modulation of plant defense via phytochrome control of jasmonate sensitivity. Proc Natl Acad Sci USA 106:4935–4940
Morgan DC, Smith H (1979) A systematic relationship between phytochrome-controlled and species habitat, for plants grown in simulated natural radiation. Planta 145:253–258
Nakamura A, Higuchi K, Goda H, Fujiwara MT, Sawa S, Koshiba T, Shimada Y, Yoshida S (2003) Brassinolide induces IAA5, IAA19, and DR5, a synthetic auxin response element in Arabidopsis, implying a cross talk point of brassinosteroid and auxin signaling. Plant Physiol 133:1843–1853
Nakamura A, Goda H, Shimada Y, Yoshida S (2004) Brassinosteroid selectively regulates PIN gene expression in *Arabidopsis*. Biosci Biotechnol Biochem 68:952–954
Navarro L, Bari R, Achard P, Lison P, Nemri A, Harberd NP, Jones JDG (2008) DELLAs control plant immune responses by modulating the balance and salicylic acid signaling. Curr Biol 18:650–655
Neff MM, Nguyen SM, Malancharuvil EJ, Fujioka S, Noguchi T, Seto H, Tsubuki M, Honda T, Takatsuto S, Yoshida S, Chory J (1999) BAS1: a gene regulating brassinosteroid levels and light responsiveness in *Arabidopsis*. Proc Natl Acad Sci USA 96:15316–15323
Nemhauser JL, Mockler TC, Chory J (2004) Interdependency of brassinosteroid and auxin signaling in *Arabidopsis*. Plos Biol 2:1460–1471
Novoplansky A (2009) Picking battles wisely: plant behaviour under competition. Plant Cell Environ 32:726–741
Nozue K, Covington MF, Duek PD, Lorrain S, Fankhauser C, Harmer SL, Maloof JN (2007) Rhythmic growth explained by coincidence between internal and external cues. Nature 448:358–361
Ohgishi M, Saji K, Okada K, Sakai T (2004) Functional analysis of each blue light receptor, cry1, cry2, phot1, and phot2, by using combinatorial multiple mutants in *Arabidopsis*. Proc Natl Acad Sci USA 101:2223–2228
Osterlund MT, Hardtke CS, Wei N, Deng XW (2000) Targeted destabilization of HY5 during light-regulated development of *Arabidopsis*. Nature 405:462–466
Park CM (2007) Auxin homeostasis in plant stress adaptation response. Plant Signal Behav 2:306–307
Pedmale UV, Liscum E (2007) Regulation of phototropic signaling in *Arabidopsis* via phosphorylation state changes in the phototropin 1-interacting protein NPH3. J Biol Chem 282:19992–20001
Peng J, Harberd NP (1997) Gibberellin deficiency and response mutations suppress the stem elongation phenotype of phytochrome-deficient mutants of *Arabidopsis*. Plant Physiol 113:1051–1058
Penninckx IAMA, Thomma BPHJ, Buchala A, Metraux JP, Broekaert WF (1998) Concomitant activation of jasmonate and ethylene response pathways is required for induction of a plant defensin gene in *Arabidopsis*. Plant Cell 10:2103–2113
Pierik R, Visser EJW, de Kroon H, Voesenek LACJ (2003) Ethylene is required in tobacco to successfully compete with proximate neighbours. Plant Cell Environ 26:1229–1234

Pierik R, Whitelam GC, Voesenek LACJ, de Kroon H, Visser EJW (2004a) Canopy studies on ethylene-insensitive tobacco identify ethylene as a novel element in blue light and plant–plant signalling. Plant J 38:310–319

Pierik R, Cuppens MLC, Voesenek LACJ, Visser EJW (2004b) Interactions between ethylene and gibberellins in phytochrome-mediated shade avoidance responses in tobacco. Plant Physiol 136:2928–2936

Pierik R, Djakovic-Petrovic T, Keuskamp DH, de Wit M, Voesenek LACJ (2009) Auxin and ethylene regulate elongation responses to neighbor proximity signals independent of gibberellin and DELLA proteins in *Arabidopsis*. Plant Physiol 149:1701–1712

Potter I, Fry SC (1993) Xyloglucan endotransglycosylase activity in Pea internodes. Effects of applied gibberellic acid. Plant Physiol 103:235–241

Potters G, Pasternak TP, Guisez Y, Palme KJ, Jansen MAK (2007) Stress-induced morphogenic responses: growing out of trouble? Trends Plant Sci 12:98–105

Quail PH, Boylan MT, Parks BM, Short TW, Xu Y, Wagner D (1995) Phytochromes – photosensory perception and signal-transduction. Science 268:675–680

Quail PH (2002) Photosensory perception and signalling in plant cells: new paradigms? Curr Opin Cell Biol 14:180–188

Reed JW, Nagpal P, Poole DS, Furuya M, Chory J (1993) Mutations in the gene for the red far-red light receptor phytochrome-B alter cell elongation and physiological-responses throughout *Arabidopsis* development. Plant Cell 5:147–157

Reed JW, Foster KR, Morgan PW, Chory J (1996) Phytochrome B affects responsiveness to gibberellins in *Arabidopsis*. Plant Physiol 112:337–342

Robert-Seilaniantz A, Navarro L, Bari R, Jones JD (2007) Pathological hormone imbalances. Curr Opin Plant Biol 10:372–379

Robson PRH, Whitelam GC, Smith H (1993) Selected components of the shade-avoidance syndrome are displayed in a normal manner in mutants of *Arabidopsis thaliana* and *Brassica rapa* deficient in phytochrome-B. Plant Physiol 102:1179–1184

Robson PRH, McCormac AC, Irvine AS, Smith H (1996) Genetic engineering of harvest index in tobacco through overexpression of a phytochrome gene. Nat Biotechnol 14:995–998

Robson PRH, Smith H (1997) Fundamental and biotechnological applications of phytochrome transgenes. Plant Cell Environ 20:831–839

Roig-Villanova I, Bou-Torrent J, Galstyan A, Carretero-Paulet L, Portoles S, Rodriguez-Conception M, Martinez-Garcia JF (2007) Interaction of shade avoidance and auxin responses: a role for two novel atypical bHLH proteins. EMBO J 26:4756–4767

Rojo E, Leon J, Sanchez-Serrano JJ (1999) Cross-talk between wound signalling pathways determines local versus systemic gene expression in *Arabidopsis thaliana*. Plant J 20:135–142

Rose JKC, Lee HH, Bennett AB (1997) Expression of a divergent expansin gene is fruit-specific and ripening-regulated. Proc Natl Acad Sci USA 94:5955–5960

Rose JKC, Braam J, Fry SC, Nishitani K (2002) The XTH family of enzymes involved in xyloglucan endotransglucosylation and endohydrolysis: current perspectives and a new unifying nomenclature. Plant Cell Physiol 43:1421–1435

Ruzicka K, Ljung K, Vanneste S, Podhorska R, Beeckman T, Friml J, Benkova E (2007) Ethylene regulates root growth through effects on auxin biosynthesis and transport-dependent auxin distribution. Plant Cell 19:2197–2212

Salter MG, Franklin KA, Whitelam GC (2003) Gating of the rapid shade-avoidance response by the circadian clock in plants. Nature 426:680–683

Sasidharan R, Chinnappa CC, Voesenek LACJ, Pierik R (2008) The regulation of cell wall extensibility during shade avoidance: a study using two contrasting ecotypes of *Stellaria longipes*. Plant Physiol 148:1557–1569

Schenk HJ (2006) Root competition: beyond resource depletion. J Ecol 94:725–739

Schmitt J, McCormac AC, Smith H (1995) A test of the adaptive plasticity hypothesis using transgenic and mutant plants disabled in phytochrome-mediated elongation responses to neighbors. Am Nat 146:937–953

Sessa G, Carabelli M, Sassi M, Ciolfi A, Possenti M, Mittempergher F, Becker J, Morelli G, Ruberti I (2005) A dynamic balance between gene activation and repression regulates the shade avoidance response in *Arabidopsis*. Genes Dev 19:2811–2815

Somers DE, Sharrock RA, Tepperman JM, Quail PH (1991) The HY3 long hypocotyl mutant of *Arabidopsis* is deficient in phytochrome-B. Plant Cell 3:1263–1274

Somers DE, Devlin PF, Kay SA (1998) Phytochromes and cryptochromes in the entrainment of the *Arabidopsis* circadian clock. Science 282:1488–1490

Tao Y, Ferrer JL, Ljung K, Pojer F, Hong FX, Long JA, Li L, Moreno JE, Bowman ME, Ivans LJ, Cheng YF, Lim J, Zhao YD, Ballare CL, Sandberg G, Noel JP, Chory J (2008) Rapid synthesis of auxin via a new tryptophan-dependent pathway is required for shade avoidance in plants. Cell 133:164–176

Teale WD, Paponov IA, Palme K (2006) Auxin in action: signalling, transport and the control of plant growth and development. Nat Rev Mol Cell Biol 7:847–859

Terzaghi WB, Cashmore AR (1995) Light-regulated transcription. Annu Rev Plant Physiol Plant Mol Biol 46:445–474

Ueguchi-Tanaka M, Ashikari M, Nakajima M, Itoh H, Katoh E, Kobayashi M, Chow TY, Hsing YIC, Kitano H, Yamaguchi I, Matsuoka M (2005) Gibberellin insensitive dwarf1 encodes a soluble receptor for gibberellin. Nature 437:693–698

van Loon LC, Geraats BPJ, Linthorst HJM (2006) Ethylene as a modulator of disease resistance in plants. Trends Plant Sci 11:184–191

Vandenbussche F, Vriezen WH, Smalle J, Laarhoven LJJ, Harren FJM, Van der Straeten D (2003) Ethylene and auxin control decreased light intensity. Plant Physiol 133:517–527

Vandenbussche F, Pierik R, Millenaar FF, Voesenek LA, Van der Straeten D (2005) Reaching out of the shade. Curr Opin Plant Biol 8:462–468

von Dahl CC, Baldwin IT (2007) Deciphering the role of ethylene in plant–herbivore interactions. J Plant Growth Regul 26:201–209

Wang D, Pajerowska-Mukhtar K, Culler AH, Dong XN (2007) Salicylic acid inhibits pathogen growth in plants through repression of the auxin signaling pathway. Curr Biol 17:1784–1790

Wang HY, Ma LG, Li JM, Zhao HY, Deng XW (2001) Direct interaction of *Arabidopsis* cryptochromes with COP1 in light control development. Science 294:154–158

Wasternack C (2007) Jasmonates: an update on biosynthesis, signal transduction and action in plant stress response, growth and development. Ann Bot 100:681–697

Weinig C (2000) Plasticity versus canalization: population differences in the timing of shade-avoidance responses. Evolution 54:441–451

Yi HC, Joo S, Nam KH, Lee JS, Kang BG, Kim WT (1999) Auxin and brassinosteroid differentially regulate the expression of three members of the 1-aminocyclopropane-1-carboxylate synthase gene family in mung bean (*Vigna radiata* L.). Plant Mol Biol 41:443–454

Zurek DM, Rayle DL, Mcmorris TC, Clouse SD (1994) Investigation of gene-expression, growth-kinetics, and wall extensibility during brassinosteroid-regulated stem elongation. Plant Physiol 104:505–513

Global Atmospheric Change and Trophic Interactions: Are There Any General Responses?

Geraldine D. Ryan, Susanne Rasmussen, and Jonathan A. Newman

Abstract Increasing atmospheric CO_2 is hypothesized to alter plant physiology and metabolism, which may have important implications for species interactions. In this chapter, we review published studies on the effects of elevated atmospheric CO_2 on plant-derived allelochemicals and the possible effects of CO_2-mediated changes on higher trophic levels such as herbivores, parasitoids, and predators. We provide a critical assessment of conventional ecological theories used to predict phytochemical responses to CO_2 and we make some suggestions as to how this field may be expanded and improved.

1 Introduction

Atmospheric carbon dioxide levels have risen by approximately 100 ppm since the onset of the industrial revolution to the present level of 385 ppm. The Intergovernmental Panel on Climate Change (IPCC) predicts that levels of CO_2 in the atmosphere will rise to between 500 and 975 ppm in the next century depending on economic growth and energy use scenarios (IPCC 2000), though recent projections suggest that these estimates may be conservative (Raupach et al. 2007). Elevated CO_2, along with increased emissions of other greenhouse gases such as methane and ozone, is primarily due to anthropogenic activities such as fossil fuel

G.D. Ryan and J.A. Newman (✉)
School of Environmental Sciences, University of Guelph, Guelph, ON, Canada N1G 2W1
e-mail: gryan@uoguelph.ca; jnewma01@uoguelph.ca

S. Rasmussen
AgResearch, 11008 Palmerston North, New Zealand
e-mail: susanne.rasmussen@agresearch.co.nz

F. Baluška and V. Ninkovic (eds.), *Plant Communication from an Ecological Perspective*, Signaling and Communication in Plants,
DOI 10.1007/978-3-642-12162-3_11,

consumption and forest clearing. Greenhouse gas emissions are projected to cause increases in surface air temperatures of between 1 and 4°C in this century, along with changes in precipitation and cloud cover. While the predictions regarding climatic changes have been debated, levels of atmospheric CO_2 are unequivocally rising. This will have important implications for plant growth and development. The effects of increasing CO_2 on plant physiology are complex and have been extensively reviewed (Bazzaz 1990; Bowes 1993; Drake and Gonzàlez-Meler 1997; Makino and Mae 1999; Medlyn et al. 2001; Woodward 2002; Nowak et al. 2004; Ainsworth and Long 2005). Plants exposed to elevated CO_2 experience enhanced photosynthesis, reduced photorespiration, decreases in stomatal conductance, decreased transpiration rates and subsequent enhanced water-use efficiency, and alterations in resource allocation, phenology and reproductive biology (as reviewed in Bazzaz 1990). In addition, decreases in plant nitrogen content, increases in the carbon to nitrogen ratio (C:N), alterations in the concentrations of defensive compounds and overall changes in plant quality are likely to directly impact herbivores. Changes in allocation to secondary metabolites can also have both direct and indirect (herbivore-mediated) effects on higher trophic levels, which may have broad implications for community structure. As we see in the next section, plant-derived allelochemicals are a critical component in shaping ecosystems in general. Thus, plant responses to increased photosynthate production under elevated CO_2 and subsequent changes in antiherbivore phytochemical production are among the critical impacts of rising CO_2 on plant growth (Bazzaz 1990).

This chapter explores the ways in which elevated CO_2 may impact resource allocation to plant allelochemicals – where an allelochemical may be defined as "any nonnutritional chemical produced by an individual of one species that affects the growth, health, behavior or population biology of another species" (Schoonhoven et al., 2005). We then examine the potential effects of these changes on higher trophic levels. Several excellent reviews have been produced on the general effects of CO_2 on ecosystems (Bazzaz 1990; Coley 1998; Coviella and Trumble 1999; Hunter 2001; Lindroth and Dearing 2005). We focus primarily on literature that addresses CO_2 effects in terms of plant secondary chemistry. The projected increase in atmospheric CO_2 is unlikely to have direct physiological effects on insect herbivores (Fajer et al. 1991) or their enemies, so effects are expected to be plant-mediated. Here, we examine the current body of literature on this topic and evaluate empirical studies in light of theoretical predictions, which consider how altered resource allocation may change the production and concentration of allelochemicals. We assess empirical studies within the framework of the prevailing ecological theory and we also discuss the limitations of such models. Where pertinent, discussions include factors associated with primary plant metabolism, especially when evaluating the effects of elevated CO_2 on host quality for herbivores. This chapter is structured to address the effects of CO_2 at increasing trophic levels starting with the effects on plant-derived allelochemicals, the implications for insect herbivores, and in turn the direct and indirect effects on herbivore enemies such as parasitoids and predators.

2 Climate Change and Plant-Derived Allelochemicals

2.1 Plant-Derived Allelochemicals as a Driving Force for Community Structure

Plants and their herbivores collectively comprise a large proportion of global species diversity, with these two groups representing almost 50% of all identified species on earth (Strong et al. 1984). The quality of host plants for herbivorous insects is dependent on a number of factors including plant metabolism (nutritional make-up and production of defenses) and morphology (physical defenses). Although the definition of plant quality depends on factors intrinsic to the herbivore in question (e.g., nutrient requirements and sensitivity to plant defenses), many generalizations can be made about the requirements of insect herbivores. In this section, we give an overview of the ways in which plant-derived allelochemicals can shape ecosystems and why predicting changes in plant chemicals are a critical part of understanding how ecosystems may function in the future. Plants produce a vast range of secondary metabolites such as alkaloids, terpenoids, saponins, flavonoids, and tannins, (Futuyama 1983) which are critically important to the reproductive fitness of plants and serve as protection against herbivores, viruses, bacteria, fungi, and competing plants.

Herbivores can encounter secondary metabolites at many levels during the process of searching for, and feeding on, plant food. Prior to feeding, herbivores may encounter plant volatile organic compounds (VOCs) in the headspace surrounding the potential host, and these compounds (e.g., monoterpenes, sesquiterpenes, aldehydes, alcohols and ketones) may provide the insect with information about the suitability of the plant as a host (Pickett et al. 1998; Bruce et al. 2005). Deciphering the effects of plant volatiles on insect herbivores can be challenging as these volatiles can act as both attractants and deterrents. The effects of specific compounds will likely be dependent on the herbivore group in question and it has been demonstrated that the peripheral receptors of phytophagous insects are tuned to the detection of specific host plant "blends" of ubiquitous VOCs, rather than any one single VOC (Fraser et al. 2003; Birkett et al. 2004).

Once a herbivore has commenced feeding, it may be exposed to a diverse array of internal plant compounds such as phenolics (e.g., tannins and flavonoids), nitrogenous compounds (e.g., alkaloids, amines, cyanogenic glycosides, and glucosinolates), terpenoids, organic acids, lipids and sulfur-containing compounds (Futuyama 1983; Howe and Jander 2008; Rasmann and Agrawal 2009). Some insects have adapted to these plant-produced defensive compounds and in some cases may exploit them for their own needs. For example, propenylbenzenes, coumarins, and polyacetylene produced in the surface wax of plants are known to be growth inhibitors for a number of taxa including fungi, bacteria and many generalist herbivores. However the same compounds have been shown to stimulate oviposition in the carrot fly *Psila rosae* (Städler and Buser 1984). Many insects can also sequester toxic plant compounds, e.g., pyrrolizidine alkaloids and cardenolides, which are stored in specialized insect tissues, and protect the insect from predators (Brower and Fink 1985; Narberhaus et al. 2005; Opitz and Müller 2009). Plant secondary chemicals can vastly alter the physiology of insect hosts and can induce complex behaviors associated with host location, oviposition, deterrence and attraction. Secondary metabolites are a crucial component in determining host-quality for insect herbivores and ingested secondary metabolites can have profound and diverse effects on feeding herbivores including the disruption of digestion, metabolism, growth, and development.

Plant-produced allelochemicals can also affect higher trophic levels either directly or indirectly (via herbivores) and thus can serve a multitude of functions in the broader ecosystem context. The induction of plant VOC release by feeding herbivores can signal the presence of herbivores to natural enemies such as predators and parasitoids which are then attracted to the damaged plant (Kessler and Baldwin 2001; Rasmann et al. 2005; Schnee et al. 2006). Damage-induced VOCs can also be perceived by neighboring plants which are alerted to the presence of herbivores and may subsequently alter their secondary chemistry (Baldwin and Schultz 1983; Karban and Shiojiri 2009). Volatiles released from different plant parts may serve different functions. For example, pollinators are attracted to emissions from flowers, while seed dispersers can recognize and orient towards volatiles released from developing fruit (Dudareva et al. 2006). The diversity of chemical groups and the types of responses they elicit are extremely complex and are beyond the scope of this chapter. However this section serves to highlight the importance of allelochemicals in community structure and function and to emphasize the importance in predicting changes in allelochemical production in a high CO_2 atmosphere.

2.2 Elevated CO_2 and Resource Allocation to Allelochemicals

Plants have a finite capacity to acquire necessary resources such as carbon, mineral nutrients and water. Plants may allocate photosynthates and nutrients to reproduction, growth, storage or defense (Ayres 1993) depending on the needs of the plant and the availability of these resources. Increased atmospheric CO_2 is expected to have a stimulatory effect on the production of photosynthates though this effect is generally stronger for C_3 plants than for C_4 plants (Stitt and Krapp 1999; Ainsworth and Long 2005). How these additional resources are allocated depends on the evolutionary history of the plant (i.e., genotypic response) and the range of strategies allowed for within the physiological constraints of the plant (i.e., phenotypic response). Changes in allocation to chemical defenses under elevated CO_2 are expected to have important implications for herbivores and to have both direct effects and indirect herbivore-mediated effects on higher trophic levels. In the search for a "general theory of plant defense" ecologists have formulated several hypotheses which have been used to predict resource allocation to allelochemicals under altered environmental conditions such as changes in light, water and nutrient inputs and sink/source relationships. These can be extended to predictions of resource allocation and allelochemical production under elevated CO_2. Among the most widely used of these are: (1) the Optimal Defense (OD) Hypothesis; (2) the Growth-Differentiation Balance (GDB) Hypothesis; and (3) the Carbon–Nutrient Balance (CNB) Hypothesis. Many of the predictions invoked by these hypotheses broadly overlap and all contain the same basic assumption, that allocation of finite plant resources constitutes a trade-off between secondary defenses (morphological and chemical defenses and cellular differentiation) and primary metabolism (photosynthesis, respiration and growth) (Cronin and Hay 1996).

The optimal defense hypothesis:

The OD (McKey 1979; Strauss et al. 2004), which incorporates aspects from the coevolution theory (Ehrlich and Raven 1964) and from the plant apparency theory (Feeny 1975, 1976; Rhoades 1979; Rhoades and Cates 1976; Cornell and Hawkins 2003) attempts to explain how the defensive needs of a plant (as defined by the plant's risk of attack by herbivores) determine the evolution of defensive secondary metabolites. The ODH comprises two central hypotheses: (1) "organisms evolve and allocate defenses in a way that maximizes individual inclusive fitness" and (2) "defenses are costly in terms of fitness" (Rhoades 1979; Stamp 2003; Strauss et al. 2002, 2004). Hypothesis 2 encompasses four subhypotheses which are (1) "organisms evolve defenses in direct proportion to their risk from predators and in inverse proportion to the cost of defense," (2) "within an organism, defenses are allocated in proportion to risk of the plant part and value of it to plant fitness, and in inverse proportion to the cost of defense," (3) "defense is decreased when enemies are absent and increased when they are present (i.e., allocation pattern of constitutive and inducible defenses)," and (4) "there is a trade-off between defense and other plant functions (growth and reproduction) such that stressed individuals are less defended" (Rhoades 1979; Stamp 2003). While some of the assumptions of the ODH are implicit in studies of plant responses to elevated CO_2, its hypotheses have not generally been formally tested in climate change scenarios. The principle of optimal defense makes different predictions about defense allocation under elevated CO_2 depending on assumptions concerning the defensive strategies of plants (Ayres 1993). The Plant Apparency Hypothesis (Feeny 1975, 1976; Rhoades 1979) predicts that plants generally have one of two defensive strategies: (1) "quantitative" (apparent plants) or (2) "qualitative" (unapparent plants). So-called "apparent" plants (e.g. long-lived woody species) are readily found by herbivores and therefore invest heavily in plant defenses that are effective against a broad range of herbivorous animals. Apparent plants produce quantitative defenses (e.g., tannins and lignins); their effects are dose dependent and they act to reduce plant tissue digestibility to herbivores. Unapparent plants on the other hand can readily escape herbivores in time and space and as a result invest much less in defensive compounds, producing low doses of overtly toxic chemicals such as glucosinolates, cyanogenic glycosides and alkaloids. If the benefits of defense increase linearly with investment in secondary chemistry (as in the case of apparent plants) and if the cost of quantitative defenses decreases under elevated CO_2, then the ODH would predict that increased CO_2 would result in increased concentrations of allelochemicals in apparent plants. On the other hand if plants are maximally protected with low concentrations of qualitative compounds (i.e., unapparent plants) then the ODH predicts no change in allelochemical production under elevated CO_2 (Ayres 1993).

Growth Differentiation Balance Hypothesis:

The GDB (Loomis 1932, 1953; Herms and Mattson 1992) is a framework for explaining resource allocation to growth vs. differentiation processes under environmental perturbations such as alterations in water and nutrient inputs. Growth can be defined as the irreversible increase in plant size due to cell division and enlargement, while differentiation refers to processes that occur after cell expansion that often (but not always) have an antiherbivore function. Processes such as deposition of lignin, leaf thickening, production of trichomes and thorns, production of secondary metabolites, and development of reproductive organs are examples of differentiation. The GDB hypothesis is based on the premise that plant growth is inversely proportional to differentiation representing a trade-off in resource allocation to different sinks. The GDB hypothesis simply states that limitations in any resource that inhibits growth more than it does photosynthesis will increase the pool of resources that can be allocated to differentiation. Empirical studies show that net photosynthesis is not as sensitive to shortages of nutrients and water, as growth processes (reviewed by Herms and Mattson 1992). Applied to climate change studies (Mattson et al. 2005; Gayler et al. 2008) the GDB hypothesis predicts that increases in atmospheric CO_2, which stimulate photosynthesis causing an increase in the C:N ratio and a decrease in plant nitrogen (Cotrufo et al. 1998), will result in a shunting of photosynthates into differentiation. Hence the production of secondary metabolites is expected to increase under elevated CO_2. As we will see, predictions based on the GDB hypothesis broadly overlap with those of the CNB (below) but the two hypotheses differ in their explanations of predicted results and only the GDB hypothesis addresses cellular development.

The Carbon-Nutrient Balance Hypothesis:

The CNB (Bryant et al. 1983) has been the most widely used of the available defense allocation models in climate change studies (Fajer et al. 1992; Roth and Lindroth 1995; Kinney et al. 1997; Hemming and Lindroth 1999; Bezemer et al. 2000; Coviella et al. 2000; Schädler et al. 2007), in part owing to the simplicity of empirical tests of its predictions. As previously mentioned the predictions of the CNB broadly overlap with those of the GDB hypothesis. However while the GDB makes predictions about changes in any resource, the CNB hypothesis addresses changes in nutrient levels only. Under the CNB framework a similar prediction can be made for elevated CO_2 whereby increased photosynthetic rates and consequential increases in the C:N ratio result in carbon products in excess of those needed for primary metabolic functions and as such result in increased carbon-based secondary metabolites (Fig. 1). A related prediction is that nitrogen limitations resulting from plant growth under elevated CO_2 will result in a decrease of N-based secondary metabolites (Karowe et al. 1997). Both the GDB and the CNB hypotheses predict that allocation to secondary metabolites under elevated CO_2 will depend critically on

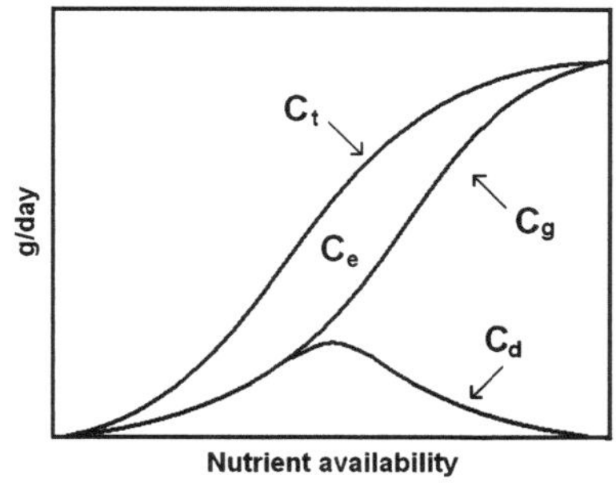

Fig. 1 The carbon–nutrient balance hypothesis: (C_t) = total available carbon, (C_g) = carbon diverted to growth, (C_e) = excess carbon, (C_d) = carbon-based allelochemicals. Nutrient limitations limit growth more than photosynthesis resulting in a net excess of carbon resources. Carbon-based plant defenses are highest where excess carbon is highest as carbon in shunted from primary metabolism to secondary metabolism. Redrawn from Stamp (2003)

other plant resource inputs such as nutrients and light, and there is much empirical evidence to support this (Johnson and Lincoln 1991; Hättenschwiler and Schafellner 1999; McDonald et al. 1999; Agrell et al. 2000; Saxon et al. 2004).

2.3 *C and N Allocation to Biosynthetic Classes of Secondary Metabolites*

We performed a literature search using the search terms: (plant) + (antiherbivory, alkaloids, cyanogenic glycosides, terpenoids, phenolics, secondary metabolites, defense) + (elevated, increased) + (climate change, CO_2) in Google Scholar. The search returned 101 relevant studies (Appendix 1) with measurements of 608 plant secondary metabolites under elevated CO_2 from 191 plants representing 102 species. Measurements of secondary metabolites under elevated CO_2 were recorded and placed into one of four chemical classes (phenolics, terpenoids, nitrogen-based compounds, and volatiles). Since some studies measured several individual allelochemicals, while others reported only totals (e.g., total phenolics), studies with a greater number of measured allelochemicals are thus overrepresented here relative to single allelochemical reports. The phytochemicals surveyed here represent constitutively produced compounds, however, the production of allelochemicals can be induced in response to damage by herbivores (see Chapter "Within-Plant Signalling by Volatiles Triggers Systemic Defences"). The dynamics of herbivore induction of plant allelochemicals may be altered under elevated CO_2. For example Bidart-Bouzat et al. (2005) found that herbivores induced increases in glucosinolate levels under elevated CO_2, but not in ambient conditions.

Figure 2 shows the proportion of measured allelochemicals in each of the four chemical classes that showed increases, decreases or no change in concentration. For all classes except phenolics the majority of studies found no change under

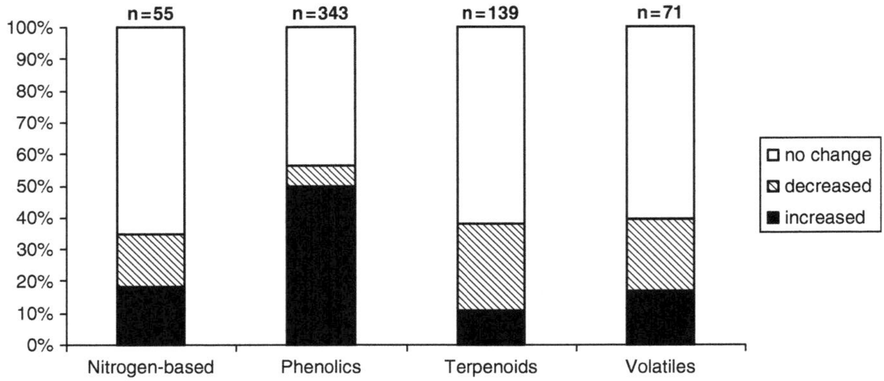

Fig. 2 The percentage of plant-derived allelochemicals that increased, decreased or showed no change under elevated CO_2 for four chemical classes: nitrogen-based compounds ($n = 55$), phenolics ($n = 343$), terpenoids ($n = 139$) and volatiles ($n = 71$). N values represent a single measured chemical compound except in cases where only the change for a specific group was given (e.g., "total phenolics"). "Elevated" CO_2 ranged from 500 to 1,400. Some terpenoids (e.g., some monoterpenes and sesquiterpenes) which can be classified as both terpenoids and volatiles, were placed in the "terpenoid" category when expressed as a concentration (mass compound/unit mass plant tissue) and were placed in the "volatile" category when expressed as an emission rate (mass compound/unit plant area/unit time). Studies which measured allelochemicals derived from plant endosymbionts, or where concentrations of plant-derived allelochemicals were measured in an interacting nonplant species (e.g., insect tissues) were excluded. In the cases where allelochemical response to CO_2 depended on an interacting factor, we attempted to characterize significant effects based on subjective "ambient" conditions in order to examine the effects of CO_2 alone (e.g., ambient temperatures)

elevated CO_2. This may be because the null hypothesis is true, or it may be the result of type II (β) statistical error due to low sample sizes (and hence a lack of statistical power). It may also be an artifact of the way in which these results are usually considered. Plant chemical concentrations are usually expressed on a dry weight basis and the accumulation of predominantly total *nonstructural* carbohydrates (TNC) under elevated CO_2 might mask changes in concentrations of carbon-based secondary or structural compounds (CBSSC). When concentrations of CBSSC were expressed on a *structural* dry weight basis, nonsignificant average increases of flavonoids and soluble phenolics became significant (Poorter et al. 1997; Peñuelas and Estiarte 1998). The two main predictions of the CNB: an increase in C-based allelochemicals and a decrease in N-based allelochemicals are not met. Under elevated CO_2 N-based compounds increased (18% of cases) more often than they decreased (16% of cases). For the carbon-based terpenoids, concentrations increased in 11% of cases and decreased in 27%. The same was true for the C-based volatile class with increases in 17% of cases and decreases in 23% of cases. Again, these decreases in carbon-based secondary metabolites under elevated CO_2 do not support the CNB. In the phenolic class however, allelochemicals increased in 50% of cases with decreases in only 7% of cases. It has been suggested that the predictive power of the CNB hypothesis is greater for certain chemical classes or subclasses over others (Reichardt et al. 1991; Lerdau and Coley 2002). For example, the CNB hypothesis may better predict allocation to stable end products and may be less effective for unstable or transient metabolic intermediates. Reichardt et al. (1991) showed that metabolite turnover determined which secondary metabolites could be predicted using the CNB framework and suggest that "dynamic" or transient intermediates (e.g. monoterpenes, sesquiterpenes, and phenolic glycosides) are less predictable than "stable" metabolic end products such as tannins and lignin. Predictions may be dependent on the biochemical pathways involved and appear to be much less conclusive for terpenes (products of the DOXP (1-deoxy-D-xylulose-5-phosphate and mevalonic acid pathways) and alkaloids (multiple pathways) (Lerdau and Coley 2002). Even within a given biochemical pathway, predictions may be more robust for stable metabolic endpoints. Figure 3 shows the breakdown of three subclasses of phenolics: tannins, phenolic glycosides and phenolic acids, from the literature search described above. Tannins are generally thought to be stable endpoints of the shikimic acid biochemical pathway while phenolic glycosides and phenolic acids are transient intermediates. The prediction that C-based secondary metabolites increase under elevated CO_2 was met only for the tannin subclass where less than 2% of measurements showed decreases in tannins while increases were shown in over 60% of cases. Results are much more varied for metabolic intermediates, which corroborates the view that predictions may be dependent on the stability of metabolites. In addition, compounds such as phenolic acids and other metabolic intermediates of the shikimic acid pathway generally represent only a small proportion of plant biomass (Koricheva 2002) and previous work has suggested that caution should be taken in applying a cost-benefit approach to compounds which are found in only very low concentrations (Gulmon and Mooney 1986).

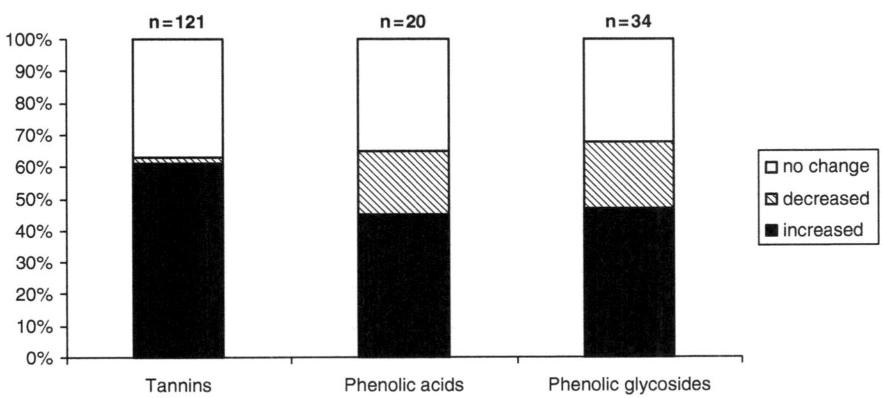

Fig. 3 The percentage of plant-derived allelochemicals that increased, decreased or showed no change under elevated CO_2 for three subclasses of phenolic compounds: tannins ($n = 121$), phenolic acids ($n = 20$) and phenolic glycosides ($n = 34$). n values represent a single measured chemical compound. Tannins are so called stable or "static" metabolic endpoints while phenolic glycosides and phenolic acids are classified as intermediate or "dynamic" metabolites

A stipulation of the CNB and GDB frameworks is that changes in allelochemical allocation are predominantly driven by nutrient limitations that cause excess photosynthates to be shunted into secondary metabolism. We therefore wished to consider simultaneous changes in nitrogen and allelochemicals. Our literature search resulted in 378 studies of carbon-based allelochemical changes for which nitrogen concentrations were measured simultaneously. Figure 4 shows the N x allelochemical contingency table for these results and the number of measurements showing the specified changes in nitrogen and allelochemicals. In 69% of cases the nitrogen concentration decreased under elevated CO_2 as predicted. Only a single study found an increase in N concentration and those that did not detect a statistical change in N concentrations generally showed a trend towards decreased nitrogen. For the CNB to be a good predictor we would expect most of the data points to consolidate in the top right cell corresponding to decreased N and increased C-based allelochemicals. However, decreased nitrogen and increased allelochemical production accounted for only 32% of all possible combinations. In only 6% of cases did decreased plant nitrogen show a simultaneous decrease in C-based allelochemicals. When only tannins were considered, 52 out of 106 cases (49% – not shown in figure) reported simultaneous decreases in nitrogen and increases in tannins. Thus, even when nitrogen concentrations are considered, the results of empirical studies of allelochemical allocation under elevated CO_2 are only weakly predicted by existing frameworks such as the CNB, except maybe for tannins.

Generally, the usefulness of the CNB hypothesis as a predictive and explanatory tool has been widely debated (Hamilton et al. 2001; Koricheva 2002; Lerdau and Coley 2002; Stamp 2003). Some suggest that the CNB has outlived its usefulness (Hamilton et al. 2001) while others suggest that the limitations of the CNB framework arise from a lack of understanding of the model's fundamental assumptions (Stamp 2003). One criticism of the CNB hypothesis is that it ignores the complexities of the biochemical machinery involved in the production of plant defenses (Hamilton et al. 2001). This has resulted in the generation of more

		[Nitrogen]			
[Allelochemical]		+	0	–	TOTAL
	+	1	18	119	138
	0	0	68	122	190
	–	0	29	21	50
	TOTAL	1	115	262	**378**

Fig. 4 Nitrogen concentration × allelochemical concentration contingency table for $n = 378$ measurements which simultaneously measured carbon-based plant allelochemicals and nitrogen concentration under elevated CO_2. Each cell represents the percentage of studies with the corresponding change in allelochemical and nitrogen concentrations. (+) = increased concentration under elevated CO_2, (0) = no change in concentration under elevated CO_2, (–) = decreased concentration under elevated CO_2. Data exclude N-based allelochemicals

mechanistic hypotheses that consider biochemical intermediates rather than simple cause and effect relationships between the composition of resource inputs and metabolic endpoints. For example, the Protein Competition Model (PCM) of phenolic allocation (Jones and Hartley 1999) predicts the allocation and concentration of phenolics in leaves of terrestrial higher plants based on biochemical pathways and regulatory mechanisms. Specifically, the model predicts that protein and phenolic synthesis compete for the common, limiting resource phenylalanine (PHE), such that protein and phenolic production are inversely correlated. Under elevated CO_2 the concentration of phenolic compounds can thus be predicted from the effects of increased carbon on leaf functions that create competing demands for proteins vs. phenolics. However, it should be noted here that the classification of defense compounds into "C-based" (e.g., tannins) and "N-based" (e.g., alkaloids) might not be justified per se, as the production of e.g. alkaloids might be equally or even more C-costly in terms of required glucose as is the production of phenolics (3.24 vs. 2.11 g of glucose per gram of alkaloid vs. phenolic; Gershenzon 1994; Schoonhoven et al. 2005). Furthermore, the production of tannins requires a large suite of enzymes (>10; Winkel-Shirley 2001), which are very N-costly, whereas the biosynthesis of the N-based tyrosine derived cyanogenic glycoside dhurrin requires only three enzymes (Nielsen et al. 2008). Furthermore, not only phenylalanine derived phenolics compete with the production of growth-related proteins, but so do all N-based secondary metabolites which require amino acids as precursors. Figure 5 shows the biochemical pathways involved in secondary metabolite production, phytohormone production (see Sect. 2.4, this chapter) and primary metabolic endpoints. The complexity of these biochemical pathways illustrates how simple predictions based on chemical composition may be misleading.

Even more confounding is the fact that classifications of compounds based on their biosynthetic origin might not be relevant in the context of defense (or differentiation) vs. growth allocation. The compound class "phenolics" in fact represents a wide array of possible structures which are all derived from the aromatic amino acid phenylalanine produced in the shikimic acid pathway. However, the major "phenolic" in most plants is lignin, a very complex and large, more or less inert polymer. Lignin is essential for xylem and cell wall development and so plays a critical role in all processes related to water transport including photosynthesis and mineral nutrient transport, as well as in structural support for land plants. Even though one might argue that xylem and secondary cell wall formation is already "differentiation" as opposed to "growth," and lignin should therefore be seen as a "secondary metabolite" it is also clear that the absence of lignin would lead to the cessation of any growth in land plants. Another example is the even larger class of isoprene derived terpenoids which comprises molecules critical for photosynthesis (carotenoids, terpenoid side chain of chlorophylls), electron carriers (e.g., side chains of plasto- and ubiquinones), membrane structures (sterols), and plant development related phytohormones (abscisic acid, gibberellins) – all essential for and not separable from plant growth. We suggest therefore that models of resource allocation should not be based on chemical structure of metabolites per se, but rather on functionality of these metabolites.

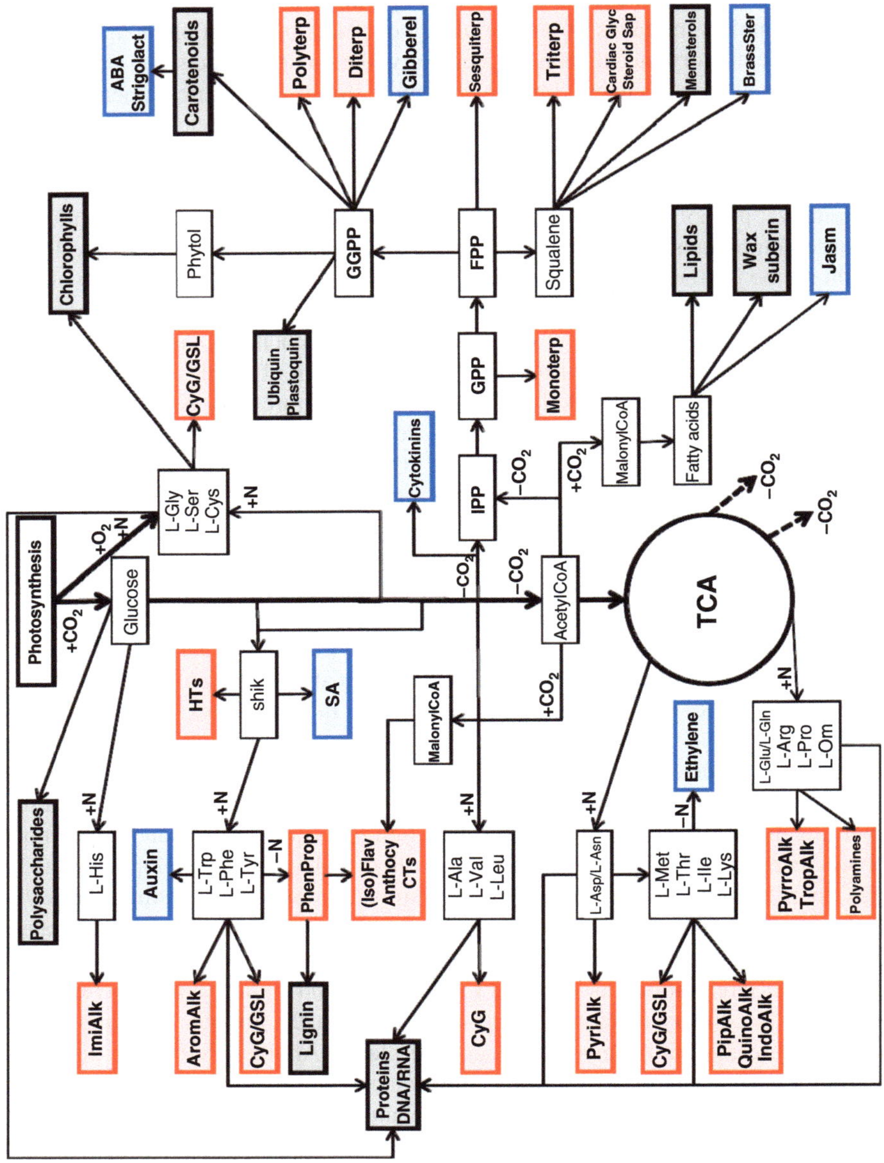

Photosynthesis
+CO2
+O2
+N
Glucose
L-Gly
L-Ser
L-Cys
CyG/GSL
Chlorophylls
Phytol
GGPP
Ubiquin
Plastoquin
ABA
Strigolact
Carotenoids
Polyterp
Diterp
Gibberel
FPP
Sesquiterp
Squalene
Triterp
Cardiac Glyc
Steroid Sap
Memsterols
BrassSter
GPP
Monoterp
IPP
Cytokinins
−CO2
+CO2
AcetylCoA
MalonylCoA
Fatty acids
Lipids
Wax
suberin
Jasm
TCA
shik
HTs
SA
Polysaccharides
L-His
ImiAlk
Auxin
L-Trp
L-Phe
L-Tyr
−N
PhenProp
(Iso)Flav
Anthocy
CTs
AromAlk
CyG/GSL
Lignin
Proteins
DNA/RNA
L-Ala
L-Val
L-Leu
CyG
L-Asp/L-Asn
PyriAlk
Ethylene
L-Met
L-Thr
L-Ile
L-Lys
CyG/GSL
PipAlk
QuinoAlk
IndoAlk
L-Glu/L-Gln
L-Arg
L-Pro
L-Orn
PyrroAlk
TropAlk
Polyamines

2.4 Phytohormones: The Molecular Link Between Physiological Responses to Elevated CO_2 and Secondary Metabolite Production?

A metaanalysis of responses of photosynthesis, canopy properties and plant production to elevated CO_2 has shown general trends of increases in light-saturated carbon uptake, diurnal C assimilation, growth, and above-ground production, and decreases in specific leaf area and stomatal conductance (Ainsworth and Long 2005). It has also been shown that, on average, stomatal density decreases under elevated CO_2 (Woodward 1987; Woodward and Bazzaz 1988), resulting in reduced transpiration rates and improved water use efficiency (Woodward 2002; Teng et al. 2006). It has been proposed that the signaling of CO_2 and the induction of these physiological responses is mediated by shifts in the biosynthesis and accumulation of phytohormones, in particular abscisic acid, ethylene, jasmonates, and cytokinins (Yong et al. 2000; Lake et al. 2002; Teng et al. 2006). Furthermore, sugars like e.g. glucose can act as signaling molecules and influence the biosynthesis of these phytohormones (Léon and Sheen 2003; Gibson 2004, 2005; Rolland et al. 2006; Bossi et al. 2009), and increased hexose levels in plants under elevated CO_2 might affect these signaling processes as well.

Phytohormones are also closely linked to the biosynthesis of a wide range of secondary metabolites involved in plant defenses. Figure 5 shows some of the pathways involved in secondary metabolite and phytohormone production. Jasmonates in particular have been linked to plant defense and immunity, and recent excellent reviews have summarized molecular mechanisms of their action as signal transducers in plant–herbivore interactions (Howe and Jander 2008; Browse 2009). Evidence for herbivore-induced production of jasmonates and for their roles in plant immunity is largely based on studies with jasmonate mutants that are compromised in resistance against a wide range of insect herbivores (see references in: Howe and Jander 2008; Browse 2009). It has also been demonstrated that jasmonates interact with other phytohormones like ABA and ethylene, as well as sugar signals (Gazzarrini and McCourt 2001), and that cross-talk between ethylene and jasmonate signaling pathways determine the activation of specific defense responses (Lorenzo et al. 2003).

Jasmonates are formed from the fatty acid linolenic acid in the octadecanoic pathway (Schaller et al. 2004). Several intermediates of this pathway as well as the resulting jasmonates have been shown to induce the biosynthesis of terpenoids in conifers (Martin et al. 2002) and tomatoes (Ament et al. 2004), terpenoid indole alkaloids in *Catharanthus rosea* (Menke et al. 1999), and terpene volatiles in

Fig. 5 Schematic overview of plant metabolic pathways (simplified and plant specific modified version of KEGG overview of biosynthetic pathways); http://www.genome.jp/kegg/pathway/map/map01010.html) from primary metabolites (*black open*) to secondary metabolites (*red shaded*), primary "end" metabolites (*black shaded*) and phytohormones (*blue shaded*). $\pm CO_2$ and N represent the regions where biochemical processes lead to the incorporation (+) or release (−) of CO_2 or N. See Appendix 2 for a list of abbreviations

Brassica oleraceae (Bruinsma et al. 2009). A study on terpene volatile induction in lima beans also reported that early and late intermediates of the octadecanoic pathway induce the accumulation of different terpenoids, resulting in a different "blend" of volatiles (Koch et al. 1999).

Only a few studies tested a relationship between elevated CO_2, phytohormone production, and plant defense responses, but it has been demonstrated that elevated CO_2 down-regulated genes involved in ethylene and jasmonate production, and that in turn resulted in the down-regulation of the expression of genes for an antiherbivorous cysteine proteinase inhibitor and enzymes of the phenylpropanoid pathway in soybean, compromising the plants' resistance against some insect herbivores (Zavala et al. 2008, 2009). We propose therefore that studies on phytohormones may provide a molecular link between physiological responses to elevated CO_2 and plant secondary metabolites and should be studied in more detail to unravel mechanisms of changes in defense metabolite production under elevated CO_2.

3 Trophic Interactions and Climate Change

3.1 Plant–Herbivore Interactions and Elevated CO_2

Changes in the quality of host plants for phytophagous insects in an atmosphere with higher levels of CO_2 will likely be driven by two predominant factors: allelochemicals and nitrogen. As discussed in Sect. 2.2 nitrogen decreases under elevated CO_2 are almost ubiquitous in plants, whereas carbon-based allelochemicals are expected to increase, albeit with many exceptions. Thus, a general prediction emerges whereby simultaneous decreases in nitrogen and increases in allelochemicals will have a net negative effect on herbivorous insects. As we will see below, this simple prediction is also subject to many exceptions and dependent on many interacting factors. Understanding the impact of CO_2-mediated changes in plant quality on insect herbivores, and how herbivory in turn will affect plant growth, is dependent on understanding how these changes affect insect performance parameters at both the individual and population levels.

Empirical studies:

For the literature search described in Sect. 2.3, a subset of studies ($n = 38$) also reported herbivore responses to CO_2-induced changes in secondary metabolites. Figures 6 and 7 below are derived from these reports. It should be noted however that correlation analyzes between insect performance parameters and allelochemical production were reported in only a minority of these studies. As such, the covariation of allelochemical concentration and insect performance presented here do not necessarily imply causation and these data should thus be interpreted with caution. Insect performance was typically measured using life history traits such as survival, body size, development time, fecundity, and growth rate. As predicted by life history theory, many of these traits typically covaried. For example decreased body size was often associated with increased development time. An overall performance score (increased, decreased or no change) was given to each study based on these traits. Again plant chemistry data must be interpreted with caution as they are based on simultaneous measurements of allelochemicals and insect performance, and as such, performance of a given insect is overrepresented here in cases where multiple allelochemicals are measured in a single study.

Figure 6a shows the contingency table derived from these data for insect performance vs. allelochemical concentration ($n = 216$). Insect performance decreased in 44% of cases (96 out of 216), was unchanged in 43% of cases (92 out of 216) and increased in 13% of cases (28 out of 216). In cases where allelochemicals increased (80 out of 216) insect performance decreased 48% of the time (38 out of 80). In some cases insect performance was negatively correlated with increases in specific groups of allelochemicals such as phenolic glycosides (Lindroth et al. 1997; McDonald et al. 1999). However, where allelochemicals decreased (19 out of 216) surprisingly more often than not insect performance also

a

[Allelochemical]		Insect performance			
		+	0	–	TOTAL
	+	9	33	38	80
	0	16	53	48	117
	–	3	6	10	19
	TOTAL	28	92	96	**216**

b

[Nitrogen]		Insect performance			
		+	0	–	TOTAL
	+	0	0	0	0
	0	2	9	3	14
	–	5	17	23	45
	TOTAL	7	26	26	**59**

Fig. 6 (**a**) Insect performance × allelochemical concentration contingency table for $n = 216$ measurements which simultaneously measured allelochemical concentration and insect performance under elevated CO_2 and (**b**) Insect performance × nitrogen concentration contingency table for $n = 59$ measurements. Each cell represents the number of measurements with the corresponding change in allelochemical concentration, N concentration and insect performance. (+) = increased concentration or performance under elevated CO_2, (0) = no change in concentration or performance under elevated CO_2, (–) = decreased concentration or performance under elevated CO_2. Insect performance was classified as negative in cases where decreases in survival, mass, growth rate or fecundity, or increases in development time or mortality were reported. The opposite was used to denote decreases in performance. Where none of the above performance parameters changed, overall insect performance was counted as unchanged

decreased (53% of cases) which suggests that factors other than allelochemicals may be strong determinants of insect performance under elevated CO_2. We examined the effect of nitrogen on insect performance in the same way (Fig. 6b). As expected, nitrogen did not increase in any of the 59 observations but decreased, as predicted, in 76% of cases. Where nitrogen decreased, insect performance decreased in 23 of 45 cases (51%). Thus, several generalities can be made from the effect of CO_2 on nitrogen and allelochemicals, and their subsequent effect on insect herbivores, however some of the conceptual and theoretical models proposed here are only weakly predictive. The variable nature of these data is likely to occur for several reasons. Firstly, the predictive frameworks explained here, and those often used in climate change biology, are often far too simplistic to capture the complexities of species interactions in heterogeneous environments. These interactions will depend critically on the environmental context in which they occur and will be subject to several interacting variables. For example, changes in the interactions between plants and insects under elevated CO_2 are dependent on light regime (McDonald et al. 1999; Agrell et al. 2000), soil nutrient status (Johnson and Lincoln 1991; Hättenschwiler and Schafellner 1999; Saxon et al. 2004), atmospheric ozone levels (Kopper et al. 2001; Kopper and Lindroth 2003), temperature (Veteli et al. 2002; Williams et al. 2003) and plant and insect genotype (Goverde et al. 2004; Saxon et al. 2004). The plant-mediated effects on insect herbivores will also depend on herbivore sensitivity to such changes. Just as plants have a range of plastic responses to elevated CO_2, herbivores too can alter or compensate for elevated CO_2-mediated changes in plant quality by altering feeding rates and efficiency.

Compensatory feeding:

In many cases, insect herbivores can compensate for decreases in the nutritional quality of plants grown under elevated CO_2 by increasing their food intake (Johnson and Lincoln 1990; Lindroth et al. 1993; Kinney et al. 1997; Mansfield et al. 1999; Agrell et al. 2000). Figure 7a shows the herbivore consumption rate vs. nitrogen concentration contingency table generated from the reports described above. This table summarizes insect plastic feeding response to elevated CO_2-induced nutrient deficiencies. In no cases did consumption decrease under elevated CO_2 suggesting that food intake rate is critical for insects feeding on elevated CO_2-grown plants. In the 30 cases where N decreased, increased consumption rates occurred in 57% (17) of observations. However, increased intake did not always correlate with insect performance in the studies examined here and compensatory feeding did not always result in full N intake compensation. For example, Hättenschwiler and Schafellner (1999) found that nun moth larvae that increased their relative consumption rates on N-deficient plants grown under elevated CO_2 still consumed on average 33% less nitrogen.

Consumption rates were unchanged in 43% of cases where plant N decreased (Fig. 7a). In some cases compensatory feeding was shown to be limited by the presence of different classes of allelochemicals such as terpenoids (Johnson and Lincoln 1990, 1991) and phenolic glycosides (Roth et al. 1998). Figure 7b shows the contingency table for allelochemicals vs. consumption. In cases where

a

[Nitrogen]		Consumption			
		+	0	–	TOTAL
	+	0	0	0	0
	0	3	4	0	7
	–	17	13	0	30
	TOTAL	20	17	0	**37**

b

[Allelochemical]		Consumption			
		+	0	–	TOTAL
	+	31	30	0	61
	0	32	19	0	51
	–	6	3	0	9
	TOTAL	69	52	0	**121**

Fig. 7 (**a**) Insect herbivore consumption × nitrogen concentration contingency table for $n = 37$ measurements which simultaneously measured nitrogen concentration and consumption rates under elevated CO_2. (**b**) Insect herbivore consumption × allelochemical concentration contingency table for $n = 121$ measurements which simultaneously measured carbon-based plant allelochemicals and consumption rates under elevated CO_2. Each cell represents the number of studies with the corresponding change in allelochemical concentrations, consumption rates and nitrogen concentrations. (+) = increased concentration under elevated CO_2, (0) = no change in concentration under elevated CO_2, (–) = decreased concentration under elevated CO_2

allelochemicals increased under elevated CO_2, compensatory feeding occurred about half of the time, where allelochemicals did not change compensatory feeding occurred in 63% of cases and where allelochemicals decreased compensatory feeding occurred in 67% of cases. This suggests that allelochemical concentration may have an effect on compensatory feeding though this is difficult to assess as cases where allelochemicals decreased were low (9 out of 121 measurements). Herbivores may also compensate for decreases in nutritive value of host plants in ways that alter food processing efficiency. For example, larvae of the red-headed pine sawfly increased their nitrogen utilization efficiency under elevated CO_2 in response to declines in nitrogen concentration in loblolly pine (Williams et al. 1994). In addition, it has been shown that insect herbivores exposed to multiple species of plants may offset the negative effects associated with elevated CO_2 by partial host shifts or by feeding on alternative plant parts (Williams et al. 1997; Agrell et al. 2005).

In agricultural systems, compensatory feeding may have implications for crop yields. Insects that increase their consumption due to high C:N ratios under elevated CO_2 can increase their exposure to pesticides resulting in increased mortality (Coviella and Trumble 2000). It is generally predicted that elevated CO_2 will have a net positive effect on agricultural yields and increased efficacy of insecticides could potentially augment this result. However, assessing the net impact of elevated CO_2 on agricultural systems is subject to several considerations. In many cases compensatory feeding has been shown to occur in response to decreases in the nutritional quality of plants under elevated CO_2 with no corresponding changes in insect performance (Johnson and Lincoln 1990; Lindroth et al. 1993; Williams et al. 1994; Docherty et al. 1996; Kinney et al. 1997; Roth et al. 1998). However, the net effect on plant productivity will depend on whether increases in plant biomass can ameliorate increases in defoliation (Hunter 2001). Other factors may also influence herbivore feeding rates under elevated CO_2. For example, Hamilton et al. (2005) found that the leaf area of soybean plants removed by insect herbivores increased by up to 57% under elevated CO_2 due to the phagostimulatory effect of sugar which increased in concentration. However, since CO_2-mediated resource allocation in plants is likely to depend critically on nutrient inputs (Johnson and Lincoln 1991; Hättenschwiler and Schafellner 1999; Saxon et al. 2004), this means that some of these effects may be buffered by controlling fertilization levels in agricultural contexts.

A challenge for climate change ecologists is the ability to broadly predict the effects of elevated CO_2 on plant–insect interactions which can in turn have profound effects on ecosystems as a whole. It may be possible that generalities can be made and models derived at specific levels of ecosystem complexity. Plant variables which are likely to be strong determinants of ecosystem responses and which should be incorporated into mechanistic models are photosynthetic machinery (C3 vs. C4), growth rates (fast vs. slow growing species), symbiotic associations (e.g., rhizobia and N_2 fixation; fungal endophytes and their antiherbivorous toxins; mycorrhizae and P uptake), agricultural vs. natural ecosystems and plant chemical profiles. Insect responses in turn will depend on levels of insect specialization, guild effects, sensitivity to changing plant quality and capacity to compensate for nutritive deficits.

3.2 The Effects of Elevated CO_2 on Higher Trophic Levels

Global atmospheric change can be reasonably expected to have consequences, not just for plants and their herbivores, but for ecosystems as a whole. Changes in primary productivity are expected to alter an array of community interactions and ecosystem functions. For example, increased lignin:nitrogen and C:N ratios in the litter of deciduous trees grown under elevated CO_2 can cause decreases in decomposition and respiration rates (Cotrufo et al. 1994). Elevated CO_2 has also been shown to affect soil microbial composition and activity (Runion et al. 1994). Changes in the biotic and abiotic structure of soils are likely to have profound effects on ecosystems via feedbacks to primary production. Plant–fungal interactions are likely to also be affected by elevated CO_2. Symbiotic fungal endophytes of grasses can impact the primary metabolic response of the host plant to elevated CO_2 which in turn is expected to affect other trophic levels. Recent studies have shown that endophyte infection can buffer the plant's nitrogen metabolism response to elevated CO_2 in both tall fescue (Newman et al. 2003) and perennial ryegrass (Hunt et al. 2005). For soil fungi, increases in the hyphal length and activity of saprophytic fungi, along with increases in mycorrhizal infection of roots, have been reported (Dhillion et al. 1996). Increasing atmospheric CO_2 may also have consequences for plant–plant communication though to our knowledge this has never been explicitly tested. Baldwin and Schultz (1983) demonstrated that herbivory can induce volatile cues in plants that signal the presence of herbivores to other plants. Nearby plants then alter their secondary chemistry in a way that makes them less palatable to herbivores, and this response is likely to depend on the genetic relatedness of plants (Karban and Shiojiri 2009). Thus, plant–plant communication may be affected by CO_2-induced changes in volatile emissions which have been demonstrated in several studies (Tognetti et al. 1998; Constable et al. 1999; Loreto et al. 2001; Kreuzwieser et al. 2002; Rapparini et al. 2004; Vuorinen et al. 2004a; Himanen et al. 2009). Figure 8 shows the number of studies that have examined the effects of CO_2-induced changes in plant defensive chemistry on several important community interactions. Studies have been dominated by the effects of elevated CO_2 on plant secondary chemistry, are common for subsequent effects on herbivores and decrease in number with higher trophic levels, highlighting the need for investigation of elevated CO_2 on broader community interactions. Recent studies suggest that changes in plant defensive chemistry under elevated CO_2 may have profound consequences for herbivore enemies.

The effects of elevated CO_2 on the production of plant allelochemicals may influence higher trophic levels either directly or indirectly (herbivore-mediated)

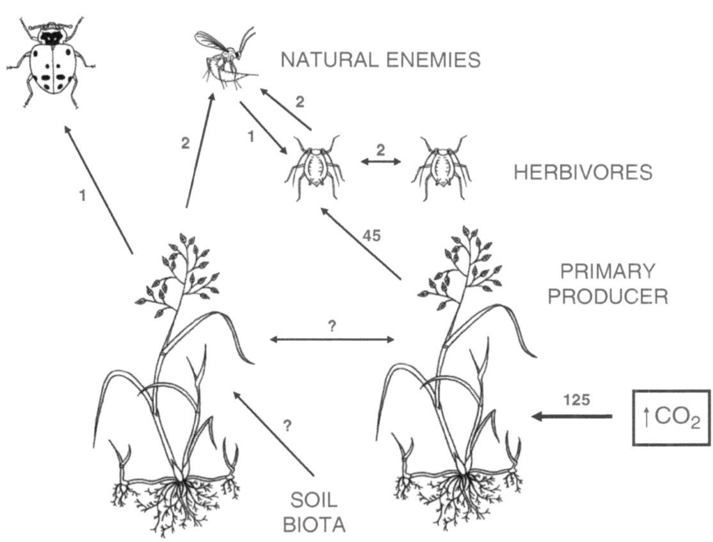

Fig. 8 The number of publications located on the effects of elevated CO_2 on plant secondary chemicals and the subsequent effects on higher trophic levels

(see Fig. 8). Higher trophic levels are attracted by herbivore-induced damage volatiles which signal to natural enemies (predators and parasitoids) that herbivores are present (direct effect). Changes in emission rates and composition of plant-derived VOCs under elevated CO_2 may alter the fitness of natural enemies through alterations in this olfactory cue and the subsequent capability of natural enemies to locate herbivores. Vuorinen et al. (2004b) found changes in the orientation behavior of a generalist predator *Podisus maculiventris* in the presence of herbivore-damaged cabbage plants grown under elevated CO_2. In the same study it was found that the specialist parasitoid *Cotesia plutellae* was unable to orient itself towards herbivore-damaged plants grown under elevated CO_2. At ambient CO_2 levels both species are able to detect herbivore presence via herbivore-induced plant volatiles. Thus herbivore-induced damage response by higher trophic levels may be diminished under elevated CO_2. This is expected to have profound effects for herbivore enemies in changing atmospheres particularly in light of the suggestion that herbivore-induced volatiles are the single most reliable prey location cue for generalist predators (Vet and Dicke 1992). Research suggests that natural enemy attraction may depend on the concentration of volatiles released. For example, transgenic *Arabidopsis thaliana* plants engineered to overexpress a terpene synthase gene were found to increase in attractiveness to the parasitic wasp *C. marginiventris* when plants were damaged by lepidopteran larvae (Schnee et al., 2006). However data on volatile emission rates under elevated CO_2 have been highly inconsistent and may covary with other plant variables sensitive to CO_2-induced changes. Compensatory feeding by herbivores on nutrient deficient plants grown under elevated CO_2 may cause increases in inducible defenses such as VOC emissions (Himanen et al. 2009) which could alter enemy behavior. However, changes in orientation behavior may be dependent on the sensitivity of herbivore enemies to changes in plant volatiles. For example, Himanen et al. (2009) found that damage-induced volatiles from *B. napus* plants increased under elevated CO_2 but there was no difference in the orientation behavior of the endoparasitoid *C. vestalis* between the treatments.

The effects of elevated CO_2 on plant defensive chemistry may also affect higher trophic levels indirectly through ingestion of plant chemicals by herbivores that in turn may alter parasitoid or predator survival. However, few studies have examined the indirect effects of CO_2-induced changes in plant chemistry on higher trophic levels. Roth and Lindroth (1995) found that the effect of parasitism on gypsy moth performance did not change under elevated CO_2. However, parasitoid mortality increased, especially on aspen grown under elevated CO_2 where sensitivity of parasitoids to the accumulation of phenolic glycosides by the gypsy moth due to compensatory feeding may have been a factor. In a similar study elevated CO_2 had little effect on the survivorship of the dipteran parasitoid *Compsilura concinnata* when its host *Malacosoma disstria* was fed on aspen grown under elevated CO_2 (Holton et al. 2003).

Elevated CO_2 may also have implications for plant-mediated production and perception of alarm responses of herbivores. For example, in potato aphids (*Aulacorthum solani*) feeding on broad bean (*Vicia faba*) grown in elevated CO_2 it was found that the ability to produce alarm pheromone in the presence of a disturbance was reduced, as was the ability of the recipient to detect the pheromone (Awmack et al. 1997). Diminished escape responses by aphids under elevated CO_2 have also been demonstrated in subsequent studies (Mondor et al. 2004). It has been hypothesized that the ability of herbivores to detect conspecific alarm signals may depend on the surrounding chemical environment produced by the plant (Dill et al. 1990). Thus CO_2-induced changes in volatile emissions may have implications for herbivore–herbivore communication.

4 Conclusions

Here we examined the effect of elevated CO_2 on plant allocation to secondary defense and subsequent effects on higher trophic levels. A great deal of empirical studies have been undertaken at the level of the plant and some general patterns have emerged from these investigations. In general carbon-based secondary metabolites have a tendency to increase, though this is likely to be pathway dependent. Thus, more mechanistic models are needed in order to help us understand some of this complexity. If we are to predict the effects of CO_2 on plant chemistry we will need to be able to determine where generalities can be made, where they cannot and what can be learned from exceptions to these generalities. Although we have focused on secondary metabolites here, the primary metabolic response will also have broad implications for bottom–up interactions in a high CO_2 atmosphere. In addition, interacting factors are likely to profoundly affect the response to elevated CO_2. We have already seen that changes in plant chemistry and nutrient status are the most common determinants of herbivore responses and subsequent responses of higher trophic levels. In general, insect herbivores are expected to have decreased performance if nitrogen concentrations decrease and plant defenses increase. However, plant responses and subsequent herbivore responses will likely depend on factors such as temperature, ozone, nutrients, water availability and light availability.

The possible role of evolution in species' responses to CO_2 and other climate change projections in the next century is often precluded from discussions of climate change impacts and the majority of studies assess only the plastic response of organisms. Thus it is generally accepted that species will either adjust to perturbations within their physiological range or become extinct. This is true for many of the long-lived woody plant species which tend to dominate the literature on elevated CO_2. However, in many cases herbivorous insects may fit the criteria for rapid evolutionary adaptation, that is: large population sizes ($>10^5$–10^6), short generation times (e.g. <1 year) and high intrinsic rates of increase (e.g. $r_m > 0.5$/generation) (Kingsolver 1996). For example, genetic changes in fruit flies have been associated with adaptation to climate warming (Umina et al. 2005; Balanyá et al. 2006). Thus there is a need for longer term studies on the community effects of CO_2. Even in the absence of adaptation per se, plants have the potential to acclimatize to changing conditions over time and studies have shown than plants can exhibit a high level of adaptive plasticity. For example, for some plants, changes in photosynthetic rates associated with elevated CO_2 may be transient (Bazzaz 1990). Also, the single-step increase in elevated CO_2 adopted in the design of most experiments is not representative of the gradual increases seen in reality and this may have implications for observed changes. For example, Klironomos et al. (2005) showed that long term gradual increases in elevated CO_2 resulted in no change in the diversity of mycorrhizal fungi relative to ambient but an abrupt step increase reduced diversity due to the sensitivity of some species to this change.

Ecosystem-wide responses to increasing CO_2 will be difficult to predict given the complexity of interactions which exist in even the most elementary ecosystems. More research is needed to even begin to formulate general principles with regard to responses of higher trophic levels. Even then, individual responses may not predict the response of communities as a whole. Differences in the physiological sensitivities of certain groups or species and the differential ability of organisms to adjust to atmospheric changes might reasonably be expected to alter relative abundances, which in turn may have consequences for ecosystem diversity and function. Figure 8 highlights the gaps in our knowledge with respect to community effects of elevated CO_2 and reveals the scarcity of studies on higher trophic levels, below ground effects and intraspecific communication. The complexity of communities necessitates the identification of those variables which are likely to be of utmost importance. Ecosystem-wide studies will benefit from technologies such as free-air carbon enrichment (FACE) and open-topped chambers (OTC) which will be of great importance in assessing effects on natural communities.

References

Agrell J, McDonald EP, Lindroth RL (2000) Effects of CO_2 and light on tree phytochemistry and insect performance. Oikos 88:259–272

Agrell J, Kopper B, McDonald EP, Lindroth RL (2005) CO_2 and O_3 effects on host plant preferences of the forest tent caterpillar (*Malacosoma disstria*). Glob Chang Biol 11:588–599

Ainsworth EA, Long SP (2005) What have we learned from 15 years of free-air CO_2 enrichment (FACE)? A meta-analytic review of the responses of photosynthesis, canopy properties and plant production to rising CO_2. New Phytol 165:351–372

Ament K, Kant MR, Sabelis MW, Haring MA, Schuurink RC (2004) Jasmonic acid is a key regulator of spider mite-induced volatile terpenoid and methyl salicylate emission in tomato. Plant Physiol 135:2025–2037

Awmack CS, Woodcock CM, Harrington R (1997) Climate change may increase vulnerability of aphids to natural enemies. Ecol Entomol 22:366–368

Ayres MP (1993) Plant defense, herbivory and climate change. In: Kareiva PM, Kingsolver JG, Huey RB (eds) Biotic interactions and global change. Sinauer Associates, Sunderland, MA, pp 75–94

Balanyá J, Oller JM, Huey RB, Gilchrist GW, Serra L (2006) Global genetic change tracks global climate warming in *Drosophila subobscura*. Science 313:1773–1775

Baldwin JT, Schultz JC (1983) Rapid changes in tree leaf chemistry induced by damage: evidence for communication between plants. Science 221:277–279

Bazzaz FA (1990) Response of natural ecosystems to the rising global CO_2 level. Annu Rev Ecol Syst 21:167–196

Bezemer TM, Jones TH, Newington JE (2000) Effects of carbon dioxide and nitrogen fertilzation on phenolic content in *Poa annua* L. Biochem Sys Ecol 28:839–846

Bidart-Bouzat MG, Mithen R, Berenbaum MR (2005) Elevated CO_2 influences herbivory-induced defense responses of *Arabidopsis thaliana*. Oecologia 145:415–424

Birkett MA, Bruce TJA, Martin JL, Smart LE, Oakley J, Wadhams LJ (2004) Responses of female orange wheat blossom midge, *Sitodiplosis mosellana*, to wheat panicle volatiles. J Chem Ecol 30:1319–1328

Bossi F, Cordoba E, Dupré P, Mendoza MS, San Román C, Léon P (2009) The *Arabidopsis* ABA-INSENSITIVE (ABI) 4 factor acts as a central transcription activator of the expression of its own gene, and for the induction of *ABI5* and *SBE2.2* genes during signaling. Plant J 59:359–374

Bowes G (1993) Facing the inevitable: plants and increasing atmospheric CO_2. Annu Rev Plant Physiol, Plant Mol Biol 44:309–32

Brower LP, Fink LS (1985) A natural toxic defense system – cardenolides in butterflies versus birds. Ann N Y Acad Sci 443:171–188

Browse J (2009) Jasmonate passes muster: a receptor and targets for the defense hormone. Annu Rev Plant Biol 60:183–205

Bruce TJA, Wadhams LJ, Woodcock CM (2005) Insect host location: a volatile situation. Trends Plant Sci 10:269–274

Bruinsma M, Posthumus MA, Mumm R, Mueller MJ, van Loon JJA, Dicke M (2009) Jasmonic acid-induced volatiles of *Brassica oleraceae* attract parasitoids: effects of time and dose, and comparison with induction by herbivores. J Exp Bot 60:2575–2587

Bryant JP, Chapin FS III, Klein DR (1983) Carbon/nutrient balance of boreal plants in relation to vertebrate herbivory. Oikos 40:357–368

Coley PD (1998) Possible effects of climate change on plant/herbivore interactions in moist tropical forests. Clim Change 39:455–472

Constable JVH, Litvak ME, Greenberg JP, Monson RK (1999) Monoterpene emission from coniferous trees in response to elevated CO_2 concentration and climate warming. Glob Chang Biol 5:255–267

Cornell HV, Hawkins BA (2003) Herbivore responses to plant secondary compounds: a test of phytochemical theory. Am Nat 161:507–522

Cotrufo MF, Ineson P, Rowland AP (1994) Decomposition of tree leaf litters grown under elevated CO_2: effect of litter quality. Plant Soil 163:121–130

Cotrufo MF, Ineson P, Scott A (1998) Elevated CO_2 reduces the nitrogen concentration of plant tissues. Glob Chang Biol 4:43–54

Coviella CE, Trumble JT (1999) Effects of elevated atmospheric carbon dioxide on insect–plant interactions. Conserv Biol 18:700–712

Coviella CE, Trumble JT (2000) Effect of elevated atmospheric carbon dioxide on the use of foliar application of *Bacillus thuringiensis*. Biocontrol 45:325–336

Coviella CE, Morgan DJW, Trumble JT (2000) Interactions of elevated CO_2 and nitrogen fertilization: effects on production of *Bacillus thuringiensis* toxins in transgenic plants. J Econ Entomol 93:781–787

Cronin G, Hay ME (1996) Within-plant variation in seaweed palatability and chemical defenses: optimal defense theory versus the growth-differentiation balance hypothesis. Oecologia 105:361–368

Dhillion SS, Roy J, Abrams M (1996) Assessing the impact of elevated CO_2 on soil microbial activity in a Mediterranean model ecosystem. Plant Soil 187:333–342

Dill LM, Fraser AHG, Roitberg BD (1990) The economics of escape behaviour in the pea aphid, *Acyrthosiphon pisum*. Oecologia 83:473–478

Docherty M, Hurst DK, Holopainen JK, Whittaker JB, Lea PJ, Watt AD (1996) Carbon dioxide-induced changes in beech foliage cause female beech weevil larvae to feed in a compensatory manner. Glob Chang Biol 2:335–331

Drake BG, Gonzàlez-Meler MA (1997) More efficient plants: a consequence of rising atmospheric CO_2? Annu Rev Plant Physiol Plant Mol Biol 48:609–639

Dudareva N, Negre F, Nagegowda DA, Orlova I (2006) Plant volatiles: recent advances and future perspectives. Crit Rev Plant Sci 25:417–440

Ehrlich PR, Raven PH (1964) Butterflies and plants: a study in coevolution. Evolution 18:586–608

Fajer ED, Bowers MD, Bazzaz FA (1991) The effects of enriched CO_2 atmospheres on the Buckeye Butterfly, *Junonia coenia*. Ecology 72:751–754

Fajer ED, Bowers MD, Bazzaz FA (1992) The effect of nutrients and enriched CO_2 environments on production of carbon-based allelochemicals in plantago: a test of the carbon/nutrient balance hypothesis. Am Nat 140:707–723

Feeny P (1975) Biochemical coevolution between plants and their insect herbivores. In: Gilbert LE, Raven PH (eds) Coevolution of animals and plants. University of Texas Press, Austin, TX, pp 3–19

Feeny P (1976) Plant apparency and chemical defense. In: Wallace JW, Mansell RL (eds) Recent advances in phytochemistry, vol 10. Plenum, New York, pp 1–40

Fraser AM, Mechaber WL, Hildebrand JG (2003) Electroantennographic and behavioural responses of the sphinx moth *Manduca sexta* to host plant headspace volatiles. J Chem Ecol 29:1813–1833

Futuyama DJ (1983) Evolutionary interactions among herbivorous insects and plants. In: Futuyama DJ, Slatkins M (eds) Coevolution. Sinauer, Sunderland, MA, pp 207–231

Gayler S, Grams TEE, Heller W, Treutter D, Priesack E (2008) A dynamical model of environmental effects on allocation to carbon-based secondary compounds in juvenile trees. Ann Bot 101:1089–1098

Gazzarrini S, McCourt P (2001) Genetic interactions between ABA, ethylene and sugar signaling pathways. Curr Opin Plant Biol 4:387–391

Gershenzon J (1994) Metabolic costs of terpenoid accumulation in higher plants. J Chem Ecol 20:1281–1328

Gibson SI (2004) Sugar and phytohormone response pathways: navigating a signalling network. J Exp Bot 55:253–264

Gibson SI (2005) Control of plant development and gene expression by sugar signaling. Curr Opin Plant Biol 8:93–102

Goverde M, Erhardt A, Stocklin J (2004) Genotype-specific response of a lycaenid herbivore to elevated carbon dioxide and phosphorous availability in calcareous grassland. Oecologia 139:383–391

Gulmon SL, Mooney HA (1986) Costs of defense on plant productivity. In: Givnish TJ (ed) On the economy of plant form and function. Cambridge University Press, Cambridge, pp 681–698

Hamilton JG, Zangerl AR, DeLucia EH, Berenbaum MR (2001) The carbon–nutrient balance hypothesis: its rise and fall. Ecol Lett 4:86–95

Hamilton JG, Dermody O, Mihai A, Zangerl AR, Rogers A, Berenbaum MR, Delucia EH (2005) Anthropogenic changes in tropospheric composition increase susceptibility of soybean to insect herbivory. Environ Entomol 32:479–485

Hättenschwiler S, Schafellner C (1999) Opposing effects of elevated CO_2 and N deposition on *Lymantria monacha* larvae feeding on spruce trees. Oecologia 118:210–217

Hemming JDC, Lindroth RL (1999) Effect of light and nutrient availability on aspen: growth, phytochemistry and insect performance. J Chem Ecol 25:1687–1714

Herms DA, Mattson WJ (1992) The dilemma of plants: to grow or defend. Q Rev Biol 67:283–335

Himanen SJ, Nerg AM, Nissinen A, Pinto DM, Stewart CN Jr, Poppy GM, Holopainen JK (2009) Effects of elevated carbon dioxide and ozone on volatile terpenoid emissions and multitrophic communication of transgenic insecticidal oilseed rape (*Brassica napus*). New Phytol 181:174–186

Holton MK, Lindroth RL, Nordheim EV (2003) Foliar quality influences tree–herbivore–parasitoid interactions: effects of elevate CO_2, O_3, and plant genotype. Oecologia 137:233–244
Howe GA, Jander G (2008) Plant immunity to insect herbivores. Annu Rev Plant Biol 59:41–66
Hunt MG, Rasmussen S, Newton PCD, Parsons AJ, Newman JA (2005) Near-term impacts of elevated CO_2, nitrogen and fungal endophyte infection on perennial ryegrass: growth, chemical composition and alkaloid production. Plant Cell Environ 28:1345–1354
Hunter MD (2001) Effects of elevated atmospheric carbon dioxide on insect–plant interactions. Agric For Entomol 3:153–159
IPCC Special Report on Emissions Scenarios (SRES), Nakicenovic N, Swart R (eds) (2000) Special report on emissions scenarios, working group III, intergovernmental panel on climate change (IPCC). Cambridge University Press, Cambridge
Johnson RH, Lincoln DE (1990) Sagebrush and grasshopper responses to atmospheric carbon dioxide concentration. Oecologia 84:103–110
Johnson RH, Lincoln DE (1991) Sagebrush carbon allocation patterns and grasshopper nutrition: the influence of CO_2 enrichment and soil mineral limitation. Oecologia 87:127–134
Jones CG, Hartley SE (1999) A protein competition model of phenolic allocation. Oikos 86:27–44
Karban R, Shiojiri K (2009) Self-recognition affects plant communication and defense. Ecol Lett 12:502–506
Karowe DN, Seimens DH, Mitchell-Olds T (1997) Species-specific response of glucosinolate content to elevated atmospheric CO_2. J Chem Ecol 23:2569–2582
Kessler A, Baldwin IT (2001) Defensive function of herbivore-induced plant volatile emissions in nature. Science 291:2141–2144
Kingsolver JG (1996) Physiological sensitivity and evolutionary responses to climate change. In: Körner C, Bazzaz FA (eds) Carbon dioxide, populations and communities. Academic, San Diego, CA, pp 3–12
Kinney KK, Lindroth RL, Jung SM, Nordheim EV (1997) Effects of CO_2 and NO_3^- availability on deciduous trees: phytochemistry and insect performance. Ecology 78:215–230
Klironomos JN, Allen MF, Rillig MC, Pietrowski J, Makvandi-Nejad S, Wolfe BE, Powell JR (2005) Abrupt rise in atmospheric CO_2 overestimates community response in a model plant—soil system. Nature 433:621–624
Koch T, Krumm T, Jung V, Engelberth J, Boland W (1999) Differential induction of plant volatile biosynthesis in the lima bean by early and late intermediates of the octadecanoid-signaling pathway. Plant Physiol 121:153–162
Kopper BJ, Lindroth RL (2003) Effects of elevated carbon dioxide and ozone on the phytochemistry of aspen and performance of an herbivore. Oecologia 134:95–103
Kopper BJ, Lindroth RL, Nordheim EV (2001) CO_2 and O_3 effects on paper birch (Betulaceae: *Betula papyrifera*) phytochemistry and whitemarked tussock moth (Lymantriidae: *Orgyia leucostigma*) performance. Environ Entomol 30:1119–1126
Koricheva J (2002) The carbon–nutrient balance hypothesis is dead; long live the carbon–nutrient balance hypothesis? Oikos 98:537–539
Kreuzwieser J, Cojocariu C, Jüssen V, Rennenberg H (2002) Elevated atmospheric CO_2 causes seasonal changes in carbonyl emissions from *Quercus ilex*. New Phytol 154:327–333
Lake JA, Woodward FI, Quick WP (2002) Long-distance CO_2 signalling in plants. J Exp Bot 53:183–193
Léon P, Sheen J (2003) Sugar and hormone connections. Trends Plant Sci 8:110–116
Lerdau M, Coley PD (2002) Benefits of the carbon–nutrient balance hypothesis. Oikos 98:534–536
Lindroth RL, Dearing MD (2005) Herbivory in a world of elevated CO_2. In: Ehleringer JR, Cerling TE, Dearing MD (eds) A history of atmospheric CO_2 and its effects on plants, animals and ecosystems. Springer, Berlin, pp 468–486
Lindroth RL, Kinney KK, Platz CL (1993) Responses of deciduous trees to elevated atmospheric CO_2: productivity, phytochemistry, and insect performance. Ecology 74:763–777

Lindroth RL, Roth S, Kruger EL, Volin JC, Koss PA (1997) CO_2-mediated changes in aspen chemistry: effects on gypsy moth performance and susceptibility to virus. Glob Chang Biol 3:279–289
Loomis WE (1932) Growth-differentiation balance vs carbohydrate–nitrogen ratio. Proc Am Soc Hortic Sci 29:240–245
Loomis WE (1953) Growth and differentiation – an introduction and summary. In: Loomis WE (ed) Growth and differentiation in plants. State College Press, Ames, IA, pp 1–17
Lorenzo O, Piqueras R, Sánchez-Serrano JJ, Solano R (2003) Ethylene response factor1 integrates signals from ethylene and jasmonate pathways in plant defense. Plant Cell 15:165–178
Loreto F, Fischbach RJ, Schnitzler JP, Cicciolo P, Brancaleoni E, Calfapietra C, Seufert G (2001) Monoterpene emission and monoterpene synthase activities in the Mediterranean evergreen oak *Quercus ilex* L. grown at elevated CO_2 concentrations. Glob Chang Biol 7:709–717
Makino A, Mae T (1999) Photosynthesis and plant growth at elevated levels of CO_2. Plant Cell Physiol 40:999–1006
Mansfield JL, Curtis PS, Zak DR, Pregitzer KS (1999) Genotypic variation for condensed tannin production in trembling aspen (*Populus tremuloides*, Salicaceae) under elevated CO_2 and in high- and low-fertility soil. Am J Bot 86:1154–1159
Martin D, Tholl D, Gershenzon J, Bohlmann J (2002) Methyl jasmonate induces traumatc resin ducts, terpenoid resin biosynthesis, and terpenoid accumulation in developing xylem of Norway spruce stems. Plant Physiol 129:1003–1018
Mattson WJ, Julkunen-Tiitto R, Herms DA (2005) CO_2 enrichment and carbon partitioning to phenolics: do plant responses accord better with the protein competition or the growth-differentiation balance models? Oikos 111:337–347
McDonald EP, Agrell J, Lindroth RL (1999) CO_2 and light effects on deciduous trees: growth, foliar chemistry, and insect performance. Oecologia 119:389–399
McKey DB (1979) The distribution of secondary compounds within plants. In: Rosenthal JA, Janzen DH (eds) Herbivores: their interactions with secondary plant metabolites. Academic, New York, pp 55–133
Medlyn BE, Barton CVM, Broadmeadow MSJ, Ceulemans R, De Angelis P, Forstreuter M, Freeman M, Jackson SB, Kellomäki S, Laitat E, Rey A, Roberntz P, Sigurdsson BD, Strassemeyer J, Wang K, Curtis PS, Jarvis PG (2001) Stomatal conductance of forest species after long-term exposure to elevated CO_2 concentration: a synthesis. New Phytol 149:247–264
Menke FLH, Champion A, Kijne JW, Memelink J (1999) A novel jasmonate- and elicitor-responsive element in the periwinkle secondary metabolite biosynthetic gene *Str* interacts with a jasmonate- and elicitor-inducible AP2-domain transcription factor, ORCA2. EMBO J 18:4455–4463
Mondor EB, Tremblay NM, Awmack CS, Lindroth RL (2004) Divergent pheromone-mediated insect behaviour under global atmospheric change. Glob Chang Biol 10:1820–1824
Narberhaus I, Zintgraf V, Dobler S (2005) Pyrrolizidine alkaloids on three trophic levels – evidence for toxic and deterrent effects on phytophages and predators. Chemoecology 15:121–125
Newman JA, Abner ML, Dado RG, Gibson DJ, Brookings A, Parsons AJ (2003) Effects of elevated CO_2, nitrogen and fungal endophyte infection on tall fescue: growth, photosynthesis, chemical composition and digestibility. Glob Chang Biol 9:425–437
Nielsen KA, Tattersall DB, Jones PR, Møller BL (2008) Metabolon formation in dhurrin biosynthesis. Phytochemistry 69:88–98
Nowak RS, Ellsworth DS, Smith SD (2004) Functional responses of plants to elevated atmospheric CO_2: do photosynthetic and productivity data from FACE experiments support early predictions? New Phytol 162:253–280
Opitz SEW, Müller C (2009) Plant chemistry and insect sequestration. Chemoecology 19:117–154
Peñuelas J, Estiarte M (1998) Can elevated CO_2 affect secondary metabolism and ecosystem function? Trends Ecol Evol 13:20–24

Pickett JA, Wadhams LJ, Woodcock CM (1998) Insect supersense: mate and host location by insects as model systems for exploiting olfactory interactions. Biochemist 20:8–13

Poorter H, Van Berkel Y, Baxter R, Den Hertog J, Dijkstra P, Gifford RM, Griffin KL, Roumet C, Roy J, Wong SC (1997) The effect of elevated CO_2 on the chemical composition and construction costs of leaves of 27 C_3 species. Plant Cell Environ 20:472–482

Rapparini F, Baraldi R, Miglietta F, Loreto F (2004) Isoprenoid emission in trees of *Quercus pubescens* and *Quercus ilex* with lifetime exposure to naturally high CO_2 environment. Plant Cell Environ 27:381–391

Rasmann S, Agrawal AA (2009) Plant defense against herbivory: progress in identifying synergism, redundancy, and antagonism between resistance traits. Curr Opin Plant Biol 12:473–478

Rasmann S, Köllner TG, Degenhardt J, Hiltpold I, Toepfer S, Kuhlmann U, Gershenzon J, Turlings TCJ (2005) Recruitment of entomopathogenic nematodes by insect-damaged maize roots. Nature 434:732–737

Raupach MR, Marland G, Ciais P, Le Quéré C, Canadell JG, Klepper G, Field CB (2007) Global and regional drivers of accelerating CO_2 emissions. Proc Natl Acad Sci USA 104:10288–10293

Reichardt PB, Chapin FS III, Bryant JP, Mattes BR, Clausen TP (1991) Carbon/nutrient balance as a predictor of plant defense in Alaskan balsam poplar: potential importance of metabolite turnover. Oecologia 88:401–406

Rhoades DF (1979) Evolution of plant chemical defense against herbivores. In: Rosenthal GA, Janzen DH (eds) Herbivores: their interaction with secondary plant metabolites. Academic, New York, pp 1–55

Rhoades DF, Cates RG (1976) Toward a general theory of plant antiherbivore chemistry. In: Wallace JW, Mansell RL (eds) Recent advances in phytochemistry, vol 10. Plenum, New York, pp 168–213

Rolland F, Baena-Gonzalez E, Sheen J (2006) Sugar sensing and signaling in plants: conserved and novel machanisms. Annu Rev Plant Biol 57:675–709

Roth SK, Lindroth RL (1995) Elevated atmospheric CO_2: effects on phytochemistry, insect performance and insect–parasitoid interactions. Glob Chang Biol 1:173–182

Roth S, Lindroth RL, Volin JC, Kruger EL (1998) Enriched atmospheric CO_2 and defoliation: effects on tree chemistry and insect performance. Glob Chang Biol 4:419–430

Runion GB, Curl EA, Rogers HH, Backman PA, Rodriguez-Kabana R, Helms BE (1994) Effects of free-air CO_2 enrichment on microbial populations in the rhizosphere and phyllosphere of cotton. Agric For Meteorol 70:117–130

Saxon ME, Davis MA, Pritchard SG, Runion GB, Prior SA, Stelzer HE, Rogers HH, Dute RR (2004) Influence of elevated CO_2, nitrogen, and *Pinus elliottii* genotypes on performance of the redheaded pine sawfly, *Neodiprion lecontei*. Can J Res 34:1007–1017

Schädler M, Roeder M, Brandl R (2007) Interacting effects of elevated CO_2, nutrient availability and plant species on a generalist invertebrate herbivore. Glob Chang Biol 13:1005–1015

Schaller F, Schaller A, Stintzi A (2004) Biosynthesis and metabolism of jasmonates. J Plant Growth Regul 23:179–199

Schnee C, Kollner TG, Held M, Turlings TCJ, Gershenzon J, Degenhardt J (2006) The products of a single maize sesquiterpene synthase form a volatile defense signal that attracts natural enemies of maize herbivores. Proc Natl Acad Sci USA 103:1129–1134

Schoonhoven LM, van Loon JJA, Dicke M (2005) Insect–plant biology, 2nd edn. Oxford University Press, Oxford

Städler E, Buser H-R (1984) Defense chemicals in leaf surface wax synergistically stimulate oviposition by a phytophagous insect. Experientia 40:1157–1159

Stamp N (2003) Out of the quagmire of plant defense hypotheses. Q Rev Biol 78:23–55

Stitt M, Krapp A (1999) The interaction between elevated carbon dioxide and nitrogen nutrition: the physiological and molecular background. Plant Cell Environ 22:583–621

Strauss SY, Rudgers JA, Lau JA, Irwin RE (2002) Direct and ecological costs of resistance to herbivory. Trends Ecol Evol 17:278–285

Strauss SY, Irwin RE, Lambrix VM (2004) Optimal defence theory and flower petal colour predict variation in the secondary chemistry of wild radish. J Ecol 92:132–141
Strong DR, Lawton JH, Southwood TRE (1984) Insects on plants: community patterns and mechanisms. Blackwell, Oxford
Teng N, Wang J, Chen T, Wu X, Wang Y, Lin J (2006) Elevated CO_2 induces physiological, biochemical and structural changes in leaves of *Arabidopsis thaliana*. New Phytol 172:92–103
Tognetti R, Johnson JD, Michelozzi M, Raschi A (1998) Response of foliar metabolism in mature trees of *Quercus pubescens* and *Quercus ilex* to long-term elevated CO_2. Environ Exp Bot 39:233–245
Umina PA, Weeks AR, Kearney MR, McKechnie SW, Hoffmann AA (2005) A rapid shift in a classic clinal pattern in *Drosophila* reflecting climate change. Science 308:691–693
Vet LEM, Dicke M (1992) Ecology of infochemical use by natural enemies in a tritrophic context. Annu Rev Entomol 37:141–172
Veteli TO, Kuokkanen K, Julkunen-Tiitto R, Roininen H, Tahavanainen J (2002) Effects of elevated CO_2 and temperature on plant growth and herbivore defensive chemistry. Glob Chang Biol 8:1240–1252
Vuorinen T, Reddy GVP, Nerg AM, Holopainen JK (2004a) Monoterpene and herbivore-induced emission from cabbage plants grown at elevated atmospheric CO_2 concentration. Atmos Environ 38:675–682
Vuorinen T, Nerg AM, Ibrahim MA, Reddy GVP, Holopainen JK (2004b) Emission of *Plutella xylostella*-induced compounds from cabbages grown at elevated CO_2 and orientation behavior of the natural enemies. Plant Physiol 135:1984–1992
Williams RS, Lincoln DE, Thomas RB (1994) Loblolly pine grown under elevated CO_2 affects early instar pine sawfly performance. Oecologia 98:64–71
Williams RS, Lincoln DE, Thomas RB (1997) Effects of elevated CO_2-grown loblolly pine needles on the growth, consumption, development, and pupal weight of red-headed pine sawfly larvae reared within open-topped chambers. Glob Chang Biol 3:501–511
Williams RS, Lincoln DE, Norby RJ (2003) Development of gypsy moth larval feeding on red maple saplings at elevated CO_2 and temperature. Oecologia 137:114–122
Winkel-Shirley B (2001) Flavonoid biosynthesis. A colorful model for genetics, biochemistry, cell biology, and biotechnology. Plant Physiol 126:485–493
Woodward FI (1987) Stomatal numbers are sensitive to increase in CO_2 from pre-industrial levels. Nature 327:617–618
Woodward F (2002) Potential impacts of global elevated CO_2 concentrations on plants. Curr Opin Plant Biol 5:207–211
Woodward FI, Bazzaz FA (1988) The responses of stomatal density to CO_2 partial pressure. J Exp Bot 39:1771–1781
Yong JWH, Wong SC, Letham DS, Hocart CH, Farquhar GD (2000) Effects of elevated [CO_2] and nitrogen nutrition on cytokinins in the xylem sap and leaves of cotton. Plant Physiol 124:767–779
Zavala JA, Casteel CL, DeLucia EH, Berenbaum MR (2008) Anthropogenic increase in carbon dioxide compromises plant defense against invasive insects. Proc Natl Acad Sci USA 105:5129–5133
Zavala JA, Casteel CL, Nabity PD, Berenbaum MR, DeLucia EH (2009) Role of cysteine proteinase inhibitors in preference of Japanese beetles (*Popillia japonica*) for soybean (*Glycine max*) leaves of different ages and grown under elevated CO_2. Oecologia 161:35–41

Appendix 1
References Used for (Figs. 11.2–11.4, 11.6, and 11.7)

Agrell J, McDonald EP, Lindroth RL (2000) Effects of CO_2 and light on tree phytochemistry and insect performance. Oikos 88:259–272

Agrell J, Anderson P, Oleszek W, Stochmal A, Agrell C (2004) Combined effects of elevated CO_2 and herbivore damage on alfalfa and cotton. J Chem Ecol 30:2309–2324

Agrell A, Kopper B, McDonald EP, Lindroth RL (2005) CO_2 and O_3 effects on host plant preferences of the forest tent caterpillar (*Malacosoma disstria*). Glob Change Biol 11:588–599

Bazin A, Goverde M, Erhardt A, Shykoff (2002) Influence of atmospheric carbon dioxide enrichment on induced response and growth compensation after herbivore damage in *Lotus corniculatus*. Ecol Entomol 27:271–278

Bezemer TM, Jones TH, Newington JE (2000) Effects of carbon dioxide and nitrogen fertilzation on phenolic content in *Poa annua* L. Biochem Syst Ecol 28:839–846

Bidart-Bouzat MG, Mithen R, Berenbaum MR (2005) Elevated CO_2 influences herbivory-induced defense responses of *Arabidopsis thaliana*. Oecologia 145:415–424

Booker FL (2000) Influence of carbon dioxide enrichment, ozone and nitrogen fertilization on cotton (*Gossypium hirsutum* L.) leaf and root composition. Plant Cell Environ 23:573–583

Booker FL, Maier CA (2001) Atmospheric carbon dioxide, irrigation, and fertilization effects on phenolic and nitrogen concentration in loblolly pine (*Pinus taeda*) needles. Tree Physiol 21:609–616

Castells E, Roumet C, Penuelas J, Roy J (2002) Intraspecific variability of phenolic concentrations and their responses to elevated CO_2 in two Mediterranean perennial grasses. Environ Exp Bot 47:205–216

Causin HF, Rufty TW, Reynolds JF (2006) Gas exchange and carbon metabolism in two *Prosopis* species (Fabaceae) from semiarid habitats: effects of elevated CO_2, N supply, and N source. Am J Bot 93:716–723

Centritto M, Nascetti PN, Petrilli L, Raschi A, Loreto F (2004) Profiles of isoprene emission and photosynthetic parameters in hybrid poplars exposed to free-air CO_2 enrichment. Plant Cell Environ 27:403–412

Chen F, Wu G, Ge F, Parajulee MN, Shrestha RB (2005) Effects of elevated CO_2 and transgenic Bt cotton on plant chemistry, performance, and feeding of an insect herbivore, the cotton bollworm. Entomol Exp Appl 115:341–350

Cipollini ML, Drake BG, Whigham D (1993) Effects of elevated CO_2 on growth and carbon/nutrient balance in the deciduous woody shrub *Lindera benzoin* (L.) Blume (Lauraceae). Oecologia 96:339–346

Coley PD, Massa M, Lovelock CE, Winter K (2002) Effects of elevated CO_2 on foliar chemistry of saplings on nine species of tropical tree. Oecologia 133:62–69

Constable JVH, Litvak ME, Greenberg JP, Monson RK (1999) Monoterpene emission from coniferous trees in response to elevated CO_2 concentration and climate warming. Glob Change Biol 5:255–267

Cornelissen T, Stiling P, Drake B (2003) Elevated CO_2 decreases leaf fluctuating asymmetry and herbivory by leaf miners on two oak species. Glob Change Biol 10:27–36

Coviella CE, Stipanovic RD, Trumble JT (2002) Plant allocation to defensive compounds: interactions between elevated CO_2 and nitrogen in transgenic cotton plants. J Exp Bot 53:323–331

Davey MP, Bryant DN, Cummins I, Ashenden TW, Gates P, Baxter R, Edwards R (2004) Effects of elevated CO_2 on the vasculature and phenolic secondary metabolism of *Plantago maritima*. Phytochemistry 65:2197–2204

Diaz S, Fraser LH, Grime JP, Falczuk (1998) The impact of elevated CO_2 on plant-herbivore interactions: experimental evidence of moderating effects at the community level. Oecologia 117:177–186

Dury SJ, Good JEG, Perrins CM, Buse A, Kaye T (1998) The effects of increasing CO_2 and temperature on oak leaf palatability and the implications for herbivorous insects. Glob Change Biol 4:55–61

Estiarte M, Penuelas J, Kimball BA, Hendrix DL, Pinter PJ Jr, Wall GW, LaMorte RL, Hunsaker DJ (1999) Free-air CO_2 enrichment of wheat: leaf flavonoid concentration throughout the growth cycle. Physiol Plant 105:423–433

Fajer ED (1989) The effects of enriched CO_2 atmospheres on plant-insect herbivore interactions: growth responses of larvae of the specialist butterfly, *Junonia coenia* (Lepidoptera: Nymphalidae). Oecologia 81:514–520

Fajer ED, Bowers MD, Bazzaz FA (1992) The effect of nutrients and enriched CO_2 environments on production of carbon-based allelochemicals in plantago: a test of the carbon/nutrient balance hypothesis. Am Nat 140:707–723

Gebauer RLE, Strain BR, Reynolds JF (1998) The effect of elevated CO_2 and N availability on tissue concentrations and whole plant pools of carbon-based secondary compounds in loblolly pin (*Pinus taeda*). Oecologia 113:29–36

Gleadow RM, Foley WJ, Woodrow IE (1998) Enhanced CO_2 alters the relationship between photosynthesis and defense in cyanogenic *Eucalyptus cladocalyx* F. Muell. Plant Cell Environ 21:12–22

Goverde M, Bazin A, Shykoff JA, Erhardt A (1999) Influence of leaf chemistry of *Lotus corniculatus* (Fabaceae) on larval development of *Polyommatus icarus* (Lepidoptera, Lycaenidae): effects of elevated CO_2 and plant genotype. Funct Ecol 13:801–810

Goverde M, Erhardt A, Niklaus PA (2002) In situ development of a satyrid butterfly on calcareous grassland exposed to elevated carbon dioxide. Ecology 83:1399–1411

Goverde M, Erhardt A, Stocklin J (2004) Genotype-specific response of a lycaenid herbivore to elevated carbon dioxide and phosphorous availability in calcareous grassland. Oecologia 139:383–391

Hall MC, Stiling P, Moon DC, Drake BG, Hunter MD (2005) Effects of elevated CO_2 on foliar quality and herbivore damage in a scrub oak ecosystem. J Chem Ecol 31:267–286

Haring DA, Korner CH (2004) CO_2 enrichment reduces the relative contribution of latex and latex-related hydrocarbons to biomass in *Euphorbi lathyris*. Plant Cell Environ 27:209–217

Hartley SE, Jones CG, Couper GC, Jones TH (2000) Biosynthesis of plant phenolic compounds in elevated atmospheric CO_2. Glob Change Biol 6:497–506

Hattas D, Stock WD, Mabusela WT, Green IR (2005) Phytochemical changes in leaves of subtropical grasses and fynbos shrubs at elevated atmospheric CO_2 concentrations. Glob Planet Change 47:181–192

Hättenschwiler S, Schafellner C (1999) Opposiing effects of elevated CO_2 and N desposition on *Lymantria monacha* larvae feeding on spruce trees. Oecologia 118:210–217

Heyworth CJ, Iason GR, Temperton V, Jarvis PG, Duncan AJ (1998) The effect of elevated CO_2 concentration and nutrient supply on carbon-based plant secondary metabolites in *Pinus sylvestris* L. Oecologia 115:344–350

Himanen SJ, Nissinen A, Auriola S, Poppy GM, Stewart CNJ, Holopainen JK, Nerg A (2008) Constitutive and herbivore-inducible glucosinolate concentrations in oilseed rape (*Brassica napus)* leaves are not affected by *Bt Cry1Ac* insertion but change under elevated atmospheric CO_2 and O_3. Planta 227: 427–437

Himanen SJ, Nerg AM, Nissinen A, Pinto DM, Stewart CN Jr, Poppy GM, Holopainen JK (2009) Effects of elevated carbon dioxide and ozone on volatile terpenoid emissions and multitrophic communication of transgenic insecticidal oilseed rape (*Brassica napus*). New Phytol 181:174–186

Holten MK, Lindroth RL, Nordheim EV (2003) Foliar quality influences tree-herbivore-parasitoid interactions: effects of elevate CO_2, O_3, and plant genotype. Oecologia 137:233–244

Huang W, He X, Chen W, Chen Z, Ruan Y, Xu S (2008) Influence of elevated carbon dioxide and ozone on the foliar nonvolatile terpenoids in *Ginkgo biloba*. Bull Environ Contam Toxicol 81:432–435

Johnson RH, Lincoln DE (1990) Sagebrush and grasshopper responses to atmospheric carbon dioxide concentration. Oecologia 84:103–110

Johnson RH, Lincoln DE (1991) Sagebrush carbon allocation patterns and grasshopper nutrition: the influence of CO_2 enrichment and soil mineral limitation. Oecologia 87:127–134

Julkunen-Tiitto R, Tahvanainen J, Silvola J (1993) Increased CO_2 and nutrient status changes affect phytomass and the production of plant defensive secondary chemicals in *Salix myrsinifolia* (Salisb.). Oecologia 95:495–498

Kainulainen P, Holopainen JK, Holopainen T (1998) The influence of elevated CO_2 and O_3 concentrations on Scots pine needles: changes in starch and secondary metabolites over three exposure years. Oecologia 114:455–460

Karowe DN, Seimens DH, Mitchell-Olds T (1997) Species-specific response of glucosinolate content to elevated atmospheric CO_2. J Chem Ecol 23:2569–2582

Kerslake JE, Woodin SJ, Hartley SE (1998) Effects of carbon dioxide and nitrogen enrichment on a plant-insect interaction: the quality of *Calluna vulgaris* as a host for *Operophtera brumata*. New Phytol 140:43–53

Kinney KK, Lindroth RL, Jung SM, Nordheim EV (1997) Effects of CO_2 and NO_3^- availability on deciduous trees: phytochemistry and insect performance. Ecology 78:215–230

Knepp RG, Hamilton JG, Mohan JE, Zangerl AR, Berenbaum MR, DeLucia EH (2005) Elevated CO_2 reduces leaf damage by insect herbivores in a forest community. New Phytol 267:207–218

Koike T, Tobita H, Shibata T, Matsuki S, Konno K, Kitao M, Yamashita N, Maruyama Y (2006) Defense characteristics of seral deciduous broad-leaved tree seedlings grown under differing levels of CO_2 and nitrogen. Popul Ecol 48:23–29

Kopper BJ, Lindroth RL (2003) Effects of elevated carbon dioxide and ozone on the phytochemistry of aspen and performance of an herbivore. Oecologia 134:95–103

Kopper BJ, Lindroth RL, Nordheim EV (2001) CO_2 and O_3 effects on Paper Birch (Betulaceae: *Betula papyrifera*) phytochemistry and Whitemarked Tussock Moth (Lymantriidae: *Orgyia leucostigma*) performance. Environ Entomol 30:1119–1126

Kreuzwieser J, Cojocariu C, Jüssen V, Rennenberg H (2002) Elevated atmospheric CO_2 causes seasonal changes in carbonyl emissions from *Quercus ilex*. New Phytol 154:327–333

Kuokkanen K, Julkunen-Tiitto R, Keinanen M, Niemela P, Tahavanainen J (2001) The effect of elevated CO_2 and temperature on the secondary chemistry of *Betula pendula* seedlings. Trees 15:378–384

Kuokkanen K, Yan S, Niemela P (2003) Effects of elevated CO_2 and temperature on the leaf chemistry of birtch *Betula pendula* (Roth) and the feeding behaviour of the weevil *Phyllobius maculicornis*. Agric For Entomol 5:209–217

Lavola A, Julkunen-Tiitto R (1994) The effect of elevated carbon dioxide and fertilization on primary and secondary metabolites in birth, *Betula pendula* (Roth). Oecologia 99:315–321

Lavola A, Julkuen-Tiitto R, de la Rosa TM, Lehto T, Aphalo PJ (2000) Allocation of carbon to growth and secondary metabolites in birch seedlings under UV-B radiation and CO_2 exposure. Physiol Plant 109:260–267

Lawler IR, Foley WJ, Woodrow IE, Cork SJ (1997) The effects of elevated CO_2 atmospheres on the nutritional quality of eucalyptus foliage and its interaction with soil nutrient and light availability. Oecologia 109:59–68

Lincoln DE, Couvet D (1989) The effect of carbon supply on allocation to allelochemicals and caterpillar consumption of peppermint. Oecologia 78:112–114

Lindroth RL, Kinney KK (1998) Consequences of enriched atmospheric CO_2 and defoliation for foliar chemistry and gypsy moth performance. J Chem Ecol 24:1677–1695

Lindroth RL, Kinney KK, Platz CL (1993) Responses of deciduous tree to elevated atmospheric CO_2: productivity, phytochemistry, and insect performance. Ecology 74:763–777

Lindroth RL, Arteel GE, Kinney KK (1995) Responses of three saturniid species to paper birch grown under enriched CO_2 atmospheres. Funct Ecol 9:306–311

Lindroth RL, Roth S, Kruger EL, Volin JC, Koss PA (1997) CO_2-mediated changes in aspen chemistry: effects on gypsy moth performance and susceptibility to virus. Glob Change Biol 3:279–289

Lindroth RL, Kopper BJ, Parsons WFJ, Bockheim JG, Karnosky DF, Hendrey GR, Pregitzer KS, Isebrands JG, Sober J (2001) Consequences of elevated carbon dioxide and ozone for foliar chemical composition and dynamics in trembling aspen (*Populus tremuloids*) and paper birch (*Betula papyrifera*). Environ Pollut 115:395–404

Lindroth RL, Wood SA, Kopper BJ (2002) Response of quaking aspen genotypes to enriched CO_2: foliar chemistry and tussock moth performance. Agric For Entomol 4:315–323

Loreto F, Fischbach RJ, Schnitzler JP, Cicciolo P, Brancaleoni E, Calfapietra C, Seufert G (2001) Monterpene emission and monoterpene synthase activities in the Mediterranean evergreen oak *Quercus ilex* L. grown at elevated CO_2 concentrations. Glob Change Biol 7:709–717

Mansfield JL, Curtis PS, Zak DR, Pregitzer KS (1999) Genotypic variation for condensed tannin production in trembling aspen (*Poulus tremuloides*, Salicaceae) under elevated CO_2 and in high- and low-fertility soil. Am J Bot 86:1154–1159

Matros A, Amme S, Kettig B, Buck-Sorlin GH, Sonnewald U, Mock HP (2006) Growth at elevated CO_2 concentrations leads to modified profiles of secondary metabolites in tobacco cv. SamsunNN and to increased resistance against infections with potato virus Y. Plant Cell Environ 29:126–137

Mattson WJ, Kuokkanen K, Niemela P, Julkunen-Tiitto R, Kellomaki S, Tahvanainen J (2004) Elevated CO_2 alters birch resistance to Lagomorpha herbivores. Glob Change Biol 10:1402–1413

Mattson WJ, Julkunen-Tiitto R, Herms DA (2005) CO_2 enrichment and carbon partitioning to phenolics: do plant responses accord better with the protein competition or the growth-differentiation balance models? Oikos 111:337–347

McDonald EP, Agrell J, Lindroth RL (1999) CO_2 and light effects on deciduous trees: growth, foliar chemistry, and insect performance. Oecologia 119:389–399

Peltonen PA, Vapaavuori E, Julkunen-Tiitto R (2005) Accumulation of phenolic compounds in birch leaves is changed by elevated carbon dioxide and ozone. Glob Change Biol 11:1305–1324

Penuelas J, Llusia J (1997) Effects of carbon dioxide, water supply, and seasonality on terpene content and emission by *Rosmarinus officinalis*. J Chem Ecol 23:979–993

Penuelas J, Estiarte M, Kimball BA, Idso SB, Pinter PJ Jr, Wall GW, Garcia RL, Hansaker DJ, LaMorte RL, Hendrix DL (1996) Variety of responses of plant phenolic concentration to CO_2 enrichment. J Exp Bot 47:1463–1467

Pritchard S, Peterson C, Runion GB, Priorj S, Rogers H (1997) Atmospheric CO_2 concentration, N availability, and water status affect patterns of ergastic substance deposition in longleaf pine (*Pinus palustris* Mill.) foliage. Trees 11:494–503

Räisänen T, Ryyppö A, Julkunen-Tiitto R, Kellomaki S (2008) Effects of elevated CO_2 and temperature on secondary compounds in the needles of Scots pine (*Pinus sylvestris* L.). Trees 22:121–135

Rapparini F, Baraldi R, Miglietta F, Loreto F (2004) Isoprenoid emission in trees of *Quercus pubescens* and *Quercus ilex* with lifetime exposure to naturally high CO_2 environment. Plant Cell Environ 27:381–391

Reddy GVP, Tossavainen P, Nerg AM, Holopainen JK (2004) Elevated atmospheric CO_2 affects the chemical quality of *Brassica* plants and the growth rate of the specialist, *Plutella xylostella*, but not the generalist, *Spodoptera littoralis*. J Agric Food Chem 52:4185–4191

Reitz SR, Karowe DN, Diawara MM, Trumble JT (1997) Effects of elevated atmospheric carbon dioxide on the growth and linear furanocoumarin content of celery. J Agric Food Chem 45:3642–3646

Rossi AM, Stiling P, Moon DC, Cattell MV, Drake BG (2004) Induced defensive response of myrtle oak to foliar insect herbivory in ambient and elevated CO_2. J Chem Ecol 30:1143–1152

Roth SK, Lindroth RL (1994) Effects of CO_2-mediated changes in paper birch and white pine chemistry on gypsy moth performance. Oecologia 98:133–138

Roth SK, Lindroth RL (1995) Elevated atmospheric CO_2: effects on phytochemistry, insect performance and insect-parasitoid interactions. Glob Change Biol 1:173–182

Roth S, McDonald EP, Lindroth RL (1997) Atmospheric CO_2 and soil water availability: consequences for tree-insect interaction. Can J For Res 27:1281–1290

Roth S, Lindroth RL, Volin JC, Kruger EL (1998) Enriched atmospheric CO_2 and defoliation: effects on tree chemistry and insect performance. Glob Change Biol 4:419–430

Sallas L, Kainulainen P, Utriainen J, Holopainen T, Holopainen JK (2001) The influence of elevated O_3 and CO_2 concentrations on secondary metabolites of Scots pine (*Pinus sylvestris* L.) seedlings. Glob Change Biol 7:303–311

Sallas L, Luomala EM, Utriainen J, Kainulainen P, Holopainen JK (2003) Contrasting effects of elevated carbon dioxide concentration and temperature on Rubisco activity, chlorophyll fluorescence, needle ultrastructure and secondary metabolites in conifer seedlings. Tree Physiol 23:97–108

Saxon ME, Davis MA, Pritchard SG, Runion GB, Prior SA, Stelzer HE, Rogers HH, Dute RR (2004) Influence of elevated CO_2, nitrogen, and *Pinus elliottii* genotypes on performance of the redheaded pine sawfly, *Neodiprion lecontei*. Can J For Res 34:1007–1017

Scholefield PA, Doick KJ, Herbert BMJ, Hewitt CNS, Schnitzler JP, Pinelli P, Loreto F (2004) Impact of rising CO_2 on emissions of volatile organic compounds: isoprene emission from *Phragmites australis* growing at elevated CO_2 in a natural carbon dioxide spring. Plant Cell Environ 27:393–401

Schonhof I, Kläring HP, Krumbein A, Schreiner M (2007) Interaction between atmospheric CO_2 and glucosinolates in broccoli. J Chem Ecol 33:105–114

Snow MD, Bard RR, Olszyk DM, Minster LM, Hager AN, Tingey DT (2003) Monoterpene levels in needles of Douglas fir exposed to elevated CO_2 and temperature. Physiol Plant 117:352–358

Staudt M, Joffre R, Rambal S, Kesselmeier J (2001) Effects of elevated CO_2 on monoterpene emission of young *Quercus ilex* trees and its relation to structural and ecophysiological parameters. Tree Physiol 21:437–445

Tognetti R, Johnson JD, Michelozzi M, Raschi A (1998) Response of foliar metabolism in mature trees of *Quercus pubescens* and *Quercus ilex* to long-term elevated CO_2. Environ Exp Bot 39:233–245

Veteli TO, Kuokkanen K, Julkunen-Tiitto R, Roininen H, Tahavanainen J (2002) Effects of elevated CO_2 and temperature n plant growth and herbivore defensive chemistry. Glob Change Biol 8:1240–1252

Veteli TO, Mattson WJ, Niemela P, Julkunen-Tiitto R, Kellomaki S, Kuokkanen K, Lavola A (2007) Do elevated temperature and CO_2 generally have counteracting effects on phenolic phytochemistry of boreal trees? J Chem Ecol 33:287–296

Vuorinen T, Reddy GVP, Nerg AM, Holopainen JK (2004) Monoterpene and herbivore-induced emission from cabbage plants grown at elevated atmospheric CO_2 concentration. Atmos Environ 38:675–682

Vuorinen T, Nerg AM, Vapaavuori E, Holopainen JK (2005) Emission of volatile organic compounds from two silver birch (*Betula pendula* Roth) clones grown under ambient and elevated CO_2 and different O_3 concentrations. Atmos Environ 39:1185–1197

Wang XW, Ji LZ, Zhang QH, Liu Y, Wang GQ (2009) Effects of elevated CO_2 on feeding preference and performance of the gypsy moth (*Lymantria dispar*) larvae. J Appl Entomol 133:47–57

Williams RS, Lincoln DE, Thomas RB (1994) Loblolly pine grown under elevated CO_2 affects early instar pine sawfly performance. Oecologia 98:64–71

Williams RS, Lincoln DE, Thomas RB (1997) Effects of elevated CO_2-grown loblolly pine needles on the growth, consumption, development, and pupal weight of red-headed pine sawfly larvae reared within open-topped chambers. Glob Change Biol 3:501–511

Williams RS, Lincoln DE, Norby RJ (1998) Leaf age effects of elevated CO_2-grown white oak leaves on spring-feeding lepidopterans. Glob Change Biol 4:235–246
Williams RS, Norby RJ, Lincoln DE (2000) Effects of elevated CO_2 and temperature-grown red and sugar maple on gypsy moth performance. Glob Change Biol 6:685–695
Williams RS, Lincoln DE, Norby RJ (2003) Development of gypsy moth larvar feeding on red maple saplings at elevated CO_2 and temperature. Oecologia 137:114–122
Wu G, Chen FJ, Ge F, Sun YC (2007) Effects of elevated carbon dioxide on the growth and foliar chemistry of transgenic Bt Cotton. J Integr Plant Biol 49:1361–1369
Ziska LH, Emche SD, Johnson EL, George K, Reed DR, Sicher RC (2005) Alterations in the productions and concentration of selected alkaloids as a function of rising atmospheric carbon dioxide and air temperature: implications for ethno-pharmacology. Glob Change Biol 11:1798–1807

Appendix 2
List of Abbreviations for Fig. 11.5

Alkaloids (red)

ImiAlk	Imidazole alkaloids
AromAlk	Aromatic alkaloids
	Quinazoline alkaloids (from anthranilate, precursor of L-Trp)
	Quinoline alkaloids (from anthranilate, precursor of L-Trp)
	Acridine alkaloids (from anthranilate, precursor of L-Trp)
	Indole alkaloids (from L-Trp)
	Quinoline alkaloids (from L-Trp)
	Tetrahydroisoquinoline alkaloids (from L-Tyr)
	Benzylisoquinoline alkaloids (from L-Tyr)
	Amaryllidaceae alkaloids (from L-Tyr)
PyrAlk	Pyridine alkaloids (from nicotinate, product of L-Asp)
PipAlk	Piperidine alkaloids (from L-Lys)
QuinoAlk	Quinolizidine alkaloids (from L-Lys)
IndoAlk	Indolizidine alkaloids (from L-Lys)
PyrroAlk	Pyrrolizidine and pyrrolidine alkaloids (from L-Orn)
TropAlk	Tropane alkaloids

Other secondary metabolites (red)

CyG	Cyanogenic glycosides
GSL	Glucosinolates
PhenProp	Phenylpropanoids (includes hydroxycinnamic acids, e.g. caffeic acid, and their esters, e.g. chlorogenic acid; also hydroxycinnamic aldehydes and alcohols)
(Iso)Flav	Isoflavonoids and flavonoids
Anthocy	Anthocyanins

CTs	Condensed tannins
HTs	Hydrolyzable tannins, e.g. ellagitannins
...terp	...terpene
CardiacGlyc	Cardiac glycosides
SteroidSap	Steroid saponins

Phytohormones (blue)

SA	Salicylic acid
ABA	Abscisic acid
Strigolact	Strigolactone
Gibberel	Gibberellins
BrassSter	Brassinosteroids
Jasm	Jasmonates

Polymeric (essential) metabolites (gray)

Ubiquin	Ubiquinones
Plastoquin	Plastoquinones
MemSterols	Membrane sterols like e.g. cholesterol

Intermediates (no fill)

Amino acids

L-His	L-Histidine
L-Trp	L-Tryptophan
L-Phe	L-Phenylalanine
L-Tyr	L-Tyrosine
L-Ala	L-Alanine
L-Val	L-Valine
L-Leu	L-Leucine
L-Asp	L-Aspartate
L-Asn	L-Asparagine
L-Met	L-Methionine
L-Thr	L-Threonine
L-Ile	L-Isoleucine
L-Lys	L-Lysine
L-Glu	L-Glutamate
L-Gln	L-Glutamine
L-Arg	L-Arginine
L-Pro	L-Proline
L-Orn	L-Ornithine
L-Gly	L-Glycine

L-Ser	L-Serine
L-Cys	L-Cysteine
Isoprenoid/terpene intermediates	
IPP	Isopentenylpyrophosphate
GPP	Geranylpyrophosphate
FPP	Farnesylpyrophosphate
GGPP	Geranylgeranylpyrophosphate

Others

Shik	Shikimate

Exploiting Plant Signals in Sustainable Agriculture

Toby J.A. Bruce

Abstract Plants respond to chemical signals (semiochemicals) that are associated with insect or pathogen attack by modifying their metabolism accordingly so that defence pathways are switched on or primed. Once the relevant semiochemicals or analogues are identified, these signals can be artificially applied to achieve similar effects. Such plant activator agrochemicals represent an entirely different approach from the one traditionally used by the agrochemical industry of deploying pesticide molecules to kill pests. These chemicals do not have direct effects on pests and diseases but upregulate plant defence genes that increase plant resistance to attack. Plant activators are compatible with integrated pest management (IPM) systems and even enhance biocontrol techniques by promoting plant attractiveness to natural enemies of plant pests, as natural enemies of pests prefer induced plants. The plant defence traits activated are often complex relying on the expression of many genes, which makes it harder for pests to adapt to them. Current practise and future prospects are reviewed in this chapter.

1 Introduction

There is a need for new preventative approaches to achieve sustainable pest management because control measures based on pesticides alone provide only short-term relief and have undesired economic and environmental consequences (Lewis et al. 1997). Natural plant signals can be used to develop alternatives for pesticides by using them to elicit plant defence or by producing crop cultivars that are more responsive to them (Chapter 'Volatile Interactions Between Undamaged Plants: Effects and Potential for Breeding Resistance to Aphids'). Plant activator agrochemicals represent an entirely different approach from the one traditionally

T.J.A. Bruce
Rothamsted Research, Harpenden, Herts AL5 2JQ, UK
e-mail: toby.bruce@bbsrc.ac.uk

F. Baluška and V. Ninkovic (eds.), *Plant Communication from an Ecological Perspective*, Signaling and Communication in Plants,
DOI 10.1007/978-3-642-12162-3_12,

used by the agrochemical industry of deploying pesticide molecules to kill pests as they do not have direct effects on pests and diseases but upregulate plant defence genes that increase plant resistance to attack. Before the discovery of chemical elicitors, induced resistance using plant immune responses was not a feasible approach for use in agriculture because of the complication of vaccinating with organisms to protect against subsequent attack (Thaler 1999a). Now, enhancement of natural plant defence processes is a real prospect although much remains to be learnt about it. The utilisation of natural plant immunity is a goal of modern agriculture (von Rad et al. 2005) and the use of elicitors to induce plant immunity is expected to become more important in the future.

There is a growing trend towards replacement of pesticides, for example with biological control agents, because of problems with insecticide resistance and consumer pressures to eliminate pesticide residues from food (van Lenteren 2000). Broad spectrum pesticides kill predators and parasitoids as well as the pests and this sometimes leads to pest resurgences later in the season. In contrast, plant activators are compatible with integrated pest management (IPM) systems and even enhance biocontrol by increasing plant attractiveness to natural enemies (Dicke and Dijkman 1992; Whitfield 2001; Stout et al. 2006; Turlings and Ton 2006; Chapter 'Volatile Chemical Interaction Between Undamaged Plants: Effects at Higher Trophic Levels'). Activators of natural plant defences have thus been proposed as useful tools within IPM strategies that aim to minimise the use of toxic products (Vallad and Goodman 2004). Reducing the release of toxins into the environment by use of alternative control techniques in crop protection would also reduce selection pressure for the development of pesticide resistance by pests thus conserving the efficacy of pesticides. This is arguably more important for agricultural sustainability if we consider sustaining consistent agricultural yields against the threat posed by attacking pests, diseases and weeds that can evolve resistance to control measures. The plant defence traits activated are often complex relying on the expression of many genes which makes it harder for pests to adapt to them (Gardner et al. 1999). In terms of sustainability, just having another weapon in the armoury for control of pests increases the options for crop protection in the future and thus adds to improve yield security.

The ability of plants to respond to chemical signals (semiochemicals) that are associated with insect or pathogen attack allows them to fine tune their metabolism according to the likelihood of exposure to biotic stress factors. For example, emission of herbivore induced volatiles from neighbouring plants can lead to activation of defence pathways that make a plant more resistant to insect attack (Farmer and Ryan 1990; Karban et al. 2000; Baldwin et al. 2006). The concentration of volatiles that is required for this may mean that under natural conditions plant–plant interactions mediated by volatile signals occur over relatively short distances. However, in an agricultural situation artificial application of plant semiochemicals can overcome these distance limitations and there are opportunities to exploit natural plant signalling processes in alternative crop protection strategies that trigger priming or induction of defence pathways in crop plants. In this context it should also be possible to use semiochemicals that are more active than the ones plants are naturally exposed to. Induced defence occurs naturally, as exposure to early season herbivores in the field enhances defence against other herbivores that arrive later in the season (Poelman et al. 2008). The challenge for exploiting induced defence in crop protection is to identify the semiochemical signals that can achieve similar effects with interventions that use semiochemical application instead of insect damage to induce defence metabolism.

2 Induction and Priming of Plant Defences

Induced plant defence traits require a signal to elicit them (Fig. 1). Such signals have the potential to immunise plants against infection by activating defence pathways (Lyon et al. 1995). Plants possess several inducible, systemic defence responses to pests and pathogens (Agrawal 1998; Vallad and Goodman 2004; Bruce and Pickett 2007). The two signalling pathways best known to regulate plant defence against biotic stress are the one regulated by salicylic acid (SA) that is often associated with systemic acquired resistance (SAR) against microbial pathogens (Sticher et al. 1997; Lucas 1999) and the other regulated by jasmonic acid (JA) that often provides resistance against herbivorous pests and necrotrophic pathogens. Although early research on induced defence signalling in plants linked the SA pathway with defence against pathogens and the JA pathway with defence against herbivores, numerous recent studies have shown a more complex picture, with varying involvement of both pathways in different pathogen and herbivore interactions depending on the species involved (Stout et al. 2006; Bruce and Pickett 2007; Zhu-Salzman et al. 2008). Furthermore, it is possible that other defence pathways exist that we do not yet know about.

Many chemical activators of induced defences against biotic attackers are known (Paré et al. 2005) and some of these have been commercialised for crop protection (Vallad and Goodman 2004; von Rad et al. 2005). However, sustained activation of defence may be costly in terms of resources, and long-term activation

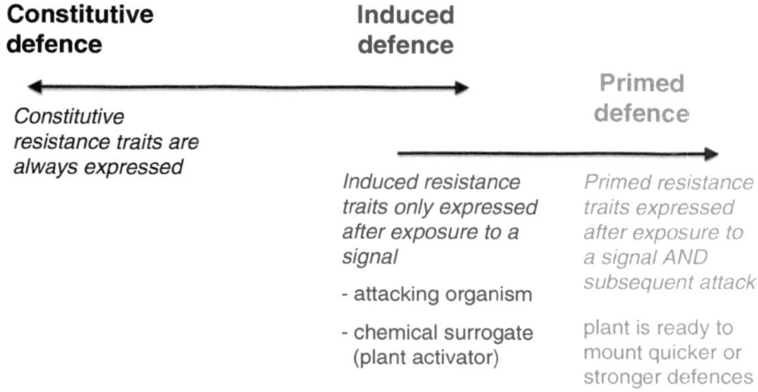

Fig. 1 The different types of plant defence traits

of induced defences can result in yield penalties (Vallad and Goodman 2004; van Hulten et al. 2006) although this was not found with JA induced tomato plants (Thaler 1999a) and the release of volatiles is not necessarily very costly (Aharoni et al. 2005). An alternative to direct activation of defence, is 'priming'.

The process of priming occurs when prior exposure to a biotic or an abiotic stimulus sensitises a plant to express a more efficient defence response to future exposure (Conrath et al. 2006; Beckers and Contrath 2007; Bruce et al. 2007). Primed plants display either faster and/or stronger activation of the various defence responses that are induced following pathogen or insect attack or exposure to abiotic stress. This was shown for example by Ton et al. (2007) in maize where exposure of plants to volatiles from caterpillar-infested plants did not activate defence genes directly but primed a subset of them for earlier and/or stronger induction upon subsequent defence elicitation. Synthetic chemicals, as well as natural stress conditions, can be used to prime defence when applied artificially to a plant (Jakab et al. 2005; Ton et al. 2005, 2007; Engelberth et al. 2007; Kessler et al. 2006; Heil and Silva Bueno 2007). Another advantage of priming is that it could avoid conflicts in plant signalling pathways (e.g. between SA and JA) because a plant may be primed for several defence pathways at once (Dr J. Ton, personal communication). Some evidence for this was obtained by van Wees et al. (2000) where JA and SA activation gave additive effects against *Pseudomonas syringae* pv. *tomato*. With priming, defence genes are not expressed immediately after treatment and so it may be possible for different suites of genes to be upregulated upon attack by different types of attackers. As plants face a diversity of different attacking organisms (Fig. 2) as well as abiotic threats it is necessary for them to have a system of defences that is suited to this (Stout et al. 2006).

There can be genetic variability in plant response to activator treatment (Vallad and Goodman 2004; Chapters 'Volatile Chemical Interaction Between Undamaged

Fig. 2 Crop plants are attacked by a variety of antagonistic pests, diseases and weeds

Plants: Effects at Higher Trophic Levels' and 'Volatile Interactions Between Undamaged Plants: Effects and Potential for Breeding Resistance to Aphids'). It is important to consider this factor otherwise there could be variable performance in terms of efficacy with different crop cultivars. However, this also represents an opportunity because there is the potential to use a plant activator in a package with selected crop cultivars that offer the best genetic potential for induced defence. Another problem is that some plant activators may be phytotoxic and this aspect should be considered when developing them.

3 Exploiting Plant Semiochemicals for Agriculture

3.1 Plant Diseases

Plant activators can be used against plant diseases and several of these have been commercialised. *S*-Methyl benzo [1,2,3]thiadiazole-7-carbothioate (BTH or benzothiadiazole) is a chemical analogue of SA and is used to induce SAR against plant diseases. It is commercially available from Syngenta as 'Bion' in Europe or 'Actigard' in the USA and has been used in the field to control a range of plant diseases reviewed by Vallad and Goodman (2004). These include bacterial spot and bacterial speck in tomatoes (Louws et al. 2001; Herman et al. 2008) and *Stemphylium* leaf blight disease in onion (Kamal et al. 2008). Other plant activators based on plant growth promoting rhizobacteria that activate induced systemic resistance (ISR) are available (Vallad and Goodman 2004; Herman et al. 2008), for example, 'BioYield Concentrate' from Bayer CropScience contains endospores of *Bacillus subtilis* and *B. amyloliquefaciens*. 'Oxycom' (Redox Chemicals Inc., Burley, ID, USA) is another commercially available product that contains reactive oxygen species, SA and other chemicals and was shown to induce SAR in tobacco (Yang et al. 2002). 'Messenger' (Eden Bioscience Co., Bothwell, WA, USA) contains harpin, an elicitor from *Erwinia* and *Pseudomonas* species that induces a hyposensitive response eventually leading to SAR. Chitosan, a deacetylated form of chitin obtained from the outer shell of crustaceans such as crabs, krill and shrimp is reported to have activity as a plant defence elicitor as well as direct fungicidal activity (Bautista-Banos et al. 2006). Less well defined activator substances are also available including plant extracts 'Bio-S' (Gebruder Schatte KG, Bad Waldsee, Germany) and 'Neudo Vidal' (W. Neudorff GmbH KG, Emmerthal, Germany) and an extract of *Pseudomonas fluorescens* ssp. *Proradix* called 'PRORADIX' (Sourcon Padena GmbH & Co. KG, Tubingen, Germany). These were found to activate systemic immunity by a complex pattern of gene activation initially involving JA-dependent genes but with a much more sustained SA-associated defence gene induction (von Rad et al. 2005). As already mentioned, a problem with using activators like BTH to cause immediate expression of plant defence genes is that there can be a cost to the plant which causes lower yield when disease is not present, thus, priming of defences would be a better approach (van Hulten et al. 2006).

Some strobilurin fungicides seem to have priming of crops as part of their mode of action (Beckers and Contrath 2007).

3.2 Insect Pests

Thaler et al. (1996) showed that application of jasmonic acid (JA) or its volatile derivative, methyl jasmonate, to the foliage of young tomato plants induced production of defence proteins similar to the ones produced with chewing herbivore feeding (*Helicoverpa zea* larvae). This was the first evidence that a jasmonate spray could induce plant defence in the field. Further field plot trials showed that JA induced tomato plants sprayed with an aqueous solution of JA with a backpack sprayer received 60% less leaf damage from herbivory than control plants (Thaler 1999a). Parasitism of lepidopteran larvae was subsequently shown to be significantly higher on treated plants (Thaler 1999b). In studies with JA treated rice Lou et al. (2005) showed that parasitism of rice brown planthopper by *Anagrus nilaparvatae* was increased in greenhouse and field experiments. In this study induction was achieved by wounding stems with a needle (200 pricks) before applying JA. This showed that the plant signalling can be induced in rice but a different delivery method of JA would be needed for large scale agricultural application. Coronalon, a synthetic 6-ethyl indnoyl iosleucine conjugate, has structural similarities to JA, and was shown in laboratory experiments to elicit plant defence responses where it had activity at lower concentrations than jasmonates (Schuler et al. 2004) although it has not yet been tested under field conditions.

Another volatile plant activator involved with plant resistance to insects is *cis*-jasmone, or (*Z*)-jasmone. Its activity was first discovered at Rothamsted when components of blackcurrant volatiles that repelled the summer form of lettuce aphid, *Nasonovia ribis-nigri*, were being identified and since then has been found to have intricate effects on interactions between pest insects and crop plants (Birkett et al. 2000; Pickett et al. 2007). It occurs naturally as a component of flower volatiles, but can also be produced by damaged plant vegetative tissues (Loughrin et al. 1995) and there is evidence that *cis*-jasmone has a role in plant defence. It is possible that *cis*-jasmone acts as an external signal, alerting recipient plants when their neighbours are being damaged by phytophagous insects and thereby enabling them to prepare their own defences prior to insect attack. The practical use of *cis*-jasmone has initially focussed on the interaction between the grain aphid *Sitobion avenae* and wheat, *Triticum aestivum*. Wheat plants sprayed with low levels of *cis*-jasmone as an aqueous emulsion are less attractive to aphids but more attractive to their parasitoids in laboratory bioassays. In the field, similarly treated plants have lower aphid infestations (Bruce et al. 2003). Field plots of wheat were sprayed hydraulically with *cis*-jasmone, at a rate equivalent to 50 g ha^{-1} in 200 l ha^{-1}, in mid May and early June in four consecutive seasons, and aphid counts were made at weekly intervals. It was consistently found that aphid infestations were reduced in *cis*-jasmone treated plots (Bruce et al. 2003) (Fig. 3).

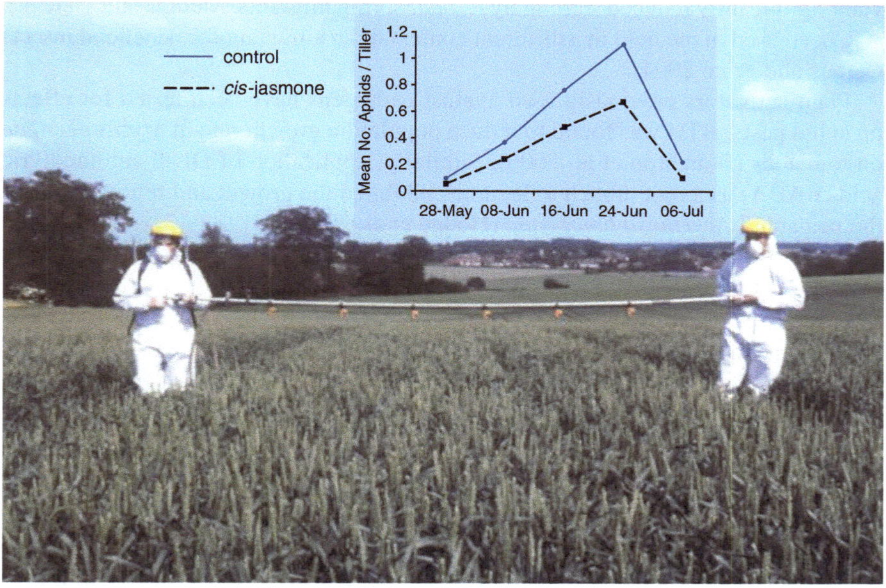

Fig. 3 Field trial with *cis*-Jasmone plant activator in Winter wheat

Barley grown next to couch grass, *Elymus repens*, is less attractive to aphids and glasshouse studies have shown that air passing by convection from *E. repens* plants to the barley plants makes them less acceptable to aphids (Glinwood et al. 2003). Further studies showed that this effect could also be achieved by exposure of barley to volatiles from thistle, *Cirsium* spp. (Glinwood et al. 2004; Chapter 'Volatile Chemical Interaction Between Undamaged Plants: Effects at Higher Trophic Levels'). However, couch grass and thistle are aggressive weeds and would not be appropriate for agricultural use as an intercrop. Thus, it would be better to identify the volatiles involved and use them directly as plant activator chemicals. The effect of exposure of barley plants to volatiles from other barley cultivars has also been investigated and it was found that volatiles from certain cultivars such as 'Frieda' could make other cultivars such as 'Hulda' significantly less attractive to aphids in subsequent choice tests (Ninkovic et al. 2002; Chapter 'Volatile Chemical Interaction Between Undamaged Plants: Effects at Higher Trophic Levels'). This is a practical proposition for agricultural use and has already been shown to work in the field as a seed bed of mixed barley gave a significant reduction in aphid acceptance compared with single cultivar plots (Ninkovic et al. 2002).

Methyl saliyclate is a plant signal associated with the winter host in bird cherry oat aphid, *Rhopalosiphum padi*. It has been successfully used to reduce cereal aphid infestations in field trials (Pettersson et al. 1994; Ninkovic et al. 2003). Methyl salicylate applied either as an aqueous emulsion or from slow-release vials significantly reduced settling by *R. padi* spring migrants. Methyl salicylate has also been used in the field in a different context as an attractant for beneficial insects (James and Price 2004).

Plant activators previously used against pathogens have been tested for effects on aphid pests. BTH was found to reduce population growth rate of *Myzus persicae* on tomatoes (Boughton et al. 2006). Similarly, application of DL-β-aminobutyric acid (BABA) as a root drench to legumes inhibited the growth and reproduction of the pea aphid *Acyrthosiphon pisum* (Hodge et al. 2005).

4 Plant Signals in 'Push–Pull' Strategies

The 'push–pull' system of IPM is different from the approaches described above in that it uses locally available companion plants instead of synthetic chemicals to deliver semiochemicals in the field (Cook et al. 2007). This approach was developed for smallholder agriculture and has been used with much success in maize and sorghum in eastern Africa (Hassanali et al. 2008; Khan et al. 2008b) against two main pest problems, stem or stalk borers and the African witchweed, *Striga*. Use of constitutive emission of relevant semiochemicals from plant sources provides an appropriate solution for smallholder African agriculture where synthetic chemicals would be too costly and logistically difficult to deliver (Fig. 4).

The system used a combination of an intercrop which releases semiochemicals that are repellent to pests (push) and a trap crop grown around the edges of the plot which releases semiochemicals that are attractive to pests of the main crop (pull).

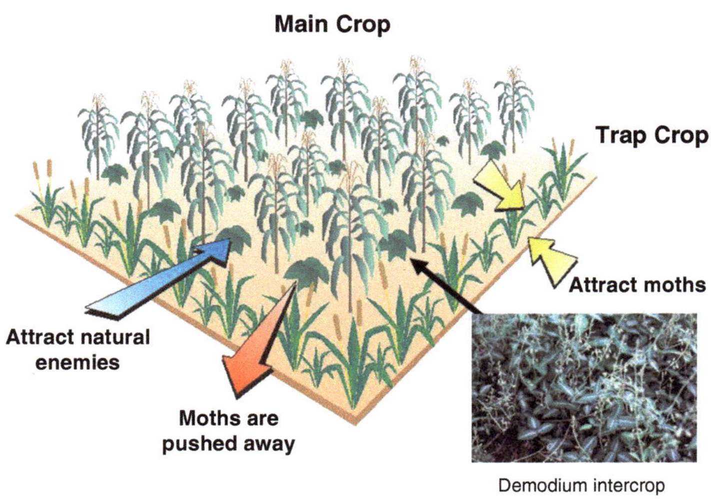

Fig. 4 Push–Pull system using companion plants to provide signals

Semiochemicals released by the intercrop also attract natural enemies of pests (Khan et al. 1997; Khan et al. 2008a). Intercrops constitutively release volatiles such as (*E*)-ocimene and (*E*)-4,8-dimethyl-1,3,7-nonatriene that are typically released from maize when attacked by chewing herbivores. These act to repel pests but attract their natural enemies. The main semiochemicals released from trap crops that make them attractive are green leaf volatiles which are produced in much larger amounts than in maize or sorghum, at nightfall when stem borer moths are active (Chamberlain et al. 2006). For stem borer control alone, the intercrop is molasses grass, but for controlling the *Striga* weed, the intercrop is silverleaf or greenleaf, two cattle forage legumes of *Desmodium* species, which also repel stem borers. The mechanism of *Striga* control consists of release of root exudate allelochemicals from *Desmodium* that induce suicidal germination of *Striga* seeds. The trap crop is preferably Napier grass, but Sudan grass can also be used. Besides controlling stem borers and the *Striga* weed, all of these companion crops are valuable as cattle forage, improving thereby livestock holdings in addition to producing a sustainable cereal harvest protected against pests and weeds.

5 Manipulating Plant Genetics to Deliver Semiochemicals

Another way of manipulating plant signals to enhance defence against pests and diseases is through altering biosynthesis of plant semiochemicals in the crop itself. This can be achieved through conventional plant breeding and marker assisted selection. However, there is scope to deliver altered volatile production in plants much faster by use of genetic engineering approaches (Degenhardt et al. 2003; Aharoni et al. 2005; Dudareva and Negre 2005). This concept has been proved in thale cress *Arabidopsis thaliana* (Brassicaceae) and the next stage will be to do this in a crop plant. Schnee et al. (2006) overexpressed a terpene synthase gene, TPS10, in *Arabidopis* and found that transformed plants were more attractive to the parasitic wasp *Cotesia marginiventris*. Another study in *Arabidopsis* showed that increasing green leaf volatile biosynthesis and emission led to increased attractiveness of plants to *C. glomerata* parasitic wasps and increased resistance to grey mould fungal infection (Shiojiri et al. 2006). A terpene synthase gene for production of the aphid alarm pheromone has been cloned into *Arabidopsis* and transformed plants were less attractive to the aphid *M. persicae* but more attractive to the aphid parasitoid *Aphidius ervi* (Beale et al. 2006) and preliminary studies at Rothamsted have demonstrated that this compound is released by transformed wheat. The rice (*E*)-β-caryophyllene synthase (OsTPS3) plays an important role in inducible volatile sesquiterpene biosynthesis and the parasitoid *A. nilaparvatae* was attracted to plants overexpressing this gene (Cheng et al. 2007). Ideally engineered plants should have synthase gene constructs that allow inducible production of semiochemicals as a continuously released signal might become habituated to by pests and would provide misleading information to natural enemies (Degenhardt et al. 2003).

6 Conclusions

Use of plant signals to alter plant metabolism has considerable potential for use in sustainable agriculture. Although much still remains to be learnt about the optimal way in which to deploy these chemicals in the field there is a growing number of studies that support the feasibility of the approach. To take this crop protection strategy into practise, there is a need for greater involvement from the agrochemical industry in the development of new commercial products. These could be plant activators themselves or seeds of plants with suitable genes for volatile production or defence response. It may even be feasible to have a package of a particular crop cultivar and activator tailored to trigger gene expression in it at appropriate times. Increasingly stringent legislation against traditional pesticides with a toxic mode of action is likely to encourage this as will increasing consumer demand for reductions in pesticide residues. In terms of agricultural sustainability, however, the plant activator approach is better seen as a supplement to pesticides rather than a complete replacement. Food security depends on ensuring that as many options as possible are kept open to reduce the losses to crop yields that are caused by pests and diseases. Overreliance on any one method is not sustainable because of the danger of breakdown of the control due to pest adaptation. Thus, plant activators should ideally be implemented as part of an IPM package alongside biocontrol with natural enemies of arthropod pest, resistant crop cultivars and use of pesticides reserved for 'fire fighting' outbreaks.

References

Agrawal AA (1998) Induced responses to herbivory and increased plant performance. Science 279:1201–1202

Aharoni A et al (2005) Volatile science? Metabolic engineering of terpenoids in plants. Trends Plant Sci 10:594–602

Baldwin IT et al (2006) Volatile signaling in plant–plant interactions: "Talking trees" in the genomics era. Science 311:812–815

Bautista-Banos S et al (2006) Chitosan as a potential natural compound to control pre and postharvest diseases of horticultural commodities. Crop Prot 25:108–118

Beale MH et al (2006) Aphid alarm pheromone produced by transgenic plants affects aphid and parasitoid behavior. Proc Natl Acad Sci USA 103:10509–10513

Beckers JM, Contrath U (2007) Priming for stress resistance: from the lab to the field. Curr Opin Plant Biol 10:425–431

Birkett MA et al (2000) New roles for *cis*-jasmone as an insect semiochemical and in plant defense. Proc Natl Acad Sci USA 97:9329–9334

Boughton AJ et al (2006) Impact of chemical elicitor applications on greenhouse tomato plants and population growth of the green peach aphid, *Myzus persicae*. Entomol Exp Appl 120:175–188

Bruce TJ, Pickett JA (2007) Plant defence signalling induced by biotic attacks. Curr Opin Plant Biol 10:387–392

Bruce TJA et al (2003) *cis*-Jasmone treatment induces resistance in wheat plants against the grain aphid, *Sitobion avenae* (Fabricius) (Homoptera: Aphididae). Pest Manag Sci 59:1031–1036

Bruce TJA, Matthes MC, Napier JA, Pickett JA (2007) Stressful "memories" of plants: evidence and possible mechanisms. Plant Sci 173:603–608

Chamberlain K et al (2006) Diel periodicity in the production of green leaf volatiles by wild and cultivated host plants of stemborer moths, *Chilo partellus* and *Busseola fusca*. J Chem Ecol 32:565–577

Cheng AX et al (2007) The rice (*E*)-beta-caryophyllene synthase (OsTPS3) accounts for the major inducible volatile sesquiterpenes. Phytochemistry 68:1632–1641

Conrath U et al (2006) Priming: getting ready for battle. Mol Plant Microbe Interact 19: 1062–1071

Cook SM et al (2007) The use of push-pull strategies in integrated pest management. Annu Rev Entomol 52:375–400

Degenhardt J et al (2003) Attracting friends to feast on foes: engineering terpene emission to make crop plants more attractive to herbivore enemies. Curr Opin Biotechnol 14:169–176

Dicke M, Dijkman H (1992) Induced defense in detached uninfested plant-leaves – effects on behavior of herbivores and their predators. Oecologia 91:554–560

Dudareva N, Negre F (2005) Practical applications of research into the regulation of plant volatile emission. Curr Opin Plant Biol 8:113–118

Engelberth J et al (2007) Insect elicitors and exposure to green leafy volatiles differentially upregulate major octadecanoids and transcripts of 12-oxo phytodienoic acid reductases in *Zea mays*. Mol Plant Microbe Interact 20:707–716

Farmer EE, Ryan CA (1990) Interplant communication – airborne methyl jasmonate induces synthesis of proteinase-inhibitors in plant-leaves. Proc Natl Acad Sci USA 87:7713–7716

Gardner SN et al (1999) Strategies to delay the evolution of resistance in pests: dose rotations and induced plant defenses. Asp Appl Biol 53:189–196

Glinwood R et al (2003) Change in acceptability of barley plants to aphids after exposure to allelochemicals from couch-grass (*Elytrigia repens*). J Chem Ecol 29:261–274

Glinwood R et al (2004) Barley exposed to aerial allelopathy from thistles (Cirsium spp.) becomes less acceptable to aphids. Ecol Entomol 29:188–195

Hassanali A et al (2008) Integrated pest management: the push–pull approach for controlling insect pests and weeds of cereals, and its potential for other agricultural systems including animal husbandry. Philos Trans R Soc Lond B Biol Sci 363:611–621

Heil M, Silva Bueno JC (2007) Within-plant signaling by volatiles leads to induction and priming of an indirect plant defense in nature. Proc Natl Acad Sci USA 104:5467–5472

Herman MAB et al (2008) Induction of plant defense gene expression by plant activators and *Pseudomonas syringae* pv. *tomato* in greenhouse-grown tomatoes. Phytopathology 98: 1226–1232

Hodge S et al (2005) Application of DL-β-aminobutyric acid (BABA) as a root drench to legumes inhibits the growth and reproduction of the pea aphid *Acyrthosiphon pisum* (Hemiptera: Aphididae). Bull Entomol Res 95:449–455

Jakab G et al (2005) Enhancing *Arabidopsis* salt and drought stress tolerance by chemical priming for its abscisic acid responses. Plant Physiol 139:267–274

James DG, Price TS (2004) Field-testing of methyl salicylate for recruitment and retention of beneficial insects in grapes and hops. J Chem Ecol 30:1613–1628

Kamal A et al (2008) Enhanced onion resistance against stemphylium leaf blight disease, caused by *Stemphylium vesicarium*, by di-potassium phosphate and benzothiadiazole treatments. Plant Pathol J 24:171–177

Karban R et al (2000) Communication between plants: induced resistance in wild tobacco plants following clipping of neighboring sagebrush. Oecologia 125:66–71

Kessler A et al (2006) Priming of plant defense responses in nature by airborne signaling between *Artemisia tridentata* and *Nicotiana attenuata*. Oecologia 148:280–292

Khan ZR et al (1997) Intercropping increases parasitism of pests. Nature 388:631–632

Khan ZR et al (2008a) Economic performance of the 'push–pull' technology for stemborer and Striga control in smallholder farming systems in western Kenya. Crop Prot 27:1084–1097

Khan ZR et al (2008b) Chemical ecology and conservation biological control. Biol Control 45:210–224

Lewis WJ et al (1997) A total system approach to sustainable pest management. Proc Natl Acad Sci USA 94:12243–12248

Lou YG et al (2005) Exogenous application of jasmonic acid induces volatile emissions in rice and enhances parasitism of *Nilaparvata lugens* eggs by the parasitoid *Anagrus nilaparvatae*. J Chem Ecol 31:1985–2002

Loughrin JH et al (1995) Volatiles emitted by different cotton varieties damaged by feeding beet armyworm larvae. J Chem Ecol 21:1217–1227

Louws FJ, Wilson M, Campbell HL, Cuppels DA, Jones JB, Shoemaker PH, Sahin F, Miller SA (2001) Field control of bacterial spot and bacterial speck of tomato using a plant activator. Plant Dis 85:481–488

Lucas JA (1999) Plant immunisation: from myth to SAR. Pest Sci 55:193–196

Lyon GD, Reglinski T, Newton AC (1995) Novel disease control compounds: the potential to 'immunise' plants against infection. Plant Pathol 44:407–427

Ninkovic V et al (2002) Mixing barley cultivars affects aphid host plant acceptance in field experiments. Entomol Exp Appl 102:177–182

Ninkovic V, Ahmed E, Glinwood R, Pettersson J (2003) Effects of two types of semiochemicals on population development of the Bird Cherry Oat Aphid, *Rhopalosiphum padi* (L.) a in barely crop. Agric For Entomol 5:27–33

Paré PW et al (2005) Elicitors and priming agents initiate plant defense responses. Photosyn Res 85:149–159

Pettersson J et al (1994) Winter host component reduces colonization by bird-cherry oat aphid, *Rhopalosiphum-Padi* (L) (Homoptera, Aphididae), and other aphids in cereal fields. J Chem Ecol 20:2565–2574

Pickett JA et al (2007) *cis*-Jasmone as allelopathic agent in inducing plant defence. Allelopath J 19:109–117

Poelman EH et al (2008) Early season herbivore differentially affects plant defence responses to subsequently colonizing herbivores and their abundance in the field. Mol Ecol 17:3352–3365

Schnee C et al (2006) The products of a single maize sesquiterpene synthase form a volatile defense signal that attracts natural enemies of maize herbivores. Proc Natl Acad Sci USA 103:1129–1134

Schuler G et al (2004) Coronalon: a powerful tool in plant stress physiology. FEBS Lett 563:17–22

Shiojiri K et al (2006) Changing green leaf volatile biosynthesis in plants: An approach for improving plant resistance against both herbivores and pathogens. Proc Natl Acad Sci USA 103:16672–16676

Sticher L, Mauch-Mani B, Métraux JP (1997) Systemic acquired resistance. Annu Rev Phytopathol 35:235–270

Stout MJ et al (2006) Plant-mediated interactions between pathogenic microorganisms and herbivorous arthropods. Annu Rev Entomol 51:663–689

Thaler JS et al (1996) Exogenous jasmonates simulate insect wounding in tomato plants (*Lycopersicon esculentum*) in the laboratory and field. J Chem Ecol 22:1767–1781

Thaler JS (1999a) Induced resistance in agricultural crops: effects of jasmonic acid on herbivory and yield in tomato plants. Environm Entomol 28:30–37

Thaler JS (1999b) Jasmonate-inducible plant defences cause increased parasitism of herbivores. Nature 399:686–688

Ton J et al (2005) Dissecting the β-aminobutyric acid-induced priming phenomenon in *Arabidopsis*. Plant Cell 17:987–999

Ton J et al (2007) Priming by airborne signals boosts direct and indirect resistance in maize. Plant J 49:16–26

Turlings TC, Ton J (2006) Exploiting scents of distress: the prospect of manipulating herbivore-induced plant odours to enhance the control of agricultural pests. Curr Opin Plant Biol 9:421–427

Vallad GE, Goodman RM (2004) Systemic acquired resistance and induced systemic resistance in conventional agriculture. Crop Sci 44:1920–1934

van Hulten M et al (2006) Costs and benefits of priming for defense in *Arabidopsis*. Proc Natl Acad Sci USA 103:5602–5607
van Lenteren JC (2000) A greenhouse without pesticides: fact or fantasy? Crop Prot 19:375–384
van Wees SCM, de Swart EAM, van Pelt JA, van Loon LC, Pieterse CMJ (2000) Enhancement of induced disease resistance by simultaneous activation of salicylate- and jasmonate-dependent defense pathways in *Arabidopsis thaliana*. Proc Natl Acad Sci USA 97:8711–8716
von Rad U et al (2005) Evaluation of natural and synthetic stimulants of plant immunity by microarray technology. New Phytol 165:191–202
Whitfield J (2001) Making crops cry for help. Nature 410:736–737
Yang KY, Blee KA, Zhang S, Anderson AJ (2002) Oxycom treatment suppresses *Pseudomonas syringae* infection and activates a mitogen-activated protein kinase pathway in tobacco. Physiol Mol Plant Pathol 61:249–256
Zhu-Salzman K et al (2008) Arthropod-inducible proteins: Broad spectrum defenses against multiple herbivores. Plant Physiol 146:852–858

Plant Volatiles: Useful Signals to Monitor Crop Health Status in Greenhouses

R.M.C. Jansen, J. Wildt, J.W. Hofstee, H.J. Bouwmeester, and E.J. van Henten

Abstract This chapter focuses on the monitoring of crop health status via the measurement of volatile organic compounds (VOCs) emitted from the plants. It includes the most important factors that affect the emission of these VOCs from crops grown in greenhouses. Since both stressors as well as nonstressors have an effect on the emission, they are covered separately. The chapter provides an overview of processes that affect the gas balance of plant VOCs in the greenhouse including the loss processes. These processes are considered as important since they contribute to the time-dynamic concentration profiles of plant-emitted VOCs. In addition, we describe the most popular techniques currently in use to measure volatiles emitted from plants, with emphasis on greenhouse application. Dynamic sampling in combination with gas chromatography coupled to mass spectrometry is considered as the most appropriate method for application at greenhouse scale. It is recommended to evaluate the state of the art in the fields concerned with this method and explore the development of a new instrument based on the specific needs for application in greenhouse practice. However, to apply such an instrument at greenhouse-scale remains a challenge, mainly due to the high costs associated with it.

R.M.C. Jansen (✉) and E.J. van Henten
Wageningen University, Farm Technology Group, P.O. Box 17, Wageningen, The Netherlands
Wageningen UR Greenhouse Horticulture, P.O. Box 644, Wageningen, The Netherlands
e-mail: Roel.Jansen@wur.nl

J. Wildt
Institute Phytosphere (ICG-III), Research Centre Jülich, Jülich 52435, Germany

J.W. Hofstee
Wageningen University, Farm Technology Group, P.O. Box 17, Wageningen, The Netherlands

H.J. Bouwmeester
Wageningen University, Laboratory of Plant Physiology, P.O. Box 658, Wageningen, The Netherlands

F. Baluška and V. Ninkovic (eds.), *Plant Communication from an Ecological Perspective,* 229
Signaling and Communication in Plants,
DOI 10.1007/978-3-642-12162-3_13,

1 Introduction

This chapter focuses on the use of plant volatiles for monitoring crop health status in greenhouses. Before going into detail, we provide a brief overview of greenhouse production systems in the world and crop health monitoring.

The definition of "greenhouse" is imprecise. In some cases, very simple plastic or shade-cloth structures are called "greenhouses." For example, there is a reported 40,000 ha. of "greenhouse" vegetable production in Almeria, Spain. Most of this production is in very simple flat-roofed structures covered with plastic. Mexico is currently producing vegetables and flowers in an estimated 2,200 ha. of more advanced, passively ventilated, high-tunnel structures. These are unheated, plastic covered metal structures, with insect netting side walls, and have computerized irrigation and fertilization systems. The high end of the greenhouse structure spectrum is found in The Netherlands (Hickman 2009). Here, more than 10,000 ha. of greenhouses are located, which are mainly used to produce vegetables, flowers, and pot plants. These greenhouses are primarily high technology metal structures covered with glass, with computer controlled environments. In such greenhouses, investments in automation will increase due to high labor costs and the poor working conditions. Furthermore, as greenhouses rapidly increase in scale (Fig. 1), growers are looking for tools to automate tasks. One task that will gradually be taken over by automation is crop inspection to assess the health status of plants.

Regular human inspections are still the primary method by which greenhouse managers assess the health status of their crops. These human inspections are indispensable. However, technological developments may help to detect emerging health problems at an early stage, which will make it easier to manage and control them. A novel approach to support the inspection of greenhouse crops is based on

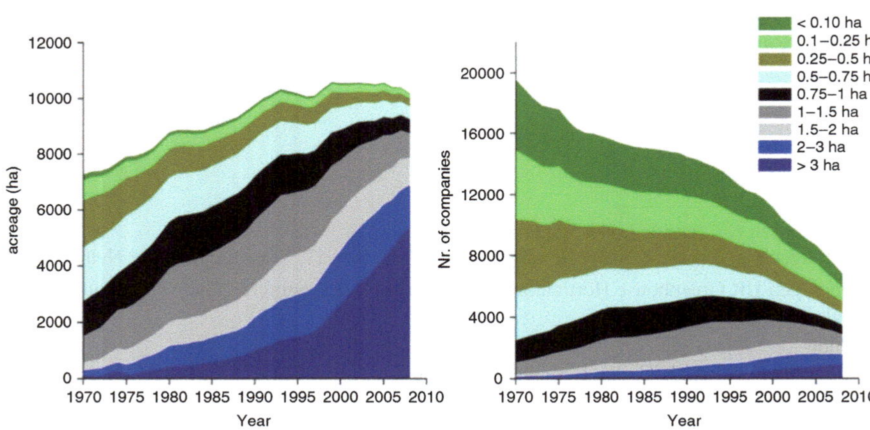

Fig. 1 Increase in scale in Dutch greenhouse horticulture
Source: Dutch Central Bureau of Statistics (CBS), available at www.cbs.nl

the measurement of volatile organic compounds (VOCs) emitted by unhealthy plants. This approach has attracted some serious interest over the last decade.

In pursuit of this interest, studies were undertaken at the laboratory-scale to pinpoint marker VOCs that can be used to indicate health problems of tomato and cucumber (Thelen et al. 2006; Laothawornkitkul et al. 2008). In addition to these laboratory studies, pilot studies were performed to verify the validity of these marker VOCs under real-world conditions (Karl et al. 2008; Markom et al. 2009).

The first section of the chapter is basically plant-oriented and covers knowledge related to the emission of VOCs from greenhouse crops. The second section of this chapter is mainly greenhouse-oriented and covers aspects related to factors that affect the gas balance of plant VOCs in the greenhouse including their loss processes. The third section of this chapter evaluates this concept from a technical point of view and reviews the most popular techniques currently in use to measure volatiles emitted from plants with emphasis on the application of these techniques in greenhouse practice. The last section of this chapter deals with trends and future possibilities and gives an outlook on the possibilities for crop health monitoring based on plant-emitted VOCs.

2 Emission of VOCs from Greenhouse Crops

The first part of this section reviews factors which affect the emission of VOCs from greenhouse crops. The second part addresses the specificity of stress-induced emissions and explains how plant-emitted volatiles can be used to characterize the stressors that contribute to plant health problems.

2.1 Factors Affecting the Emission of VOCs from Crops Grown in Greenhouses

An extensive overview of factors affecting VOC emissions from crop and noncrop plants has appeared in the literature (e.g., Kesselmeier and Staudt 1999). In this chapter we explain those factors that affect the emission of plant volatiles from crop plants grown in greenhouses. These factors are divided into two categories. The first category includes factors that have been shown to correlate with plant health. Since plant health is generally associated with responses to stresses, we termed these factors as “stressors”. The second category includes factors that do not show this correlation. These factors were termed as “nonstressors”.

The term “stressor” is extensively used in this chapter. However, this term is subjective and used with various meanings in different situations (see Gaspar et al. 2002). Since the aim of greenhouse systems is to produce, we define stressors as those factors that adversely affect crop productivity.

2.1.1 Stressors Affecting the Emission of VOCs from Crops Grown in Greenhouses

Crops grown in greenhouses might be challenged with numerous stressors. However, the number of stressors that generally occur in such systems is limited, primarily due to monoculture and environmental control. The effects of these generally occurring stressors in terms of plant volatile emissions are described below. The factors are divided into two types, biotic and abiotic stressors.

Biotic stressors are those factors that are caused by biological sources. Two biotic stressors that generally occur in greenhouses are herbivore infestation and pathogen infection. The herbivore-induced emission of VOCs has been widely studied over the past few decades. Most of these studies were performed at the laboratory scale. In these studies, numerous plant species were subjected to insect-, mite-, and snail-species to study the plant-response in terms of volatile emissions. Usually, these herbivores were applied on aerial parts of the plants (e.g., Wei et al. 2007) but herbivores were also applied on the root zone of plants (e.g., Rasmann and Turlings 2007). In general, these studies reported a significant increase in the types and amounts of VOCs emitted after herbivore infestation.

In contrast to the large amount of studies that report about herbivore-induced emission, only few studies reported about pathogen-induced emissions. These limited amount of studies include the increased emission of VOCs from silver birch (*Betula pendula*) upon a fungal infection (Vuorinen et al. 2007), from tomato plants (*Lycopersicon esculentum*) upon a viral infection (Deng et al. 2004), and from tobacco plants (*Nicotiana tabacum*) upon a bacterial infection (Heiden et al. 2003). In these studies, the aerial parts of the plants were infected. However, also root infections may result in increased emissions of certain VOCs. Preliminary experiments using cucumber plants (*Cucumis sativus*) inoculated with the root pathogen *Pythium aphanidermatum* did indicate such systemic plant response (Jansen et al. 2007).

Abiotic stressors are those factors which are caused by non-biological, environmental forces. Water deficiency, nutrient deficiency, and air pollution are abiotic stressors that may occur in practice, and have a negative effect on crop health (Peet 1999). Crops in greenhouses might be monitored for the presence of nutrient deficiency based on volatile emissions since several studies have indicated an effect of fertilization rate on volatile emission. For instance, Gouinguené and Turlings (2002) reported that the emission of volatiles was minimal when corn plants (*Zea mays*) were grown under low nutrition, even when results were corrected for plant biomass. Crops in greenhouse systems might also be monitored for water deficiency based on VOCs emitted from drought-stressed plants since several studies demonstrated an increase in the amount and types of plant volatile emitted after drought (e.g., Ebel et al. 2006). Emitted substances after drought were characterized by alcohols and aldehydes, probably as a result of the gradual collapse of the cellular structure of the plant leaves during the drying process. Finally, crops might be monitored for air pollution damage by plant-emitted volatiles since several researchers (e.g., Wildt et al. 2003), have demonstrated that harmful ozone concentrations induced an increased emission of several VOCs from a number of plant species including sunflower (*Helianthus annuus*) and Scots pine (*Pinus sylvestris*).

2.1.2 Nonstressors Affecting the Emission Rate of VOCs from Greenhouse Crops

Temperature and light are well described factors that affect the VOC emission rate from plants. These factors can be stressful for plants during periods of excessive temperature and/or light. However, due to climate control, these excessive temperature- and/or light-conditions are generally avoided in modern production systems and thus regarded as "nonstress". The effects of temperature and light in terms of volatile emissions are described below.

Temperature increases the emission rates of most VOCs exponentially up to an optimum by enhancing the biosynthetic enzymatic activities, by raising the VOC vapor pressure and by decreasing the resistance of emission pathways (Niinemets et al. 2004).

Among the studies that have examined the light dependency of plant volatile emission, there have been mixed findings, with evidence that some emissions are mainly temperature controlled (Loreto et al. 2000), while others are also significantly affected by light (e.g., Schuh et al. 1997). However, most of the literature suggests that dependencies on temperature are much stronger than those on light. Interestingly, similar chemical classes of VOCs might respond quite differently to light. When the concentration of the sesquiterpene α-copaene was examined at greenhouse scale, a clear diurnal emission pattern was evident, with an increase during the day and a decrease at night. However, the concentration of another sesquiterpene (β-caryophyllene) remained constant (Jansen et al. unpublished data). A similar observation was obtained during laboratory-scale studies in which volatiles from tomato plants were analyzed (Maes and Debergh 2003). The investigators suggested that α-copaene requires light for its biosynthesis and/or emission. In summary, temperature and light might have a strong effect on the emission of certain VOCs. Since temperature and light fluctuate in horticulture practice, these two factors have to be taken in account when correlating the concentration of such volatiles to any plant-health issue.

Besides fluctuations in temperature and light, other "nonstressors" generally occur in greenhouses that affect the emission of VOCs. Such "nonstressors" include elevated CO_2 concentrations (Loreto et al. 2006), phenological events such as budding, flowering, and fruit setting (Peñuelas and Llusià 2001), and activities of greenhouse workers such as pruning and fruit picking (Jansen et al. 2009a).

2.2 *Specificity of Stress-Induced Emissions*

A monitoring system that detects plant health problems at an early stage would enable a grower to take early action. The opportunity to identify the stressor would be of great value to such system as it would allow to decide on the proper control measure such as the release of natural enemies of white fly (*Bemisia tabaci*) in case the stressor was identified as "white fly infestation".

To identify a stressor through the measurement of plant-emitted VOCs requires the emission of highly specific chemical substances upon the onset of stress, or a highly specific time course of the stress-induced change in VOC emissions.

The emission of highly specific substances seems unlikely since it is well established that emission of many of the same substances is induced upon different biotic and abiotic stressors. For example, most of the substances reported upon pathogen infection of tomato plants were also reported upon herbivore infestation of tomato plants (Deng et al. 2004; Kant et al. 2004). Same substances were also induced when different plant species were challenged with a similar stressor. For example, herbivore damage of the plant species cucumber, apple, lima bean, corn, potato, tobacco, and cotton all induced an increase in the emission of (*E*)-β-ocimene (Paré and Tumlinson 1999). Chemical substances which are frequently reported after a stress-induced change in VOC emissions – independent of the stressor and independent of the plant species – include (*Z*)-3-hexenol, methyl salicylate, (*E*)-β-ocimene, linalool, (*E*)-β-farnesene, (*E*)-4,8-dimethylnona-1,3,7, triene, and (*E*,*E*)-4,8,12-trimethyl-1,3,7,11-tridecatetraene. This list is certainly not complete. But, to the best of our knowledge, no chemical substance has ever been exclusively ascribed to one particular stressor. Therefore it is improbable that stressors can be identified based on plant-emitted VOCs only.

Several researchers addressed the time course of stress-induced volatile emissions from plants (e.g., Kunert et al. 2002). These studies demonstrated that the emission of certain substance can increase directly after the onset of stress followed by rapid returns to low emission rates while increased emissions of other substances were delayed for some hours up to several days after the onset of stress. The time period between the first response and the delayed response in terms of increased volatile emissions might indicate the stressor exposed to the plant. For instance, this time period was dissimilar for tobacco plants in response to different strains of *Pseudomonas syringae* (Huang et al. 2003). However, to the best of our knowledge, no time course has ever been exclusively ascribed to one particular stressor. Therefore it is improbable that stressors can be identified based on the time course of stress-induced volatile emissions only. But, how stress-induced changes in VOC emissions might be used to characterize the stressor is explained below.

The first way to characterize the stressor is based upon the chemical substances present in the mixture of the plant-emitted VOCs upon the onset of the stress. These substances are to a large extent related to the plant-structure that emits these VOCs. Previous studies suggest an arbitrary classification of confined substructures and the entire plant as emitting structure.

There are several examples in which the emission from confined plant substructures changes upon stress. A first example is the release of VOCs from local plant tissue after damage of involved cell-membranes due to, e.g., herbivore infestation (Wei et al. 2007) or pathogen infection (Wildt et al. 2003). Damage of these cell-membranes will result in the local emission of several C_6-alcohols and

C_6-aldehydes at the site of damage. The emissions of these C_6-compounds originate from the oxidative cleavage of C_{18}-fatty acids in the presence of oxygen and enzymes such as lipoxygenases (Hatanaka 1993; Fall et al. 1999; Matsui 2006). These C_6-compounds thus characterize stressors in which damage of cell-membranes (that contain fatty acids) plays an important role. Also nonstressors may damage cell-membranes. For instance, shoot removal was responsible for the detection of C_6-compounds at greenhouse-scale.

A second example of plant sub-structures that emit VOCs during stress is the local emission of VOCs from damaged trichomes due to, e.g., herbivore infestation (Loughrin et al. 1994), or pathogen infection (Jansen et al. 2009b). These trichomes are outgrowths of the plant epidermis and collectively constitute the pubescence of the plant surface. A local damage of these plant sub-structures will result in the local emission of terpenes that are stored in them. These terpenes thus characterize stressors in which damage of trichomes plays an important role.

Also "nonstressors" may damage trichomes. For instance, fruit picking resulted in the damage of trichomes and a subsequent increase in the concentration of mono- and sesquiterpenes at greenhouse-scale (Jansen et al. 2009a).

Plants are attacked at different parts in different ways by the multitude of stressors. As a result, it can be expected that some types of plant sub-structures are involved, while others are not, depending on the stressor. As a consequence, the chemical substances associated with the particular type of sub-structure might thus be used to characterize the stressor that harms the plant.

The emission of methyl salicylate can be cited as an example in which the entire shoot can be regarded as emitting structure. The emission of this volatile phytohormone is generally believed to increase, but only after a certain period following the local inoculation of, or local application of herbivores (Röse et al. 1996; Shulaev et al. 1997; Kant et al. 2004).

Instantaneous damage to plants, e.g., the punching of holes within a short time period, did not result in increased emission of methyl salicylate from tomato plants (Deng et al. 2005b). Probably, a stressor needs be continuously sustained in order to increase the emission of methyl salicylate and/or other stress-associated VOCs. It is also believed that chemical signals derived from the stressor, e.g., derived from herbivore secretions, are required to increase the emission of methyl salicylate and/ or other stress-associated VOCs (Arimura et al. 2005). Thus methyl salicylate might be used to characterize stressors in which continuously sustained damage is, and/or chemical signals are involved (see Chapter "Volatile Interaction Between Undamaged Plants: A Short Cut to Coexistence"). In summary, the emission from confined plant sub-structures changes upon stress. An example of the mechanisms underlying a fungal infection induced change in emission is given in Fig. 2.

Besides the chemical substances present in the mixture of the plant-emitted VOCs, the time course of the stress-induced change in VOCs emissions may also characterize the stressor.

For example, severe *B. cinerea* infections resulted in a large increase in emissions a few hours after inoculation while mild infections resulted in a small increase in emissions several hours after inoculation (Jansen et al. 2009a). The importance of

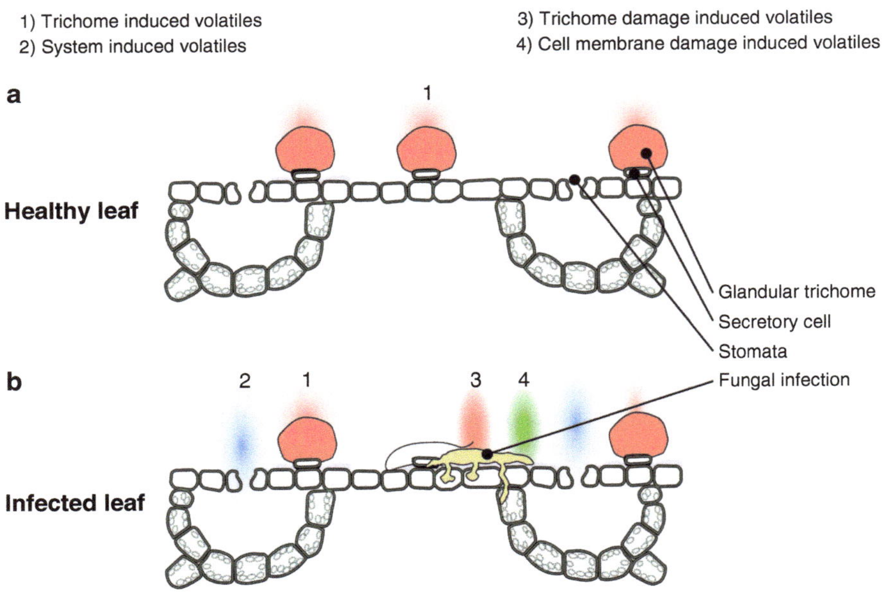

Fig. 2 Mechanisms underlying a fungal infection induced change in emission of volatiles from plants

the duration and the intensity of damage as the main factor with respect to stress-induced changes in emissions was also elegantly demonstrated using MacWorm, a robotic device designed to reproduce tissue damage caused by herbivore attack (Mithöfer et al. 2005). Besides local emissions, systemic emissions also depend on the duration and intensity of damage. For instance, the emission of systemically emitted volatiles from Brussels sprouts depends on the duration of caterpillar feeding (Mattiacci et al. 2001).

As mentioned before, the opportunity to identify the stressor would be of great value to a plant health monitoring system. This section explains how plant-emitted volatiles can be used to characterize the stressors that contribute to plant health problems. This might be sufficient since the diversity of stressors that occur per greenhouse is often limited, primarily due to monoculture and environmental control.

3 Factors that Affect the Gas Balance of Plant VOCs in the Greenhouse

This section covers the aspect related to factors that affect the gas balance of plant VOCs in the greenhouse. The crop inside of the greenhouse is probably the most important source of plant volatiles in a greenhouse. However, the gas balance of plant VOCs in the greenhouse might also be affected by the unpredictable transfer of typical plant VOCs from outside to inside of the greenhouse. Greenhouse ventilation is likely to be the most important source in that respect.

Loss processes are regarded as important aspects of the gas balance since they contribute to the time-dynamic concentration profiles of plant-emitted volatiles. On the one hand, a slow loss will cause the accumulation of VOCs in a greenhouse and thus promote the detection. On the other hand, a fast loss of VOCs enables the detection of short time dynamics, which might be required since VOC emissions during stress sometimes appear as a burst followed by rapid returns to low emission rates (e.g., Beauchamp et al. 2005; Deng et al. 2005a).

The first loss process for plant-emitted VOCs is the removal of these VOCs by greenhouse ventilation. Ventilation involves removing air from inside the greenhouse and replacing it with external air. This ventilation may be natural – caused by wind and temperature forces – or mechanical, accomplished by using fans.

The second loss process considers the degradation due to gas-phase reactions. In lower atmosphere, the major degradation processes for plant VOCs are reactions with hydroxyl radicals (OH), nitrate radicals (NO_3) and ozone (O_3) leading to a number of breakdown products (Atkinson and Arey 2003). Such oxidative breakdown of compounds not only affects the concentration of VOCs in the air surrounding unstressed plants but also the concentration of VOCs in the air surrounding stressed plants. For example, it was recently demonstrated that exposure of plants to moderately enhanced O_3 levels resulted in the partial degradation of VOCs emitted upon herbivore infestation (Pinto et al. 2007).

The third process leading to removal of VOCs from greenhouse air is the sorption on air-contact surfaces such as the floor of the greenhouse. Many researchers have shown that material surfaces interact with VOCs (e.g., Jorgenson 1999). Most of this work involved relatively simple test chamber experiments where material surfaces were exposed to VOCs and the concentration in the test chamber was monitored (Huang et al. 2006). The material surfaces in a greenhouse are a complex mixture of materials such as glass, steel, plastics, and concrete. Therefore it is difficult to estimate the effect of sorption on air-contact surfaces beforehand.

The fourth process to be taken into account is the solution of VOCs in water bodies occurring on cold greenhouse surfaces due to condensation. This water originates from plant transpiration and the amount mainly depends on greenhouse climate and plant size. The Henry's Law constant is a key parameter to estimate the maximum amount of VOCs that can be dissolved into water. This Henry's Law constant for chemical air–water partitioning is defined as the ratio of a chemical partial pressure in air to its mole fraction in water at equilibrium. However, care should be taken since Henry's Law assumes no further chemical breakdown of chemical compounds when dissolved in water.

The fifth process for losses of VOCs is absorption on the plant cuticle. Welke et al. (1998) proved that the plant cuticle can adsorb many VOCs and the amount absorbed is correlated with the concentration in air. This adsorption process might be relevant if the absorbed compounds are metabolized and the uptake potential remains.

The sixth possible loss process for VOCs is the uptake of these compounds due to absorption through the stomata. Uptake of VOCs through stomata requires a lower concentration of the compounds in the stomatal cavity than in the greenhouse air. This concentration difference is important since gasses move along the concentration gradient between the inside and the outside of the leaf. The stomatal cavity is covered by water. Therefore, compounds should be solved in this water and thereafter metabolized in plant tissues to maintain the continuous uptake potential. So, uptake through the stomata of a certain VOC depends on the water solubility of this compound and its metabolism. This loss process might thus be particularly relevant for polar VOCs such as alcohols.

4 Techniques to Measure the Emission of VOCs from Plants at Greenhouse Scale

Several excellent papers are available that review the techniques currently in use to measure the emission of VOCs from plants (Chapter "Exploiting Plant Signals in Sustainable Agriculture"; Tholl et al. 2006; Ortega and Helmig 2008). However, none of these papers describe how these techniques can be applied to monitor crop health at greenhouse scale. This section is intended to fill this knowledge gap.

In general, the measurement of plant emission consists of three steps: (1) collection of the plant-emitted VOCs, (2) separation of the plant-emitted VOCs in the mixture, and (3) identification, and/or quantification of the separate VOCs. These three steps are explained below.

4.1 Collection of the Plant-Emitted VOCs

In the first step, a fraction of the compounds emitted from the plants is collected. This sampling step is in general combined with the pre-concentration of the VOCs in the air to achieve the detection limits of commonly applied analytical instruments. Several reviews deal with the preconcentration of VOCs in air (e.g., Harper 2000; Dettmer and Engewald 2002). Therefore, we briefly mention the basic concepts and focus on appropriate methods for preconcentrating plant-emitted VOCs with emphasis on the application of these methods in greenhouse practice.

Two methods are generally applied to preconcentrate the VOCs present in air. The first method is based on the dynamic preconcentration of VOCs. This method is referred to as dynamic because the air is actively pumped through a cartridge packed with a material that traps the compounds of interest. The second method is based on the static pre-concentration of VOCs. In this case, a material is exposed to the air, in which the trapping of VOCs mainly depends on mass diffusion processes. In both cases, the selection of the material is crucial in order to trap the VOCs of interest. There are a huge number of different materials available for the pre-concentration of plant-emitted VOCs in air (supplementary material in Tholl et al. 2006). For most materials, e.g., the porous polymer Tenax [poly-(2,6-diphenyl-*p*-phenylene oxide] and carbon-based adsorbents, the preconcentration depends on adsorption. For a few other materials, e.g., polydimethylsiloxane, the preconcentration depends on absorption. The appropriate material – or combination of materials – should meet the following criteria: (1) homogeneous and inert surface to avoid artifact formation, irreversible adsorption, and catalytic effects during sampling and desorption; (2) complete and fast adsorption or absorption of the volatile organic compounds of interest; and (3) low affinity with water.

This inventory is not meant to be a complete list of criteria but rather to demonstrate the range of different aspects to consider. It is therefore obvious that care should be taken in the selection of materials since the pre-concentration step offers the opportunity to reduce the required sensitivity of the detector. It is recommended to investigate available materials in order to improve the efficiency of this step. Derivatization techniques might be employed to improve the properties of these materials in order to increase the efficiency of air sampling (see Deng et al. 2005a).

4.2 Separation of the Plant-Emitted VOCs in the Mixture

Before identification and/or quantification of the plant-emitted volatiles, the mixture of compounds is often separated. Gas chromatography (GC) is then the method of choice in most applications. This method is a type of chromatography in which the mobile phase is a carrier gas, usually an inert gas such as helium, and the stationary phase is a layer of a polymer on an inert solid support, inside a glass or metal column. The properties of this column should be selected with care since they have a large effect on the ability to separate plant-emitted volatiles.

4.3 Identification and Quantification of the Plant-Emitted VOCs

After separation, a detector is used for the identification and/or quantification of the individual VOCs present in the sample. A key-specification of any detector is their limit of detection (LOD). This LOD is generally defined as the lowest quantity of a substance that can be distinguished from the absence of that substance within a stated confidence limit, i.e., where it can be assured that a certain substance is present. The limit of quantification (LOQ) should be considered if besides detection also quantification of the concentration is required for the task of crop health monitoring. This LOQ is the minimum concentration that can be quantitatively determined with satisfying certainty. The LOQ is normally defined as ten times the standard deviation for blank samples, and is thus approximately three times higher than the LOD. In this chapter, we use two units of measure to approximate the LODs per instrument: absolute amounts in nanograms (ng) or picograms (pg), and the concentrations in air defined as nanograms per liter of air (ng L^{-1}) or picograms per liter of air (pg L^{-1}).

There are various types of detectors available in the market to identify and quantify plant-emitted VOCs. The most popular detectors in use are the flame ionization detector and the mass spectrometer (MS). Electronic noses are also widely used to detect plant-emitted VOCs in air (Kunert et al. 2002). More recently, biosensors have emerged as a promising tool to identify and quantify low levels of VOCs in ambient air. These four types of instruments are briefly described below.

4.3.1 Flame Ionization Detector

This technique involves the detection of ions. The response of the detector is determined by the number of carbon atoms hitting the detector per unit time. This makes the detector sensitive to the mass rather than the concentration, which is useful because the response of the detector is not greatly affected by changes in the carrier gas flow rate. Flame ionization detectors (FID) have been commonly used to measure VOCs emitted from plants (e.g., Greenberg et al. 1994). It offers a stable response, a wide dynamic concentration range, and a high sensitivity with limits of detection (LOD) in the order of picograms to nanograms (Tholl et al. 2006).

4.3.2 Mass Spectrometer

The MS and its applications are extensively covered in a variety of journals and books (e.g., McMaster 2008). Therefore we only briefly mention its operating principle and focus on aspects related to the application of this instrument for the identification and/or quantification of plant-emitted VOCs at greenhouse scale.

MSs measure the mass of charged molecules. Often the MS is combined with a chromatographic column (GC). This combination has become the method of choice for quantification and identification of plant-emitted VOCs at laboratory scale. It offers a high selectivity and resolution, good accuracy and precision, a high sensitivity, and a wide dynamic concentration range. Most current GC-MS instruments can achieve LODs in the low femtogram range. However, GC-MS LODs for realistic analytes are often in the picogram to nanogram range.

Conventional GC-MS systems are delicate instruments usually restricted to laboratory use. As a consequence, air samples collected in the greenhouse should be transferred to the laboratory for further analysis. The disadvantage of this transfer is the time delay between sampling and analysis. This time delay is undesirable in case the detection of plant health problems require an immediate act, e.g., in case of the detection of a highly transmittable disease. Air samples should therefore preferably be analyzed on-site. More robust GC-MS systems have therefore appeared on the market and have been applied, for example, to detect air contaminants in field settings (e.g., Smith et al. 2005) and to monitor a biogas tower reactor for the presence of potentially toxic VOCs (Matz et al. 1998).

4.3.3 Electronic Nose

The term electronic nose (e-nose) first appeared in the literature around the late 1980s. Before this time, these sensors were referred to as gas sensors. Many aspects of electronic noses have been already reviewed in detail (e.g., Arshak et al. 2004) and thus we mention only those aspects relevant to the detection of plant-emitted VOCs in greenhouse air.

E-nose instruments are good at addressing the chemical integrity of a sample, which means to determine whether the sample is the same as or different than a certain standard. In general, they are not useful for the identification and quantification of individual components (Gardner and Bartlett 1999). However, the identification of the volatiles being emitted may not be needed if the comparison and recognition of patterns in the volatile profile are sufficient for crop health monitoring through the analysis of plant-emitted volatiles. Such a profile can be obtained through the use of sensor arrays. This converges with research on volatile based inspection of potato tubers based on e-nose systems which rely on the recognition of fingerprints of volatiles released from them. For instance, a prototype device incorporating three metal oxide sensors was able to discriminate between sound tubers and the same tubers with one *Erwinia carotovora*-infected tuber added (de Lacy Costello et al. 2000). De Lacy Costello et al. (2003) recognized the problems associated with air humidity and low air temperatures. However, these authors claimed that the system was able to differentiate between sound and infected tubers when operating at 4°C and 85% relative humidity while the sampling time necessary to allow discrimination was reduced to 10 s.

A combination of the marker-compound-approach with the e-nose technique can result in e-nose systems that have the ability to quantify VOC concentration in air as demonstrated for the differentiation of fresh and rancid butter based on volatiles (Hofmann et al. 1997). This development seems to be quite promising. The remaining drawback of e-noses based on sensor arrays is that the threshold of determination of most of these systems is in the low ppm-range. However, this drawback can be overcome by utilization of pre-concentration techniques. Such a combination of a gas-chromatographic system equipped with a pre-concentration unit and e-nose was successfully applied to detect plant emitted volatiles in a small cuvette (Kunert et al. 2002). They reported LODs for relevant VOCs in the low nanogram levels.

4.3.4 Biosensor Technology

A biosensor is a particular type of chemical sensor that uses the recognition properties of biological components in the sensitive layer. Since its inception, biosensors have been expected to play a significant analytical role in medicine, agriculture, food safety, homeland security, and environmental and industrial monitoring (Luong et al. 2008). However, despite the large amount of biosensors developed in research laboratories, the commercialization of biosensor technology is still in its infancy.

Nevertheless, steady improvements of well known basic principles have resulted in improved sensitivity, reliability, and stability of traditional enzymatic biosensors. Also, new affinity sensors such as transmembrane sensors and sensors utilizing whole cells or cell networks have significantly improved. For example, The Centre for Environmental Biotechnology at the University of Tennessee developed a proof-of-concept for a whole cell bioluminescent bioreporter for the detection of VOCs (Vijayaraghavan et al. 2006). These bioluminescent bioreporters generate visible light in response to specific chemical or physical agents in their environment. Measurements were obtained at vapor phase concentrations of $<1\ \mu g\ L^{-1}$. Despite the lag in response and lack of correlation between concentration and bioluminescence it was hypothesized that the bioreporter can produce qualitative as well as quantitative results.

Today even whole animals or certain organs of animals are used in biosensors. For example, Schütz (2001) developed a biosensor to detect volatiles emitted from artificially and herbivore damaged potato plants (*Solanum tuberosum*). This biosensor, based on the intact antennae of the Colorado potato beetle (*Leptinotarsa decemlineata*), was also able to detect volatiles emitted from potato plants infected with *Phytophthora infestans*, the causal agent of the late blight disease (Schütz 2001). Sensitivity and dynamic range can compete with the performance of GC-MS instruments (LOD <1 ng L^{-1}) while the response, dead and adaptation time, are shorter by a factor 10.

5 Trends and Future Possibilities

So far, most of the research related to plant health monitoring through plant-emitted VOCs is undertaken at the laboratory scale to pinpoint marker VOCs that can be used to indicate certain health problems (e.g., Laothawornkitkul et al. 2008). Recently, experimental evidence demonstrated that the detection of plant damage based on plant-emitted VOCs is also feasible at greenhouse scale (Jansen et al. 2009a). A characteristic of their experimental system was the rather small-scale with 60 plants grown at a floor area of 42 m^2. However, commercial greenhouses are much larger in size. For example, at present, the majority of commercial greenhouses in Western European countries, such as The Netherlands, range between 10^3 and 10^4 m^2 (Henten 2006).

Proper experiments can be done to determine whether plant-emitted volatiles can be detected in these full-scale greenhouses. However, this will be a time consuming and costly operation since the effects of various greenhouse characteristics must be evaluated. A potential cost reduction can be attained through the use of model-based predictions. For that reason, mass-transfer models are increasingly being used to bridge the gap between experimental measurements and real world applications. Such model approach is considered to be cost saving in translating the results obtained in a small-scale greenhouse into the potential application of plant health monitoring in full-scale greenhouses.

This chapter indicates the potential of monitoring crop health status at greenhouse scale on the basis of volatiles emitted from the plants. It reflects on how technological developments in the field of analytical chemistry can be used in an agricultural setting. Most of these developments are driven by research in which the detection of trace level amounts of volatile contaminants in food, air, or water is the subject. Approaches to detect these contaminants are based on highly sensitive instruments including FID, MS, electronic noses and biosensors. It is recommended to evaluate the status of these instruments and to explore the development of new instruments that meet the specific needs for application in greenhouse practice.

At this moment, we consider gas chromatography coupled to mass spectrometry as the best method for monitoring the health status of crops on the basis of plant-emitted VOCs at high-input greenhouse facilities. This preference is based on its favorable combination of high selectivity and resolution, good accuracy and precision, wide dynamic concentration range, high sensitivity, and the current commercialization of robust GC-MS systems.

A disadvantage associated with GC-MS application is the large and complex data generated by this instrument. As a consequence, experienced and skilled analysts are often required to process these data in order to extract the concentrations of the chemical compounds of interest. However, developments in computer science technology and software will increase the opportunity to automatically process GC-MS data at an affordable price which will promote the efficient application of this instrument outside the laboratory. Another disadvantage related to the application of GC-MS is the relatively high cost of purchase. However, improvements in the field of mass spectrometry will likely results in affordable systems.

In conclusion, plants emit different types and amounts of volatiles during their decline in health status. Probably, it will be difficult if not impossible to identify the stressors based on VOC emissions only. However, plant-emitted volatiles can be used to characterize the stressors that contribute to plant health problems. There are instruments available (GC-MS) meeting the required technical specifications to detect these VOCs at greenhouse scale. Only due to the high costs, we are years away from having this kind of instruments in horticultural practice. But, the ongoing expansion and intensification of greenhouse production and the concern among consumers about the potential intake of pesticide residues on fruits and vegetables will support the prospected application of plant health monitoring in a commercial setting.

Acknowledgments Our laboratories have been financially supported by the European Science Foundation Scientific Program "Volatile organic compounds in the Biosphere-Atmosphere System" (VOCBAS grants 814 and 974). We thank the Royal Netherlands Society of Plant Pathology (KNPV) for the financial support to write this chapter.

References

Arimura G, Kost C, Boland W (2005) Herbivore-induced, indirect plant defences. Biochim Biophys Acta 1734:91–111

Arshak K, Moore E, Lyons GM, Harris J, Clifford S (2004) A review of gas sensors employed in electronic nose applications. Sensor Rev 24:181–198

Atkinson R, Arey J (2003) Gas-phase tropospheric chemistry of biogenic volatile organic compounds: a review. Atmos Environ 37:197–219

Beauchamp J, Wisthaler A, Hansel A, Kleist E, Miebach M, Niinemetts U, Schurr U, Wildt J (2005) Ozone induced emissions of biogenic VOC from tobacco: relationship between ozone uptake and emission of LOX products. Plant Cell Environ 28:1334–1343

de Lacy Costello BP, Ewen RJ, Gunson HE, Ratcliffe NM, Sivanand PS, Spencer-Phillips PTN (2003) A prototype sensor system for the early detection of microbially linked spoilage in stored wheat grain. Meas Sci Technol 14:397–409

de Lacy Costello BPJ, Ewen RJ, Gunson HE, Ratcliffe NM, Spencer-Phillips PTN (2000) The development of a sensor system for the early detection of soft rot in stored potato tubers. Meas Sci Technol 11:1685–1691

Deng C, Li N, Zhu W, Qian J, Yang X, Zhang X (2005a) Rapid determination of C_6-aldehydes in tomato plant emission by gas chromatography-mass spectrometry and solid-phase microextraction with on-fiber derivatization. J Sep Sci 28:172–176

Deng C, Qian J, Zhu W, Yang X, Zhang X (2005b) Rapid determination of methyl salicylate, a plant signaling compound, in tomato leaves by direct sample introduction and thermal desorption followed by GC-MS. J Sep Sci 28:1137–1142

Deng C, Zhang X, Zhu W, Qian J (2004) Investigation of tomato plant defence response to tobacco mosaic virus by determination of methyl salicylate with SPME-capillary GC-MS. Chromatographia 59:263–267

Dettmer K, Engewald W (2002) Adsorbent materials commonly used in air analysis for adsorptive enrichment and thermal desorption of volatile organic compounds. Anal Bioanal Chem 373:490–500

Ebel RC, Mattheis JP, Buchanan DA (2006) Drought stress of apple trees alters leaf emissions of volatile compounds. Physiol Plant 93:709–712

Fall R, Karl T, Hansel A, Jordan A, Lindinger W (1999) Volatile organic compounds emitted after leaf wounding: on-line analysis by proton-transfer-reaction mass spectrometry. J Geophys Res 104:15963–15974

Gardner JW, Bartlett PN (1999) Electronic noses: principles and applications. Oxford University Press, Oxford, p 245

Gaspar T, Franck T, Bisbis B, Kevers C, Louve J, Hausman JF, Dommes J (2002) Concepts in plant stress physiology. Application to plant tissue cultures. Plant Growth Regul 37:263–285

Gouinguené SP, Turlings TCJ (2002) The effects of abiotic factors on induced volatile emissions in corn plants. Plant Physiol 129:1296–1307

Greenberg JP, Lee B, Helmig D, Zimmerman PR (1994) Fully automated gas chromatograph-flame ionization detector system for the *in situ* determination of atmospheric non-methane hydrocarbons at low parts per trillion concentration. J Chromatogr 676:389–398

Harper M (2000) Sorbent trapping of volatile organic compounds from air. J Chromatogr 885:129–151

Hatanaka A (1993) The biogeneration of green odour by green leaves. Phytochemistry 34:1201–1218

Heiden AC, Kobel K, Langebartels C, Schuh-Thomas G, Wildt J (2003) Emissions of oxygenated volatile organic compounds from plants Part I: emissions from lipoxygenase activity. J Atmos Chem 45:143–172

Hickman GW (2009) Greenhouse vegetable production statistics. (http://cuestaroble.com/Documents/Sample%20Pages%20September%202009%20Greenhouse%20Vegetable%20Production%20Statistics.pdf)
Hofmann T, Schieberle P, Krummel C, Freiling A, Bock J, Heinert L, Kohl D (1997) High resolution gas chromatography/selective odorant measurement by multisensor array (HRGC/SOMSA): a useful approach to standardise multisensor arrays for use in the detection of key food odorants. Sens Actuators B Chem 41:81–87
Huang H, Haghighat F, Blondeau P (2006) Volatile organic compound (VOC) adsorption on material: influence of gas phase concentration, relative humidity and VOC type. Indoor Air 16:236–247
Huang J, Cardoza YJ, Schmelz A, Raina R, Engelberth J, Tumlinson JH (2003) Differential volatile emissions and salicylic acid levels from tobacco plants in response to different strains of *Psedumonas syringae*. Planta 5:767–775
Jansen RMC, Hofstee JW, Wildt J, Verstappen FWA, Bouwmeester HJ, Posthumus MA, van Henten EJ (2009a) Health monitoring of plants by their emitted volatiles: trichome damage and cell membrane damage are detectable at greenhouse scale. Ann Appl Biol 154:441–452
Jansen RMC, Miebach M, Kleist E, van Henten EJ, Wildt J (2009b) Release of lipoxygenase products and monoterpenes by tomato plants as an indicator of *Botrytis cinerea*-induced stress. Plant Biol 11:859–868
Jansen RMC, Postma J, Verstappen FWA, Bouwmeester HJ, Posthumus MA, van Henten EJ (2007) Changes in volatile emission of cucumber plants after inoculation with the root pathogen *Pythium aphanidermatum*. In: International workshop on PR proteins and induced resistance against pathogens and insects. Doorn, The Netherlands, p 68
Jorgenson RB (1999) Chamber testing of adsorption of volatile organic compounds (VOCs) on material surfaces. Indoor Air 9:2–9
Kant MR, Ament K, Sabelis MW, Haring MA, Schuurink RC (2004) Differential timing of spider mite induced direct and indirect defenses in tomato plants. Plant Physiol 135:483–495
Karl T, Guenther A, Turnipseed A, Patton EG, Jardine K (2008) Chemical sensing of plant stress at the ecosystem scale. Biogeosciences 5:1287–1294
Kesselmeier J, Staudt M (1999) Biogenic volatile organic compounds (VOC): an overview on emission, physiology and ecology. J Atmos Chem 33:23–88
Kunert M, Biedermann A, Koch T, Boland W (2002) Ultrafast sampling and analysis of plant volatiles by a hand-held miniaturised GC with pre-concentration unit: kinetic and quantitative aspects of plant volatile production. J Sep Sci 25:677–684
Laothawornkitkul J, Moore JP, Taylor JE, Possell M, Gibson TD, Hewitt CN, Paul ND (2008) Discrimination of plant volatile signatures by an electronic nose: a potential technology for plant pest and disease monitoring. Environ Sci Technol 42:8433–8439
Loreto F, Barta C, Brilli F, Nogues I (2006) On the induction of volatile organic compound emissions by plants as consequence of wounding or fluctuations of light and temperature. Plant Cell Environ 29:1820–1828
Loreto F, Nacetti P, Graverini A, Mannozzi M (2000) Emission and content of monoterpenes in intact and wounded needles of the Mediterranean Pine, *Pinus pinea*. Funct Ecol 14:589–595
Loughrin JH, Manukian A, Heath RR, Turlings TCJ (1994) Diurnal cycle of emission of induced volatile terpenoids by herbivore-injured cotton plants. Proc Natl Acad Sci USA 91:11836–11840
Luong JHT, Male KB, Glennon JD (2008) Biosensor technology: technology push versus market pull. Biotechnol Adv 26:492–500
Maes K, Debergh PC (2003) Volatiles emitted from *in vitro* grown tomato shoots during abiotic and biotic stress. Plant Cell Tissue Organ Cult 75:73–78
Markom MA, Shakaff AYM, Adom AH, Ahmad MN, Hidayat W, Abdullah AH, Fikri NA (2009) Intelligent electronic nose system for basal stem rot disease detection. Comput Electron Agr 66:140–146
Matsui K (2006) Green leaf volatiles: hydroperoxide lyase pathway of oxylipin metabolism. Curr Opin Plant Biol 9:274–280

Mattiacci L, Rocca BA, Scascighini N, D'Alessandro M, Hern A, Dorn S (2001) Systemically induced plant volatiles emitted at the time of "danger". J Chem Ecol 27:2233–2252
Matz G, Loogk M, Lennemann F (1998) On-line gas chromatography-mass spectrometry for process monitoring using solvent-free sample preparation. J Chromatogr 819:51
McMaster MC (2008) GC/MS a practical user's guide. Wiley, Hoboken, p 180
Mithöfer A, Wanner G, Boland W (2005) Effects of Feeding *Spodoptera littoralis* on lima bean leaves. II. Continuous mechanical wounding resembling insect feeding is sufficient to elicit herbivory-related volatile emission. Plant Physiol 137:1160–1168
Niinemets U, Loreto F, Reichstein M (2004) Physiological and physiochemical controls on foliar volatile organic compound emissions. Trends Plant Sci 9:180–186
Ortega J, Helmig D (2008) Approaches for quantifying reactive and low-volatility biogenic organic compound emission by vegetation enclosure techniques - Part A. Chemosphere 72:343–364
Paré PW, Tumlinson JH (1999) Plant volatiles as a defense against insect herbivores. Plant Physiol 121:325–331
Peet MM (1999) Greenhouse crop stress management. Acta Horticulturae 481:643–655
Peñuelas J, Llusià J (2001) The complexity of factors driving volatile organic compound emissions by plants. Biol Plant 44:481–487
Pinto DM, Blande JD, Nykänen R, Dong W-X, Nerg A-M, Holopainen JK (2007) Ozone degrades common herbivore-induced plant volatiles: does this affect herbivore prey location by predators and parasitoids? J Chem Ecol 33:683–694
Rasmann S, Turlings TCJ (2007) Simultaneous feeding by aboveground and belowground herbivores attenuates plant-mediated attraction of their respective natural enemies. Ecol Lett 10:926–936
Röse USR, Manukian A, Heath RR, Tumlinson JH (1996) Volatile semiochemicals released from undamaged cotton leaves. A systemic response of living plants to caterpillar damage. Plant Physiol 111:487–495
Schuh G, Heiden AC, Hoffmann T, Kahl J, Rockel P, Rudolph J, Wildt J (1997) Emissions of volatile organic compounds from sunflower and beech: dependence on temperature and light intensity. J Atmosph Chem 27:291–318
Schütz S (2001) Der Einfluß verletzungsinduzierter Emissionen der Kartoffelpflanze (*S. tuberosum*) auf die geruchliche Wirtspflanzendfindung und -auswahl durch den Kartoffelkäfer (L. *decemlineata*). Ein Biosensor für die Diagnose von Pflanzenschäden. Habilitationsschrift. Justus Liebig Universität, Gießen, p 236
Shulaev V, Silverman P, Raskin I (1997) Airborne signalling by methyl salicylate in plant pathogen resistance. Nature 385:718–721
Smith PA, Sng MT, Eckenrode BA, Leow SY, Koch D, Erickson RP, Lepage CRJ, Hook GL (2005) Towards smaller and faster gas chromatography-mass spectrometry systems for field chemical detection. J Chromatogr 1067:285–294
Thelen J, Harbinson J, Jansen RMC, van Straten G, Posthumus MA, Woltering EJ, Bouwmeester HJ (2006) The sesquiterpene α-copaene is induced in tomato leaves infected by *Botrytis cinerea*. J Plant Interact 1:163–170
Tholl D, Boland W, Hansel A, Loreto F, Röse U, Schnitzel JP (2006) Practical approaches to plant volatile analysis. Plant J 45:540–560
van Henten EJ (2006) Greenhouse mechanization: state of the art and future perspective. Acta Hortic 710:55–69
Vijayaraghavan R, Islam SK, Zhang M, Ripp S, Caylor S, Bull ND, Moser S, Terry SC, Blalock BJ, Sayler GS (2006) A bioreporter bioluminescent integrated circuit for very low-level chemical sensing in both gas and liquid environments. Sens Actuators B 123:922–928
Vuorinen T, Nerg AM, Syrjälä L, Peltronen P, Holopainen JK (2007) *Epirrita autumnata* induced VOC emission of silver birch differ from emission induced by leaf fungal pathogen. Arthropod Plant Interact 1:159–165

Wei J, Wang L, Zhu J, Zhang S, Nandi O, Kang L (2007) Plants attract parasitic wasps to defend themselves against insect pests by releasing hexenol. PLoS ONE 2:e852

Welke B, Ettlinger K, Riederer M (1998) Sorption of volatile organic chemicals in plant surfaces. Environ Sci Technol 32:1099–1104

Wildt J, Kobel K, Schuh-Thomas G, Heiden AC (2003) Emissions of oxygenated volatile organic compounds from plants. Part II: emissions of saturated aldehydes. J Atmosph Chem 45:173–196

Index

A

Abiotic stressor, 232, 234
Adaptation, 79
Adaptive plasticity, 160, 169
Aggression, 5
Alarm calls, 12, 15
Alkaloids, 182, 183, 187, 189, 191
Allelobiosis, 76, 92, 94, 95, 114, 116–121
Allelochemicals, 76, 180
Allelopathy, 4–5, 61–70, 76, 88
Ambrosia dumosa, 8, 9, 12
Antirrhinum majus, 79
Ants, 147
Aphid resistance, 113–122
Artemisia californica, 78
Artemisia vulgaris, 78
Auxin, 79

B

Barley, 81, 92, 116–121
Biogenic volatile organic compounds (BVOCs), airbone, 77
Biomass allocation, 80
Biosensor, 240–243
Biotic stressor, 232, 234
Blue light, 162, 163
Bruchins, 30
Burst, 237

C

Caeliferins, 32
Cakile edentula, 80
Carbon dioxide
  agriculture, 195–196
  atmospheric increase, 179
  compensatory feeding, 194–196
  decomposition, 196
  fungi, 196–197
  herbivory, 192–196
  nitrogen, 188
  parasitoids, 198
  pesticides, 195–196
  phenolics, 185–188
  pheromones, 198–199
  phytohormones, 191–192
  predators, 198
  soil, 196
  terpenoids, 185–186
  VOCs, 185–186, 198
Carbon–nutrient balance hypothesis (CNB), 184, 186–188
Cell wall, 168
  expansin, 168
  XTH, 168
Centaurea maculosa, 12
Chemical interaction, 76
cis-Jasmone, 28
Climate change, 179–180
Clonality, 12
CO_2, 233
Coevolution, 145, 150
Coexistence, 75
Collection of plant-emitted VOCs, 238–239
Communication, 83
  biological, 2–3
  evolution, 7
  general definition, 2
  true, 3, 12
Competition, 76, 160
Constitutive defence, 141

Crop health monitoring, 230, 231, 239
Crosstalk, 51
Cues, 4, 15
Cyanogenic glycosides, 189–191
Cytokinin, 79

D
Defense, 99–108, 171–172
DELLA proteins, 165, 166, 171
Dhurrin, 189
(E)-4,8-Dimethylnona-1,3,7-triene (DMNT), 34
DOXP pathway, 187, 190, 191
Dynamic headspace collection, 22

E
Ecotypic variation, 170
Electric fields
Electronic nose, 240, 241, 243
Energy transmission, 5
Ethylene (ET), 44, 47, 49, 78
Evolution, 75
Exotic species, 65, 69, 70

F
Female choice, 14, 15
Fitness, 3, 50, 52, 53
Flame ionization detector (FID), 240, 243
Fruit picking, 233, 235
Fungal infection, 232, 235, 236

G
Gas chromatography (GC), 229, 239, 240, 242, 243
Gas chromatography-Fourier transformed infrared spectroscopy (GC-FTIR), 27
Gas-phase reactions, 237
GC-FTIR. *See* Gas chromatography-Fourier transformed infrared spectroscopy
GC-MS, 240, 242, 243
Genetic engineering, 223
Germination, 78
Greenhouse, 229–243
Greenhouse gases, 179–180
Group selection, 9, 12–13
Growth-differentiation balance hypothesis (GDB), 184, 188
Growth rate, 82

H
Handicap principle, 8, 12
Herbivore-induced volatiles, 141
Herbivores, 77, 100, 103–106
(Z)-3-Hexenol, 234
Hormones, 165, 167–169
  auxin, 166–167, 171
  brassinosteroids, 167
  ethylene, 168, 171, 172
  gibberellin, 165–166

I
Individual selection, 8–12
Induced defence, 106, 137, 217
Induced resistance, 100, 106, 107
Induced systemic resistance (ISR), 43–48, 54
Information, 2
In-situ hybridization, 26
Integrated pest management, 215, 216, 222, 224
Intercropping, 88
ISR. *See* Induced systemic resistance

J
Jasmonic acid (JA), 33, 45, 47, 48, 51

K
Kin
  recognition, 80
  selection, 12–13

L
Ladybird, 93, 95
Larrea tridentate, 9
Leaf temperature, 82
Light, 233
Lima bean, 101–104, 106, 107
Liquid-liquid extraction, 24
Local emissions, 234–236

M
Mass spectrometer, 240–241
Mass spectrometry (MS), 26
Mate choice, 13
Methyl benzoate, 79
Methyl jasmonate (MeJA), 81
Methyl salicylate (MeSA), 81, 235
Mevalonic pathway, 187, 190, 191
Modularity in plants, 6, 7
Monoculture, 79
Monoterpenoid, 78
MS. *See* Mass spectrometry
Multiple stress, 169, 171
Mutualism, 129, 131
Myrmecophyte, 130

N
Natural enemies, 88
Nicotiana attenuate, 78

Nonstressors, 231, 233, 235
Nuclear magnetic resonance (NMR), 26

O
Optimal defense hypothesis, 183–184
Ovule competition, 13
Oxidative cleavage, 235
Oxylipins, 28

P
Parasitoid, 94, 95
Pathogen-associated molecular pattern (PAMP), 45
Pathogens, 77, 100, 102, 103, 105, 106
PDMS materials, 25
Phenolics, 185–187, 189–190
Phenylalanine, 189–190
Photoreceptors, 162–164
  cryptochromes, 162–163
  phototropins, 163–164
  phytochromes, 164
Phytochrome interacting factors (PIF), 164, 165
Phytohormones, 28, 189–192
Phytotoxic effects, 78
Phytotoxin, 63, 65
PIF. *See* Phytochrome interacting factors
Plant activators
  BABA, 222
  benzothiadiazole (BTH), 219
  bion, 219
  bio-S, 219
  'bioYield Concentrate', 219
  chitosan, 219
  cis-jasmone, 220
  jasmonic acid (JA), 220
  messenger, 219
  methyl jasmonate, 220
  methyl salicylate, 219
  neudo vidal, 219
  oxycom, 219
  PRORADIX, 219
Plant communication, 100–102
Plant defence 21, 25, 27, 28, 34, 36, 37, 49, 50, 101–102, 105–106, 115, 137, 146, 217–222
Plant diversity, 89
Plant growth-promoting rhizobacteria (PGPR), 44–45, 47, 52–54
Plant hormones, 78, 100
Plant–plant interaction, 79
Pollen competition, 13
Porapak Q, 23
Pre-concentration, 238, 239, 241
Priming, 47–54, 102, 103, 106, 217–219
Protein competition model (PCM), 189
Proton-transfer-reaction mass spectrometry (PTR-MS), 35
Pruning, 233
PTR-MS. *See* Proton-transfer-reaction mass spectrometry
Push-pull, 222–223

R
Reactive oxygen species (ROS), 63
Real-time mass spectrometry (MS), 35
Resistance, 100–104, 106, 107
Resistance breeding, 113–122
R:FR ratio, 161, 164, 170, 171
Root biomass, 79, 80
Root competition, 8, 12
Root elongation, 78
Root-emitted volatile, 79
Root grafting, 13, 15
Root growth, 79
Root territoriality, 9, 10

S
Sagebrush, 102–104, 108
Salicylic acid, 45, 47, 50, 51, 53
Salvia leucophylla, 78
SBSE. *See* Stir bar sorptive extraction
Self/non-self recognition, 15
Sexual signaling, 13
Shade avoidance, 160–162
Shikimic acid pathway, 189–190
Sibling rivalry, 13
Signal
  ability to ignore, 5
  biological, 3
  costs, 8, 9
  general definition, 2
  naturally selected, 3, 4
Signal evolution, 150
Signalling, 99–108
Solid phase extraction (SPE), 24
Solid phase microextraction (SPME), 23
Sorption, 237
Sound, 7
SPE. *See* Solid phase extraction
Specificity, 231, 233–236
SPME. *See* Solid phase microextraction
Stir bar sorptive extraction (SBSE), 25
Stomata, 238

Stomatal opening, 82
Stressors, 231, 232–233, 234–236, 243
Structure-activity relationships, 32
Super Q, 23
Symbiosis, 129
Systemic acquired resistance (SAR), 45, 48, 50, 54
Systemic emissions, 236
Systemic signaling, 104

T
Talking trees, 100–103
Temperature, 233, 237, 241
TENAX TA, 23
Terpene, 79
Terpenoids, 185, 186, 189–190
Time course of stress-induced volatile emissions from plants, 234, 235
Toxins, 5, 9, 11, 15
Trade-off, 171
Transcription factors
  MYB72, 46, 48
  MYC2, 48, 49
  WRKY, 49
Trichomes, 235
(E,E)-4,8,12-Trimethyltrideca-1,3, 7, 11-tetraene (TMTT), 34
Trophic levels, 76

V
Vacuum distillation, 24
Vapour phase extraction (VPE), 25
Ventilation, 237
Volatile, 77
Volatile organic compounds (VOCs), 101–107, 181, 182, 185, 186, 198
Volatile signal, 79
Volicitin, 30

W
Weeds, 92
Within-plant signaling, 99–108

MIX
Papier aus verantwortungsvollen Quellen
Paper from responsible sources
FSC® C105338

If you have any concerns about our products,
you can contact us on
ProductSafety@springernature.com

In case Publisher is established outside the EU,
the EU authorized representative is:
Springer Nature Customer Service Center GmbH
Europaplatz 3, 69115 Heidelberg, Germany

Printed by Libri Plureos GmbH
in Hamburg, Germany